Bibliographic information published by the Deutsche Nationalbibliothek

The Deutsche Nationalbibliothek lists this publication in the Deutsche Nationalbibliografie;
detailed bibliographic data are available on the internet at http://dnb.dnb.de

Publisher	PMC Media House GmbH Espenschiedstraße 1 55411 Bingen am Rhein Germany Fax (office) +49 40 228679 503 Email: office@pmcmedia.c om Internet: www.pmcmedia.com
Cover Design	Leon Zaayman
Printing	TZ-Verlag & Print GmbH, Roßdorf, Germany

3rd edition 2017
ISBN: 978-3-96245-151-6

A PMC Media House publication

PMCMedia
International Publishing

Foreword

The goal of a railway is to transport people and freight from one place to another safely, rapidly, regularly, on time and comfortably. This document is directed to the track infrastructure department whose contribution to the above goal is to ensure that the infrastructure is reliable, available, maintainable, affordable and safe; denoted by the acronym RAMAS. RAMAS is achieved through regular, effective and affordable maintenance of the track infrastructure.

The maintenance strategy and methods plays a major role in the RAMAS of the railway. This strategy and methods refer to deciding where maintenance needs to be carried out, what maintenance must be carried out, prioritising of the available resources for the maintenance and the focus of this document, the use of appropriate mechanised maintenance machinery to ensure that RAMAS can be achieved.

This book provides the basic principles of track maintenance with a focus on mechanisation.

Leon Zaayman
leonz@kimba.co.za

August 2017
3rd Edition

Table of Contents

Chapter 1
Introduction

The maintenance of railway infrastructure is a vast subject and consists of various components that must be maintained from time to time. Figure 1 provides a typical layout of some of the most important parts of railway infrastructure. All of these parts require maintenance and most of them will have some form of mechanisation used in the maintenance thereof.

This manual however focusses only on the maintenance of the track (item 1) together with all its individual components, the formation (item 2), drains (item 3) and the overhead traction equipment (item 8). As the title of this manual suggests, the focus is on those components that can be mechanised. Under mechanised maintenance it is understood the use of heavy on track mechanised maintenance machines with some reference to smaller machines as well.

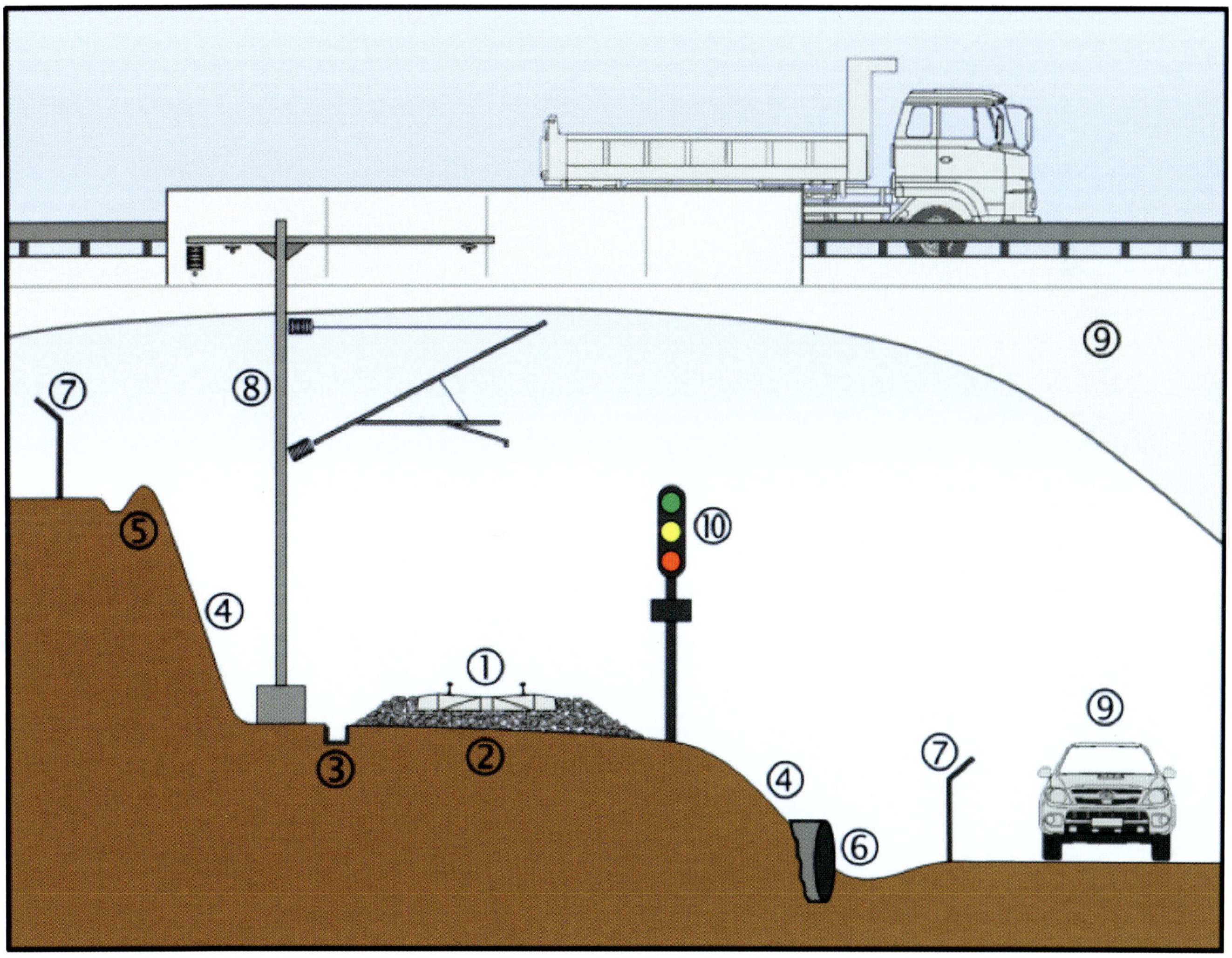

Figure 1: Typical Rail Track Infrastructure

Legend:

1. Rail, sleepers and ballast
2. Subballast and formation
3. Drains
4. Cuttings and embankments
5. Berms and cut-off drains
6. Culverts
7. Fencing and firebreaks
8. Overhead traction equipment (OHTE)
9. Service roads, level crossings, bridges and tunnels
10. Signalling and telecommunication

Why the specific focus on mechanised maintenance? As railway transport advanced over the past 60 years or so, the technology in terms of train traction, length, axle loading, speed, line congestion, economic importance, social importance etc. has demanded a far more technological approach from the maintenance of the track on which the trains run as well. It requires higher safety levels due to greater catastrophic results in the case of track failure in terms of human fatalities, economic loss of infrastructure, goods and rolling stock. Track material has advanced at the same pace and so has maintenance strategies, methods and machinery advanced as well.

There are many research studies and published books on track material and design. However, track design and construction is likely to be 30% or less of the lifecycle cost of the track whereas for the life of the track it will be maintained where more than 70% of the lifecycle cost will be spent.

There are not many literature studies focussing on track maintenance with even less on the mechanisation thereof. This manual attempts to address that shortcoming going back to basic principles with a practical approach. It was identified as a shortcoming based on the facts mentioned above; track maintenance has become very advanced and mechanisation machines have kept up with the latest technology has to offer. It extensively uses laser and computer technologies and many different machines with specialised features have been developed to carry out almost every possible track maintenance activity at the highest possible production, quality and accuracy.

However, sometimes the high production is not required but quality may not be compromised. Unless the end user of the machines knows what is available and what the criteria should be to select the correct mix of machines and features for the intended application, an informed decision cannot be made.

Also, not all machines are technologically equal. Some machine processes and technologies have been studied in detail and European Standards have been formulated to ensure quality, reliability and safety. This manual will also delve into some of the proven technologies and processes and will refer to some European Standards that refer to these. Future machine owners and end users need to take heed of these and ensure that over and above selecting the right type and mix of machinery that they also select the original equipment manufacturer (OEM) that uses standards and technologies that have been proven.

Mechanised track maintenance has become a true science and this manual will assist maintenance specialists to understand mechanised track maintenance technology and assist to develop selection criteria not purely based on price.

Chapter 2
Historic Overview

1. HISTORY OF RAILWAYS AND TRACK MATERIAL

1.1. Introduction

What would be the simplest definition of what we call a railway today? Maybe, “a specially constructed track which guides the wheels of a vehicle running on it in such a way that they cannot leave the track for the benefit of rolling efficiency”?

The earliest evidence of a “railway” that complies with this loose definition was the 6-kilometre Diolkos trackway, which transported boats in Greece during the 6th century BC and was functional for over 600 years.

The Diolkos trackway consisted of ruts cut into stone. These ruts were not created by repeated use and wear of wheels over an extended period of time. Documented research evidence shows that the ruts have been deliberately constructed to guide wheels.

The next resemblance to a railway track seems to have been related to mining in Germany in the 12th century referred to as a plankway. Mine passageways were usually wet and muddy, and moving wheelbarrows of ore along them was extremely difficult. Improvements were made by laying timber planks so that wheeled containers could be dragged along by manpower.

By the 16th century the difficulty of keeping the wagons running straight had been solved by a vertical pin running in a slot between the two narrowly separated planks (Figure 2).

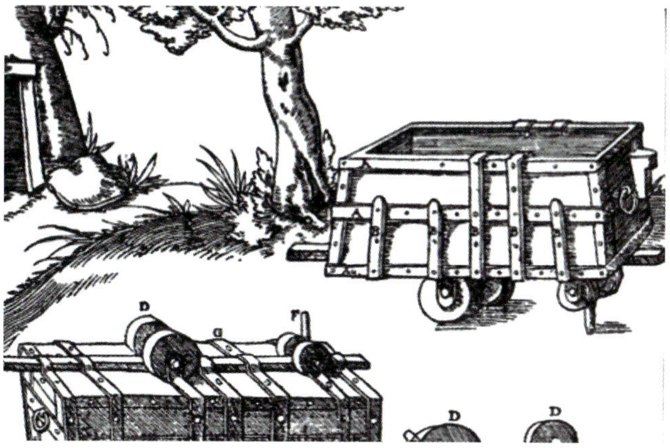

Figure 2: Plankways used in the 12th Century Mines
(Source https://commons.wikimedia.org)

1.2. The History of Rails

But getting a bit closer to what we generally perceive to be a railway was called a wagonway used by German mines as early as the 1550’s (Figure 3). These railed roads consisted of wooden logs acting as rails over which horse-drawn wagons or hand pushed carts moved with greater ease than over dirt roads. The wagonways even had turnouts of sorts. Also notice that flanged wheels were already in use though flanged on both sides.

Figure 3: Wagonway of 1550’s
(Source https://commons.wikimedia.org)

This method of guiding wheels clearly gave rise to the term railway and tramway both meaning the same thing, rail comes from the French word "raille" and tram from the Germanic word "traam", both meaning a length of timber or plank.

From the early 1700's, the wooden rails were covered with a thin flat sheathing or "plating" of iron to extend the life of the wood rails and reduce friction (Figure 4). This design was widely used on pre-steam railways and referred to as a strap rail.

Figure 4: Strap Rail
(Source https://commons.wikimedia.org)

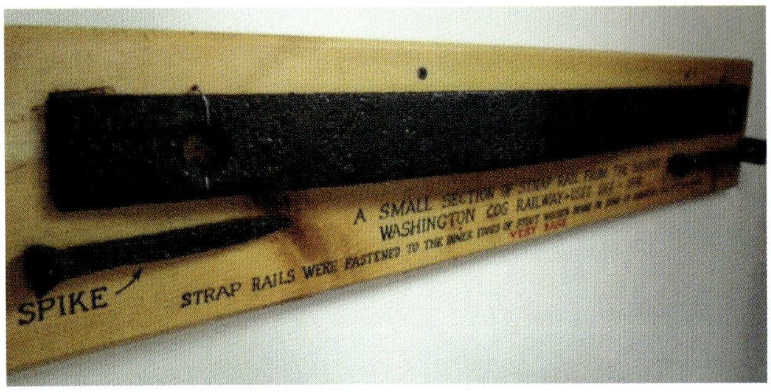

Figure 5: Flat-wheeled Wagon on L-Shaped Flangeway
(Source https://commons.wikimedia.org)
(The Little Eaton Gangway, Derbyshire (1795–1908)

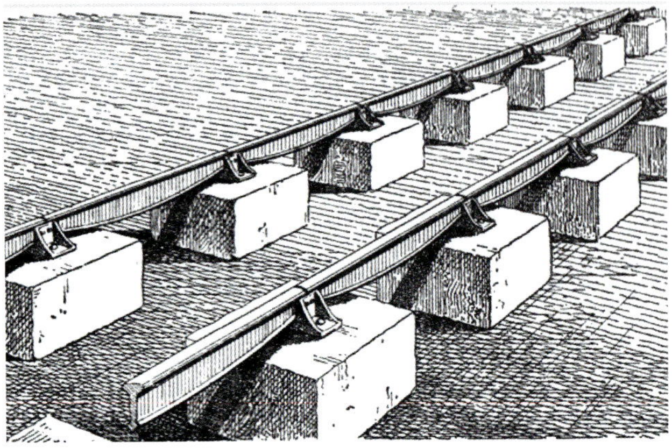

Figure 6: Jessop's Fish-Bellied Edge-Rail
(Source https://commons.wikimedia.org)

The steel straps caused rapid wear on the wooden wheels of the wagons which led to the introduction of iron wheels in the 1730's.

The rolling effect of steel on steel caused the straps to come loose which were prone to curl up and actually penetrate the wagons causing injuries. These were referred to as snakeheads.

It can be assumed that axle loads kept on increasing and the wooden rails were just not able to carry the loads. In 1767 British Coalbrookdale Iron Works introduced cast iron rails for the first time.

This was the era of the flangeway which was a cast iron flanged plate in the shape of an "L" (Figure 5). They were used with flat wagon wheels which could operate on ordinary roads as well. This type of rail gave rise to terms like "plateways" and "plate-rail", names which are still used today in the term "platelayer" applied to the men who lay and maintain the permanent way of a railway.

And talking of the term permanent way – in the earlier days of railway construction, contractors often laid a temporary track to transport materials to the construction site; when the construction was completed, the temporary track was removed and the "permanent way" installed. There is however also another version saying that at this time laws were starting to be promulgated which gave railways permanent right of way. In the 1770's the word "Railway" was used for the first time.

Not too long after the development of the L-shaped flange rail, the edge-rail was introduced by William Jessop in 1789 which resembles the modern rail we use today. In the same year Jessop also developed the first workable iron points. This rail was referred to as "fish-bellied rail" (Figure 6) due to its shape and was made of cast iron. The rails were used with flanged wheels with their flanges running on the inside edge of the rail, as we do today still, hence the description edge-rail. Flanged wheels on edge-rail negotiated curves better than the flat wagon wheels on the L-shaped flange rail.

The fish bellied rail and the L-shaped flanged rail were produced by the same company and coexisted for many decades, as a matter of fact, the L-shaped flanged rail was used until the 1940's but it definitely was the edge rail that became popular because of its better vehicle behaviour around curves.

Once again we can assume that there were demands for higher axle loads and track speeds but cast iron rails were brittle and broke easily and they could only be made in short lengths which would soon settle with an uneven longitudinal level. Therefore in 1820 John Birkinshaw introduced wrought iron rails in longer lengths as rolling techniques improved.

In 1836 Charles Vignoles in Britain suggested the use of a flat-bottomed rail. His name has become associated with rails as Vignoles rail, a term still being used today however, it appears that he should not be credited with such an honour. Flat bottomed rails have already been used in America by Stevens in 1831 before Vignoles and started looking like the rails we are still using today.

The year 1857 must have been a breakthrough year in railway technology when the first steel rails were made by Robert Mushet. Steel is a malleable alloy of iron and carbon which is much stronger than the iron castings used in the day and therefore would have lasted longer and would have been able to carry heavier axle loads.

As from 1858 onward rail profiles have not changed much as steel rolling technology improved and the current profile provided good performance. What surely has changed is the crown profile as trains became faster and heavier and steering and wheel/rail interaction became better understood. What have also changed were the steel alloys that were used to reduce rail wear and increase strength. Today of course heat treated rails are more common than alloy rails with high chromium.

1.3. The History of Sleepers

The word "sleeper" dates back from about 1789 and comes from Newcastle (where many local terms are based on old Norwegian words), which has the same root in Old Norse as the word "slab".

Sleepers made of large stone blocks were popular at the time as it provided a clear area for the horses since at the time carriages were drawn by horses, still a long way off from the commercial use of the steam locomotive. See again Figure 6.

Timber sleepers as we know them today may have originated at the same time as steam locomotion in the early 1800's but could not be confirm by the author.

In 1896 concrete sleepers were used for the first time on a main railway line by the Reading Company in America, as recorded by AREA (American Railway Engineering and Maintenance-of-Way Association). The early concrete sleepers had mixed success as they were prone to cracking and appeared to have the same design as the stone blocks used in the past but this time from concrete and with a gauge bar in-between them. They are referred to as pot sleepers and they resemble what we know as a twin block sleeper which is still in track and even produced still today.

The first general introduction of concrete sleepers on main lines took place in the late 1930's. Their design resembled timber sleepers and was fitted with chairs and bull-head rails. Concrete was probably introduced due to timber becoming scares at the time and may even be related to World War II.

European railways introduced pre-stressed concrete sleepers as we know them today which are capable of carrying heavier rail sections and long welded rails.

Steel sleepers have also been in use since the early 1900's. They are generally found on older and light traffic lines. Just like all other sleepers they have both advantages and disadvantages; a disadvantage of course would be corrosion.

Once again, there were various designs during an evolutionary process of sleeper design until it looked like the concrete sleepers we know today. Even today the research is continuing with various designs to provide better performance and less maintenance.

1.4. The History of Ballast

Early railway engineers aimed at complete rigidity of track by installing track on large heavy stone blocks laid directly on level ground. This quickly proved unsatisfactory and engineers realised that some form of resilient base under the track was essential to spread the load, to retain the track in its required position and that will allow lifting and packing of the track to keep the track top "level". The natural ground is rarely strong enough to accept the loading from locomotives without excessive settlement due to plastic deformation amplified by poor drainage.

This resilient base is the ballast bed. The date when the use of ballast stone under the track was used the first time was probably sometime in the later 1700's. The term "ballast" also stems from this period in Newcastle, UK.

Sailing ships carrying coal from Newcastle at the time were made of wood and designed to carry heavy loads. On the empty return the ships would be unstable due to their design. Ballast is a shipping term still used today to stabilise boats. Today water is used as ballast but at the time this was not possible with the wooden boats. Boats were therefore laden with gravel as ballast to maintain stability. This "ballast" was dumped by the quay side and probably became mountains of gravel over time. It probably was not by design but rather by convenience of the availability of material that the stone was used to provide a bed for the wagonways that carried the coal. One can also assume that the track started sinking into the muddy ground due to the continuous use and something was required to lift it. This association of the word "ballast" remains to the day.

In the 1840's true ballast of crushed stone as we know it today began to be widely used.

2. HISTORY OF TRACK MAINTENANCE

Track work by gangs of labourers with forks, shovels and tamping picks is extremely strenuous, time and labour-consuming and of limited accuracy, as a matter of fact, under certain circumstances it can be very damaging to the track and even reduce its life expectancy.

The result is that mechanised methods for track maintenance have been considered much further back than most people realise. Figure 7 shows a photo of an overhead maintenance "machine" which must have been for the tramways and was horse drawn. Figure 8 is some kind of a railbound ballast shoulder renewal machine that dates back to around the 1930's.

The oldest patent for a mechanised method of tamping found by the author dates as far back as 1866 by John W Oakley though only published in 1886 (Figure 9). Tamping is the most regular maintenance activity on track so it is no surprise that attempts towards mechanisation started a very long time ago.

Figure 7: Wagon Designed for Overhead Maintenance on Trams (Source https://commons.wikimedia.org)

Figure 8: Ballast Shoulder Cleaning Machine (With courtesy of www.railwaywondersoftheworld.com)

This US-patent by Evans showed a manually driven rail-bound mechanism with a number of tools pushing the ballast under the sleepers. There were quite a few others patents filed in the following 40 years but the most important historical patent for modern tamping machines was however filed in Switzerland in 1934 by Scheuchzer for the first time. His mechanism comprised of two motions, one is the vibration of the tool by an eccentric shaft which penetrates into the ballast and the second motion is squeezing the ballast underneath the sleeper by closing the tools via two spindles. There is no information on how extensively the above inventions were used but they were certainly not produced in great numbers.

It was not until the introduction of widespread mechanised track maintenance that the foundations were created for a good track geometry and consequently for a rise in the travelling speed. It basically started in 1945 when the first tamping machine was produced in reasonable numbers, the Matisa Standard which initiated the mechanisation of ballast tamping. This machine was based on the Scheuchzer patent. The company Matisa still exists today.

In 1953 Plasser & Theurer of Austria was established and filed patents for an improved tamping mechanism. For the first time hydraulic pressure was used for the vibration of the tines. The vibration is required to liquefy the ballast stones and allow them to "flow" without stressing and crushing the ballast and for crush-free penetration of the tines into the ballast.

For the closing action of the tines, these assemblies are cylinders which apply a force on the tine arm to squeeze ballast underneath the sleeper. This action replaces the spindles previously used and revolutionised the industry, giving Plasser & Theurer the competitive edge to become the world leader in its field. This principle known as asynchronous tamping is the norm today with all tamping machines by all manufacturers.

The first machine produced by Plasser & Theurer in 1953 was the HGL tamping machine (Figure 11) which probably tamped less than 4 sleepers per minute and lifting and aligning had to be done manually ahead of the machine.

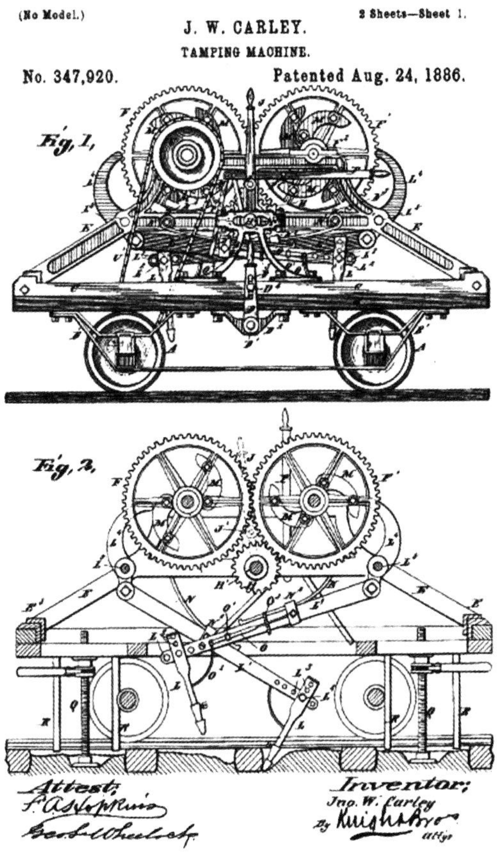

Figure 9: Tamping Patent by Oakley 1866
(US Patent 347920 A)

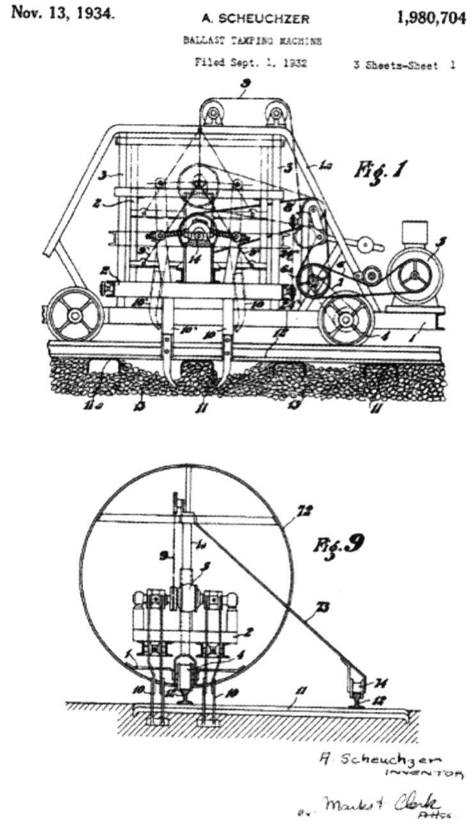

Figure 10: Tamping Patent by Scheuchzer 1934
(US Patent 1980704)

From the early 1950's the advancement in mechanised track maintenance technology improved in leaps and bounds. It was demanded by railway lines becoming increasingly heavier for supporting higher axle loading and speeds. There was also demand for every other component of the track structure to be maintained by mechanised means. The demand was for machines to produce ever increasing production and quality. Today many thousands of highly sophisticated mechanised machines are maintaining every major railway in the world carrying out almost every possible maintenance activity on the track. Mechanised systems produce quality, production and unit cost of maintenance that no manual maintenance activity can equal. Yet, some machines do it better than others, some processes have been more researched than others and the quality longevity and maintainability of some machines are just better than others.

Figure 11: Plasser & Theurer's first Tamping Machine (HGL)

Chapter 3
Track Components

1. INTRODUCTION

The objective of this chapter is to identify the different components (Figure 12) of the track, define their functions and design criteria and to emphasise why each of these should be maintained. This chapter will also identify how each of the components should be maintained while considering the advantages of mechanised maintenance. These mechanised methods are discussed in more detail in following chapters.

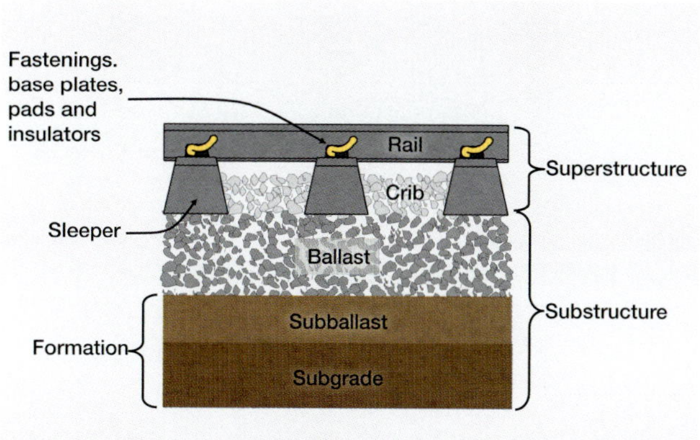

Figure 12: Track Components

2. RAILS

2.1. General Description

It is the rails that guide the train wheels and with steel wheels on steel rail, the rail is exposed to some of the toughest conditions faced by all steel products. If the rail fails, people's lives will be in danger which makes maintenance to the rail one of the most important track maintenance activities.

The rail is the most costly component in the railways asset register. A very high level of knowledge of rail deterioration and maintenance is therefore essential for lifecycle cost management.

Rails are exposed to wear on the crown (due to wheel and rail interaction and rail profiling), wear of the gauge face (due to wheel flange contact in tight curves), plastic flow (due to contact stresses between rail and wheel exceeding the steel strength) and defects (which can be internal or on the running surface due to rolling contact fatigue, dynamic loading, etc.).

Refer to Figure 13 and Figure 14 for general terminology that is used with regards to a description of rails.

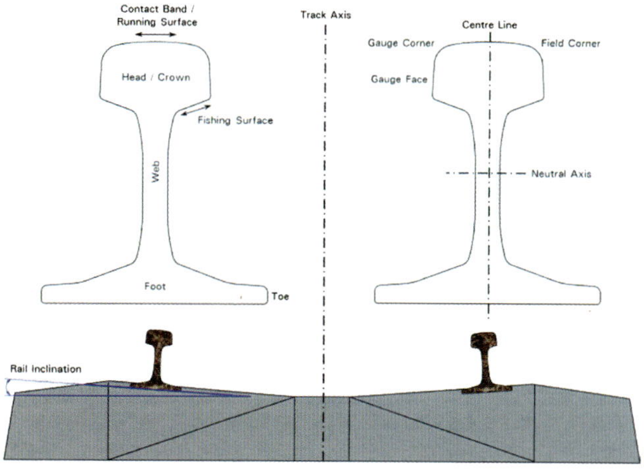

Figure 13: Rail Terminology

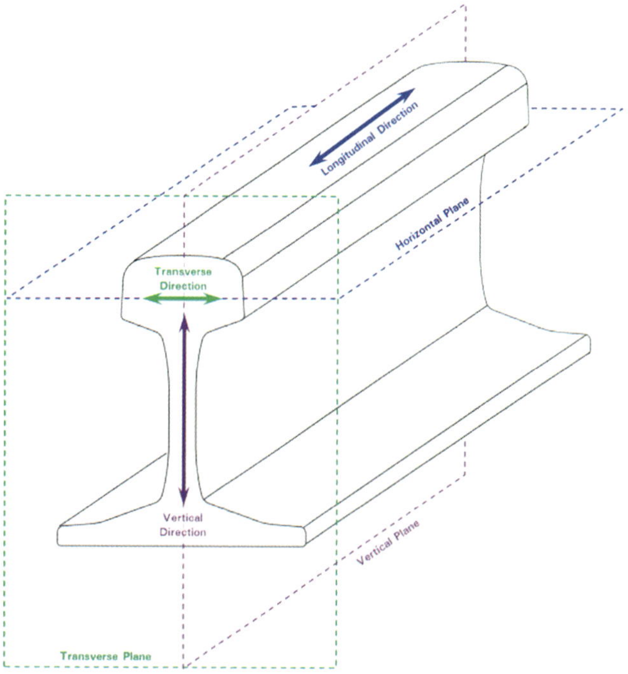

Figure 14: Directions Related to the Rail

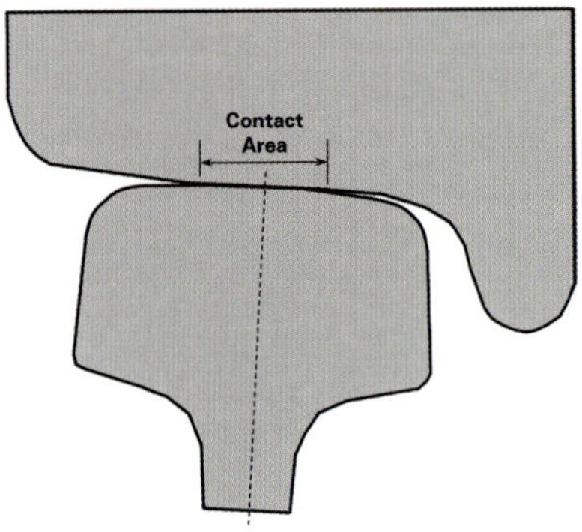

Figure 15: Rail Wheel Contact Area or Band

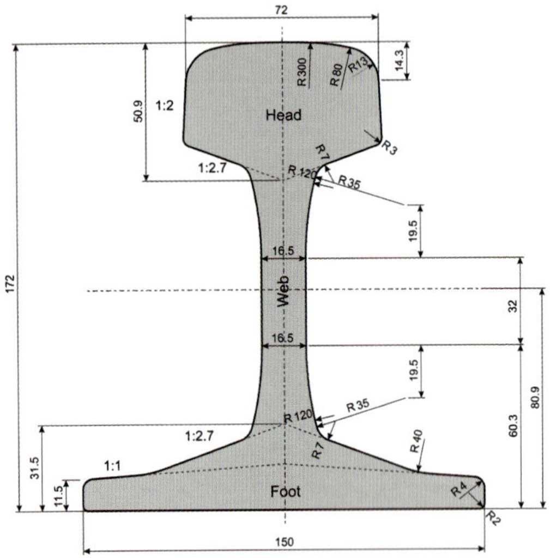

Figure 16: A Typical Complex Rail Profile (60E1) from the EN13674 Standard

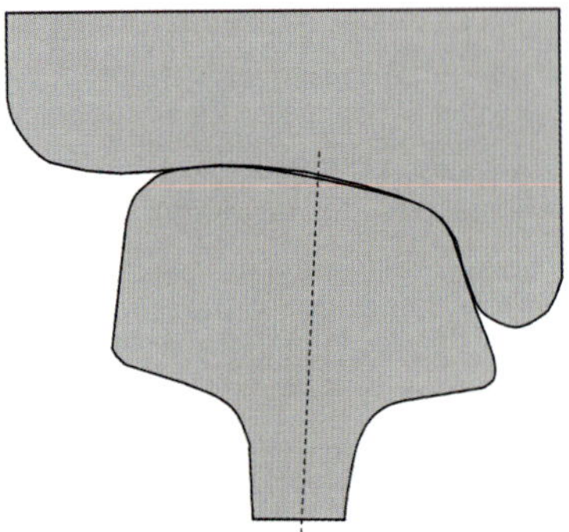

Figure 17: Hollow Worn Wheel on Worn Rail Profile in a Curve

Figure 18: Flat Wheel Crown with Side Wear Developing

2.2. Rail Profiles

Rail is identified by its weight per running metre and is generally available for freight and passenger railways from about 30 kg per metre rail to approximately 75 kg per metre. The actual rail profile is further identified by other standards such as UIC (Union of International Railways), AREMA (American Railway Engineering and Maintenance-of-Way Association), etc. The most used standard for rail is EN13674 which provides standards for steel grades, rail profiles, chemical compositions, hardness, testing methods, etc. The chemical composition and heat treatment is then a further classification which determines the rail's hardness and wear characteristics (see paragraph 2.3). See Figure 16 for a typical complex rail profile (60E1) which is widely used.

Modern rails are also referred to as broad footed rail. The rail weight, profile and hardness will determine the application of the rail, for example, heavy haul, light passenger, etc.

On many low density railway lines lighter rails are used from about 30 kg to 50 kg per metre and in some cases the profile is unique to a specific country. Standard grade (hardness) is generally used. On high speed lines and heavy haul applications heavier rail profiles of even 70kg/m are common.

Besides the standard profiles mentioned above, there are also a number of special rail profiles for example rail in switches, grooved rails, switch blades, etc.

On local railway lines (tramways) grooved rails are used and crane rails are designed so that they can support heavy loads.

All rails are branded with milling and pressing marks which are found on the web and which provides the manufacturing company, year of rolling, profile and steel quality.

The crown profile is very important from a wheel and rail interaction point of view since this is what provides the tracking of the rail wheel, especially around curves (Figure 15). However, over time and lack of preventative maintenance, the rail profile will deteriorate in many different ways, such as side wear in curves and flattening of the rail crown (Figure 18).

The rail/wheel contact area becomes undefined and the wheel tends to hunt across the rail causing accelerated wear on both rail and wheel, while tracking and steering of the wheel is adversely affected, to the discomfort of passengers and the danger of cargo shifting and damage.

The wheel profile is equally complex and its conicity is matched with the rail profile to optimise this very complex interaction to reduce friction and wear of both wheels and rail to an absolute minimum (Figure 15) and to aid steering around curves. However, similar to the rail profile, if the wheel profile is not maintained, hollow wear of the wheels develop (Figure 19) and flanges become sharp (Figure 20) due to the contact with the rail crown in curves. It becomes increasingly easy for the wheel flange to climb over the rail and cause a derailment. Refer to CHAPTER 4 on the fundamentals of rail/wheel interaction.

Figure 19: Hollow Wear

Figure 20: Sharp Wheel Flange due to Wear in Curves

Managing rolling stock and wheel profiles is not part of this manual; however, managing rail profiles is and is made simple using an infrastructure measuring vehicle equipped with rail profile scanning equipment. Refer to CHAPTER 9 paragraph 4.2 for information on rail profile scanning.

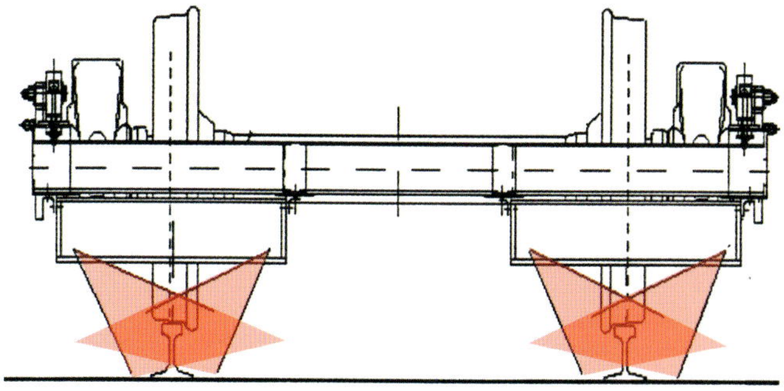

Figure 21: Rail Profile Scanning Equipment

Reports from the infrastructure measuring vehicle can then be used to determine the rail profiling and rail replacement programme. Rail profiling reduces surface defects (rolling contact fatigue) and improves bogie tracking which in turn reduces flange contact and internal rail stresses. A dramatic increase in rail life can be achieved when rail profiling, track geometry, lubrication, tracking and speed are managed optimally. Rail profiling consists of rail grinding, rail planing or rail milling which forms a very important part of preventive track maintenance to maintain the rail profile and prolong the useful life of the rail. These are complex processes which must be managed by experts in this field. Refer to CHAPTER 16, paragraph 3 and BS EN 13231-3.

Figure 22: EM120 Infrastructure Measuring Vehicle

However, if the grinding programme was not maintained, the rail profile can deteriorate beyond the economic intervention by a grinding machine. In these cases it may be required to use the rail planing machine or mobile rail milling machine which are corrective methods. Refer to CHAPTER 16, paragraph 4 and paragraph 5.

2.3. Rail Steel Properties

Rails are subject to the process of wear, work hardening and fatigue under traffic loading. Continuous welded rail is also subjected to high compression or tensile stresses due to temperature variances.

2.3.1. Mechanical Properties

- **Tensile Properties** – The conventional stress-strain diagram for mild steel in Figure 23 illustrates the position of all the following definitions:

  Ultimate tensile strength: Also called tensile or ultimate strength is the maximum tensile stress [measured as a force per unit area, Pascal (Pa)] before the rail will break if stretched; typically as could be found on a cold night on continuous welded rail.

  Stress-strain: Stress is a physical quantity that expresses the internal forces that neighbouring particles of a continuous material exert on each other, while strain is the measure of the deformation of the material (Wikipedia).

  Stress: Stress is defined as the force across a "small" boundary per unit area of that boundary, for all orientations of the boundary (Wikipedia).

  Strain: A measure representing the displacement between particles in a material relative to a reference length. It is dimensionless represented by Epsilon (ε) expressing a decimal fraction, a percentage or in parts-per notation (Wikipedia).

Proportional limit (linear range): Up to this amount of stress, stress is proportional to strain (Hooke's law), so the stress-strain graph is a straight line, and the gradient will be equal to the elastic modulus of the material (Wikipedia).

Elastic limit or yield strength (point): The stress at which a material begins to deform plastically, i.e. the material cannot return to its original shape. Prior to the yield point the material will deform elastically.

Upper and lower yield points: Some metals, such as mild steel, reach an upper yield point before dropping rapidly to a lower yield point. The material response is linear up until the upper yield point (Wikipedia).

Apparent and actual fracture strength: Fracture strength in this context is the tensile stress at which a material breaks. Ductile materials have a fracture strength lower than the ultimate tensile strength. If a ductile material reaches its ultimate tensile strength in a load-controlled situation, it will continue to deform, with no additional load application, until it fractures. If the stress-strain curve is plotted in terms of actual (true) stress and true strain the curve will always slope upwards as actual stress is corrected for the decrease in cross-sectional area. The true stress on the material at the time of fracture is known as the breaking strength. This is the maximum stress on the true stress-strain curve.

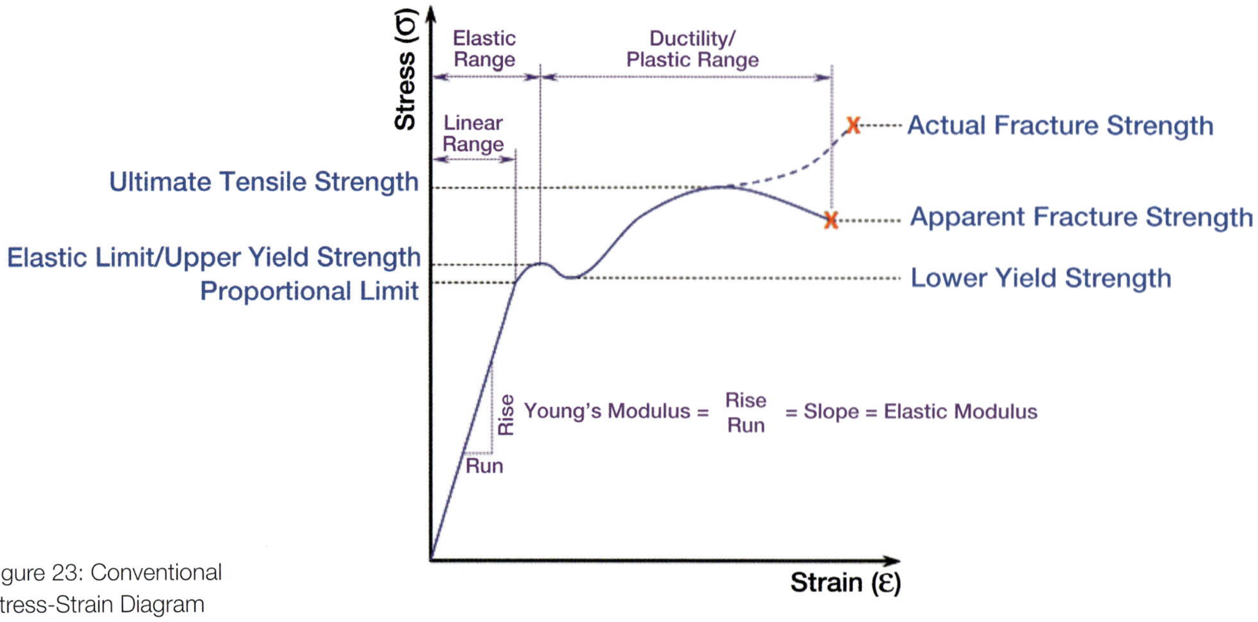

Figure 23: Conventional Stress-Strain Diagram

Ductility: The ability of material to plastically deform before fracture. The absence of plastic deformation will be brittle fracture which would not be suitable as rail material. Brittle materials will fracture at the ultimate tensile stress.

- **Hardness Properties** – The hardness identifies the rail's resistance to indenting. The most common method for determining hardness is the Brinell harness classification. This method uses a 10 mm tungsten carbide ball which is pressed at a certain load into the test piece. The size of the indent determines the Brinell hardness number (HBW). There is a strong correlation between hardness and material strength, therefore the higher the hardness the more wear resistant the rail will be. However, wear resistance cannot be increased at the expense of sufficient ductility and toughness.

- **Fatigue Properties** – Rail contact surface fatigue is the progressive weakening of the metal resulting in structural damage that occurs due to the cyclic loadings by rolling wheels. The nominal stress levels that causes fatigue may be much less than the rail material's ultimate tensile stress limit or yield stress limit. If the fatigue causing stress levels exceeds a certain threshold, fine cracks start developing on the rail surface known as rolling contact fatigue. Harder rails have a greater fatigue resistance.

  Rails are also subjected to fatigue due to cyclic bending of the rail between their supports on the sleepers and will increase with lighter rail profiles and longer sleeper spacing.

- **Toughness** – Toughness (ductility) is the ability of the rail to absorb energy and plastically deform without fracturing. Toughness is an important rail property to resist crack growth and sudden fracture in the presence of defects.

A good rail material will be one which is hard for wear resistance and will exhibit high tensile strengths (N/mm^2) to resist fatigue but should also exhibit toughness to absorb energy without fracturing. Some of the requirements for ideal rail material are in conflict with one another. To find the right balance without compromise, the desired properties can be achieved through alloys of steel and heat treatment of the steel altering its microstructure.

2.3.2. Alloy Steels

Steel is essentially an alloy of iron and carbon. Iron in its purest form is called ferrite and contains up to 0.025% carbon. Ferrite is soft and has a low strength. Increased carbon content gives steel strength and hardness, but the higher the carbon, the lower the ductility which is unsuitable for rails. To improve the ductility and strength of rail at lower carbon contents, elements may be added to the steel. In rail steel the more important elements are:

Silicon: Low contents improve tensile strength and hardness.
Manganese: Most steels contain manganese. It is mainly added to steel as a deoxidising and desulphurising agent. Up to 1.5% makes steel more forgeable.
Chromium: Makes the steel easier to harden and, if the carbon content is high, it also enables it to withstand severe abrasion and wear.

Figure 24 shows the chemical composition of typical rails used around the world.

Element	HCOB* Grade 700	UIC A Grade 900A	CrMn Grade 1100	HH
Carbon	0.55–0.68%	0.60–0.80%	0.65–0.80%	0.60–0.80%
Manganese	0.70–0.90%	0.80–1.30%	0.80–1.30%	0.80–1.30%
Chromium			0.70–1.30%	
Silicon	0.10–0.35%	0.10–0.50%	0.30–0.90%	0.10–0.50%
Phosphorus	0.06% max	0.05% max	0.03% max	0.05% max
Sulphur	0.06% max	0.05% max	0.03% max	0.05% max
Hardness (Brinell)	240	280	340	400
Strength:				
Yield	410 MPa	449 MPa	600 MPa	800 MPa
Ultimate Tensile	880 MPa	885 MPa	1080 MPa	1200 MPa

Figure 24: Rail Steel Element Composition

UIC A: This is standard rail steel and is like all rail steels pearlitic. It contains a relatively low carbon percentage resulting in a medium hardness of 280 Brinell with an ultimate tensile strength of 885 MPa.

Chrome Manganese (CrMn): CrMn rail steel is harder than UIC Grade A rails making this alloy more wear-resistant. Its high yield strength resists plastic deformation or metal flow under normal running conditions. In addition the high tensile strength (1080 MPa) of the rail makes it suitable for heavy haul applications for its expected extended fatigue life of the rail. The elongation property of CrMn steel is slightly lower than Grade A rail steels making it less ductile and more brittle. As a result, its toughness and its capability to withstand shock loads are reduced. The rail is therefore more prone to brittle crack propagation and simultaneously more sensitive to stress raisers in the form of nicks and marks.

Head Hardened Rails (HH): To overcome the disadvantages of CrMn rails, head hardened rails have been developed which are of a similar composition to UIC A steel but of which the crown has been heat-treated to give it exceptional hardness and tensile strength. Voest-Alpine Schienen GmbH developed a method to produce head-hardened rails (HSH – Head Special Hardened). It differs from other hardening methods in that it is fully integrated into the production process of the naturally hardened rails. When a certain rail temperature is reached, the rail head is quenched in a hardening bath so that only the rail head is immersed in the cooling agent. After a pre-selected time of around 2.5 minutes a highly wear resistant fine pearlite structure has formed over the entire cross-section of the rail head. The rail is then lifted from the bath and laid on the lifting beam cooling be where it is further processed like a naturally hard rail. HSH rails are especially suited for high-speed lines and heavily used tracks because during the entire service life of the rail, even with heavy wear, the rail still remains in contact with a high-strength rail material. The service life is substantially higher than that of highly wear-resistant rails of standard quality.

2.3.3. Heat-Treated Rails

Heat treatment in this regard refers to the controlled cooling of the steel. According to Becker (see references) the most efficient way to describe steels and their properties is to start with the iron/iron-carbide phase diagram (Figure 25). A quick glance at the UIC grades of rail steel shows carbon content is on average about 0.7% whilst the manganese content is 1.0%. Although the manganese content is higher than the norm for this diagram, it is fair to say that the phase diagram is still valid for these steels.

The iron/iron-carbide phase diagram shows that if an 0.83% carbon steel is heated to about 723°C the phase at room temperature changes to a phase called austenite which is an unstable, solid solution of carbon in iron at a temperature of between 723°C to 910°C. Upon slow cooling, austenite changes to pearlite below 700°C. Pearlite is a composition of alternative lamella of ferrite and cementite. The latter is a hard and brittle phase and, as said earlier, ferrite is soft and ductile.

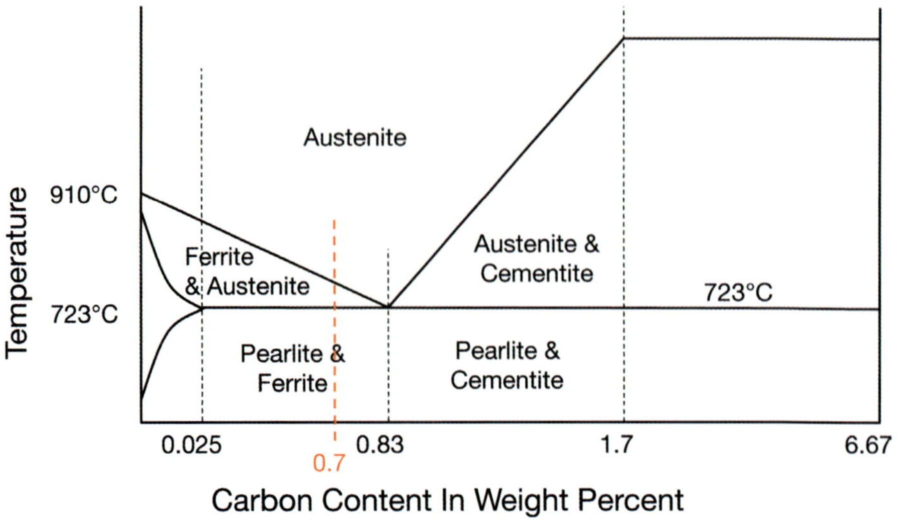

Figure 25: Iron/Iron-Carbide Phase Diagram (Becker J)

The end result is soft phase strengthened by the cementite which means that it is strong and fairly hard. Rail steel, however, has slightly lower carbon content, resulting in a microstructure that consists of ferrite and pearlite.

However, this structure only applies in the case of slow or equilibrium cooling conditions. In case of fast i.e. non-equilibrium cooling conditions, other structures will be present.

During rail welding the cooling afterwards is normally very fast because the rest of the rail acts as a heat sink. Under these conditions a different diagram is applicable to establish what structures would be present. In this case the Time Temperature Transformation (TTT) diagram is more appropriate to explain the theory.

Under non-equilibrium cooling cycles two additional phases other than pearlite are possible, namely Bainite and Martensite. Bainite is of moderate hardness (±350 Hv) while martensite is extremely hard and brittle (800 Hv) and usually results in failure as a result of cracks that form after transformation. Only a small fraction of martensite present could render the rail unsuitable and dangerous.

During welding one has to aim as far as possible at maintaining the microstructural composition of the parent material i.e. a pearlitic structure. Various steel alloys exhibit different cooling rates at which a pearlitic structure will be formed.

The TTT diagram shows the time temperature transformation curves for UIC A and CrMn steel. Curve 1 represents a cooling rate in atmosphere. As can be seen from the diagram, the cooling curve passes through the UIC A transformation curve within the critical cooling period between 800°C and 500°C. In this case full pearlitic transformation will take place.

On the other hand, cooling curve 1 does not pass through the CrMn transformation curve. The chromium in CrMn steel reduces the cooling rate at which pearlite can form. Pearlite is unable to form at cooling rates in excess of approximately 1°C/sec. If these cooling rates are exceeded then martensite and bainite are formed. For this reason the weld of CrMn rails are post-heated to slow down the cooling rates. The slower cooling rate is illustrated by cooling curve 2. Therefore, a disadvantage of CrMn rails is that its cooling rate must be slowed down after welding, complicating the process, making it slower and potentially more expensive. See more about the cooling of CrMn rails after flash butt welding in CHAPTER 17.

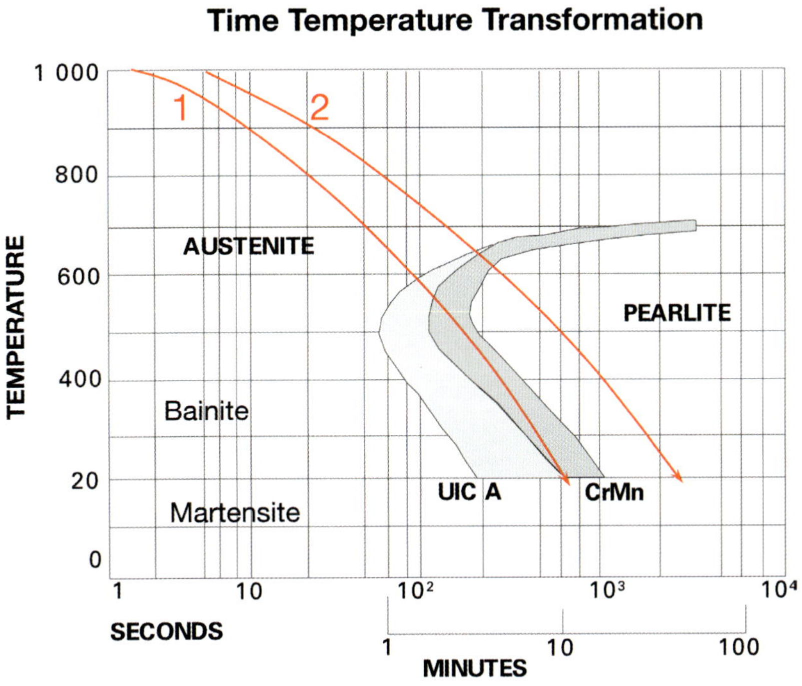

Figure 26 : Time Temperature Transformation Diagram

Figure 27: Classic Maintenance Neglect on Fish-Plated Joints – Wide Gaps, Missing Bolts

Figure 28: Classic Maintenance Neglect on Fish-Plated Joints – Battered Rail Ends

2.4. Fish-Plate, Splice Bar or Joint Bar

A fishplate, splice bar or joint bar is a metal bar that is bolted to the ends of two rails to join them together in a track. The name is derived from fish, a wooden bar with a curved profile used to strengthen a ship's mast. The top and bottom edges are tapered inwards so the device wedges itself between the top and bottom of the rail when it is bolted into place and prevents lateral and vertical movement of the rail ends while allowing longitudinal movement for expansion and contraction of the rails.

Joint bars are matched to a specific rail profile to ensure fishing angles and radii match. Each rail section has a specific spacing of holes from the end of the rail as well as dimension above the base for the installation of the joint bar. Incorrect installation will result in an inadequately supported joint and will promote rail defects such as head and web separations and bolthole breaks.

When joints are installed, it may only be done during the rail neutral temperature. The expansion gaps must be correctly set using shims set at a specific temperature.

Thereafter joints should be removed, cleaned and lubricated at least once per year. This very seldom happens, with the result that joints deteriorate exponentially until it becomes a constant source of maintenance and eventually fails in many ways such as battered ends, dipped joints, large gaps, oval bolt holes etc. See Figure 27 and Figure 28 for typical examples of common maintenance neglect. The joint is considered to be the weakest point of the track.

The holes for fishplates should be drilled with the appropriate equipment such as those supplied by Robel Bahnbaumaschinen (Figure 29) and a template for the hole positions should preferably be used.

The sleepers on either side of fish-plated joints must be of the same type and extra care must be afforded when these sleepers are tamped.

Over the years a great variety of designs have evolved, e.g. angle fishplates, U-type fishplates and Z-type fishplates. Joints can either be suspended, where the rail joints are placed over the space between two sleepers, or supported, where the rail ends are placed over an extra wide sleeper (Figure 31).

On main lines, rail joints should only be installed if the prevailing conditions do not allow for the rails to be welded together continuously.

Fish plate joints must be provided at un-ballasted bridges and may also be used to protect sets and curves.

2.5. Temporary Rail Joints

For connecting rail ends temporarily in track renewal sections (e.g. prior to welding them together continuously) or in the event of rail breakages, or as a safety measure where internal rail defects have been identified during ultrasonic rail testing, a temporary rail joint can be installed. This kind of connection consists of two flat fishplates and screw clamp/s.

Various designs have been developed but one of the most user-friendly and safe even for high speed designs is by Robel (Figure 32).

Figure 29: Rail Drilling Machine

2.6. Insulating Joints/Block Joints

An insulating joint (commonly called a block joint) is used to insulate one rail from the next for either signalling or to insulate non-electrified track from electrified track thereby creating a track circuit section.

Insulating joints are generally pre-manufactured and have very specific specifications that must be adhered to. There are different systems on the market.

Figure 30: Template for Rail Drilling

Figure 31: Rail Joint over Extra Wide Sleeper

Figure 32: Robel Rail Clamp

2.7. Compromise Joints or Junction Rails

Compromise joints or junction rails are used for joining rails of different rail profiles, spanning more than one mass if the use of junction exothermic welds is not possible. They are constructed such that the bars align the running surface and gauge sides of different rails sections (Figure 33). There are two types of joints:

- Directional compromise joints are used where a difference in the width of the head between two sections requires the offsetting of the rail to align the gauge side of the rail.
- Non-directional compromise joints are used where the difference between sections is only in the heights of the head.

2.8. Continuously Welded Rail (CWR)

Rail joints are a major source of track maintenance effort, due to its maintenance requirements, in absence of which very high dynamic forces are transferred to the sleepers, ballast and formation causing damage, dipped and battered joints etc. It is therefore general practice to eliminate this source of maintenance by welding the rail sections into continuously welded rail (CWR).

CWR also benefits carrying of the return current, noise reduction and greater passenger comfort. However, Rails may only be continuously welded, if the fastening method guarantees their good and lasting fastening down onto the sleepers with a high creep resistance and if the track is ballasted according to the regulations.

CWR started as early as 1820 when John Birkenshaw proposed to weld together his 6 metre long wrought iron rails in an attempt to eliminate the problems associated with the great number of joints in the railways then being constructed. However, the first rails were welded in 1899 using the thermite welding process developed by Goldschmidt (as far as can be established, this was only on the tramways). The first continuous welded rail was introduced in Germany in 1924. It is common practice on all mainlines world-wide today.

The minimum radius of continuously welded curves is 600 metres provided that the sleeper spacing does not exceed 700 mm and if concrete sleepers are used the minimum radius may be reduced to 400 metres provided that the ballast depth is at least 300 mm and the sleeper spacing is not more than 600 mm (Transnet standard).

Where a curve consisting of jointed track is joining unto a section of continuously welded rails, a transition length of jointed track of at least 80 sleepers between where the curve ends/begins and the continuously welded rails begins/ends must be applied.

Figure 33: Compromise Joint (Source www.rrpicturearchives.net)

Rails are welded together using either the exothermic or flash butt welding process to form one continuous rail that may be several kilometres long.

2.8.1. Exothermic (Aluminothermic or Thermite) Welding

Despite the economy and easy application of the thermite welding method (Figure 34), it does require specially trained personnel because the quality of the welding depends directly upon the skill and sense of responsibility of the operator.

During the thermite welding process, the two rails that are to be welded together are enclosed in a sand mould with a defined gap between the faces and pre-heated to a temperature which is suitable for both the materials and the process. The aluminothermic mixture of metal oxide and aluminium, called thermite, is ignited in a highly refractory mangnesite crucible using a special igniter. After the exothermic reaction, lasting only a few seconds, approximately equal volumes of molten steel and fluid aluminium oxide (the slag) are separated at a high temperature. The molten steel is allowed to flow into the gap between the rails. This reaction can be illustrated by:

After removal of the mould the rail profile is ground to within set standards using a handheld grinding machine as shown in Figure 35.

The advantage of this process is that it is cost effective to carry out in small numbers and does not burn off any of the rail material but it is a slow process and is not recommended where production welding is required or would be a benefit.

Thermite welds also suffer from a large heat affected zone of which the hardness is softer than the mother material and is therefore prone to accelerated wear and breakage. On some lines more than half of all rail breaks occur in thermite welds. It is therefore important that the weld is carefully prepared according to good practice, which is not always the case. Refer to a classic case in Figure 36 where the maximum distance to fish-bolt holes have not been adhered to, which is a risk area for cracks to develop.

The sleepers on either side of exothermic welds must be of the same type and extra care must be afforded when these sleepers are tamped.

Metal oxide	**+**	**aluminium**	**⟶**	**metal**	**+**	**aluminium oxide**	**+**	**heat**
Fe_2O_3	**+**	**$2Al$**	**⟶**	**$2Fe$**	**+**	**Al_2O_3**	**+**	**heat**

Figure 34: Thermite Welding

Figure 35: Rail Profile Grinding Machine

Figure 36: Poor Welding Practice

2.8.2. Flash Butt Welding

Flash butt welding has been used by railways since around 1930 in welding depots.

This process charges the two rails to be welded with very high amperage at low voltages. The resistance of the current flow between the two rails causes rapid heating of the two rail ends. Based purely on a time controlled system, the heated rail ends are forced together at a high pressure during which all impurities are forced out. Flash butt welding does not require any additional welding material as the rail itself is used as a welding compound. The result is a near flawless weld of which the strength exceeds that of the rail material (Figure 37).

Initially a drawback was that flash butt welding was only used in stationary depots which then required the arduous transport of the long welded rails, thus setting constructional limits for the production lengths.

Plasser & Theurer developed the first mobile flash butt welding machine in 1973 using a welding head, the K355 (Figure 38), developed at the J O Paton Institute in Kiev, Russia. The first application of mobile flash butt welding was carried out on the Rome to Florence Direttissima line where 14,000 welds were completed between November 1974 and October 1976 at a production rate of approximately 10 welds per hour. Thereafter followed countless welds the world over using mobile flash butt welding machines.

In 2009 a new EN standard was published which required additional requirements from flash butt welding. Plasser & Theurer was again the first company to produce a welding head that fully complied with the standard. Refer to CHAPTER 17 for more information on the flash butt welding process and machines available.

All railway lines still using fish-plated rail joints can benefit substantially from welding their rail into continuous welded rail using a mobile flash butt welding machine.

2.8.3. Other Methods of Welding

There are also other less popular forms of welding such as pressure gas welding which will not be discussed.

Arc welding is also used for building up rail at wheel spin marks, damaged crossing points etc.

2.9. Transportation of Long Welded Rails

Rails are available from rolling mills in various countries in various standard lengths from about 12 metres up to 120 metres. The longer the rail the fewer welds are required and the less risk of defects is built in. However, longer rails do have logistical challenges such as the availability of ships capable of handling such long rails and the handling, stocking and rail transport of these long rails. It does, however, positively affect the production rate of re-railing since fewer welds need to be made.

Rails are generally transported in their originally manufactured lengths to a welding depot from where they are welded into long rails of approximately 240 metres. These welding depots are either established depots with stationary machines or a flash butt welding machine can be set up in a temporary stationary depot as was typically done in Botswana (see Figure 39).

Figure 37: Flash Butt Weld

Figure 38: Flash Butt Welding Machine

From the welding depot the long rails must be transported to the installation site. It is often not possible to weld the short rails in track due to production requirements, the large number of rails and the fact that the track is often required to be available for traffic in-between rail replacements (long distance re-railing).

To transport the long rails safely, specialised equipment is required to ensure that the rails cannot move during transit and that the minimum bending stresses occur during unloading at the installation site. An example of such a specialised rail train is seen in Figure 40. Refer also to CHAPTER 19 for a detailed description of a rail carrying train.

2.10. Rail Installation

Rail will generally have a shorter life expectancy than sleepers and will probably be replaced a number of times before complete track renewal takes place. Rail handling is a dangerous maintenance activity for which mechanised methods are suggested such as using a rail threader (Figure 41).

Assuming that rails will be welded into continuous welded rail, the rail must be destressed and the closure weld made during the neutral temperature range.

Figure 39: Temporary Stationary Welding Depot using a Mobile Flash Butt Welding Machine

2.11. Rail Transposing

It is common practice to remove rail from main lines where the rail is nearing its wear limits and install them on branch/secondary lines of less importance. Also, on curves it is also common to transpose the high rail and low rail when the side wear on the high rail has reached a defined measure.

There are however a number of factors to consider which includes the existing gauge corner wear, crown wear, ultrasonic testing history of the section where the rails were lifted, battered rail ends, rail kinks, rail running surface condition, etc. Standards should be available that specifies the acceptable tolerances for the transposing of rails.

Figure 40: Robel Rail Train

Figure 41: Robel Rail Threader

2.12. Rail Stress, Destressing and Neutral Temperature

Steel will expand when the temperature increases and will shrink when the temperature decreases. Every material has a different response to temperature, which is characterised by its linear thermal expansion coefficient ≤αL). The thermal expansion coefficient represents the amount that the material expands/shrinks (ΔL = change in length) per each degree increase/decrease (ΔT = change in temperature). The change in length (Figure 42) can be calculated using:

$$\Delta L = \alpha L \times \Delta T \times Lo$$

This is not a problem in fish-plate jointed tracks because the gaps in the joints provides for expansion and contraction of the rail whereas with continuous welded rail (CWR) these thermal expansion gaps of the joints are no longer there.

When the rail was laid, it was stress free at that temperature when the closure weld (the last weld to form CWR) was made. When the temperature increases the rail will be under compressive stress. If the compressive stress in the rail exceeds the track's resistance to lateral displacement, the track will buckle, also called a temperature kick-out (Figure 43). The required stress for this to happen would be lower where the ballast is insufficiently stabilised or compacted, where there is insufficient ballast in the cribs or insufficient ballast at the sleeper shoulders. Where kick-outs have occurred, the rail must be cut and destressed for at least 500 metres on either side of the kick-out.

At low temperatures the restrained CWR rails will cause tensile stresses in the rail. If there are any internal rail cracks or rail surface stress raisers, this would be where the rail can break (Figure 44) on a cold night.

Compressive forces can also be caused by heavy axle loads on downgrades as a result of rail creep and/or rolling action of the wheels (causing compressive forces at the one end and tensile stresses at the other). Rail creep may also occur on steep upgrades due to the high tractive forces of modern locomotives.

Therefore, after certain maintenance activities such as ballast cleaning, sleeper replacement, etc. or when rail creep is observed or where rail stress is measured as a preventive maintenance activity, the rails must be **destressed.** During ballast cleaning special attention should be taken against kick-outs when rail temperature falls outside of the neutral temperature. It is imperative to have a good geometric position of the track in level and line and the ballast bed according to regulations before destressing starts.

Destressing of rail involves the cutting of the rail using a rail cutter that will ensure perfectly square cutting of the rail ends (see Figure 45) and loosening all the fastenings over the affected distance. Rollers are used to roll out the rail to remove the stress.

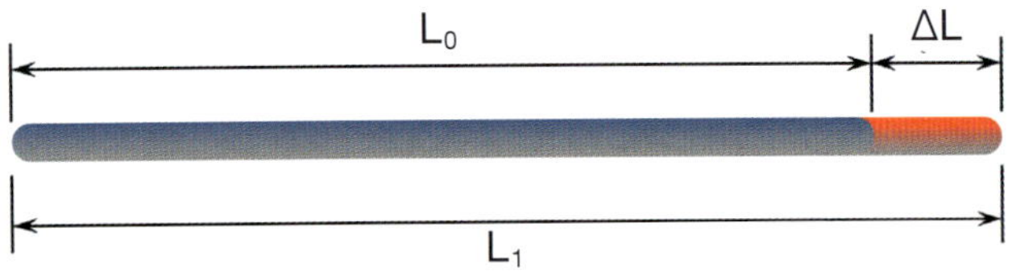

Figure 42: Change in Length of Rail due to Increase in Temperature

Figure 43: Temperature Kick-Out under Compressive Stress

Figure 44: Rail Break under Tensile Stress (Source https://commons.wikimedia.org)

The rail is then fastened to the sleepers and welded during the rail's "destressing temperature" range, also called the **"neutral rail temperature".** Specific procedure must be followed during rail temperature measurements.

The neutral temperature range is roughly midway between the extremes experienced at that location where the rail stress should be close to zero. This temperature would be different for different geographic areas and should be available in a maintenance standard for the specific area in which the line is situated. If the neutral temperature range is impractically short, rail tensors (Figure 46) can be used to artificially stress the rail to what the tensile stress would be at that temperature below the neutral temperature. This requires careful calculation of the gap for the closure weld.

In tunnels, further in than 60 m from the mouth, no attention needs to be paid to the temperature.

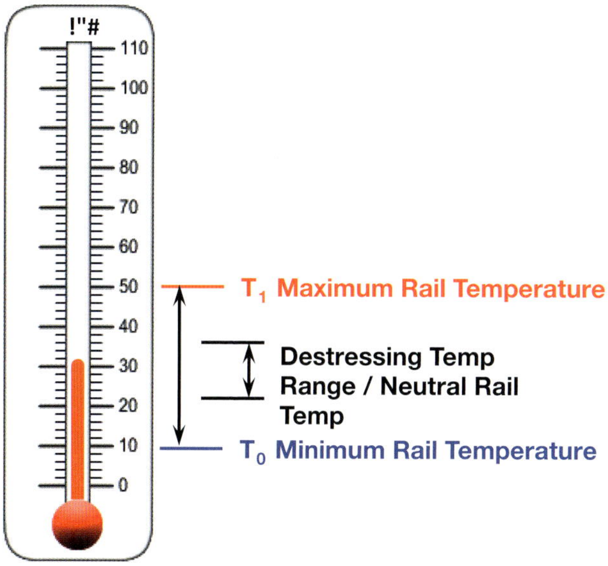

Figure 47: Destressing Temperature Range/Neutral Temperature

Figure 45: Rail Cutter

Figure 46: Rail Tensor

Where rail tensors are not used, rail maintenance activities such as destressing, rail replacement, cutting out of rail defects etc. must be planned in such a way that the closure weld, the last weld, can be carried out during the neutral rail temperature to leave the rail in a stress free state. The problem is that this neutral rail temperature is reached for a very short period of time of between 0 and 3 hours every day, depending on the location, season and climatic conditions.

The closure weld between the two sections of CWR is traditionally carried out using the Thermite welding process. Despite great strides in Thermit welding material and practice, a Thermite weld will still have up to a third of the fatigue life of a flash butt weld and prone to breaking during tensile stresses once fatigue has set in, thereby increasing the track lifecycle cost.

However, the latest flash butt welding machines are capable of welding a closure weld even at temperature below the neutral temperature without the need for additional tensor equipment using software to compensate for the temperature and the burn-off of the rail during welding. Refer to CHAPTER 17 for more information on the flash butt welding process.

2.13. Bending of Rails

Before rails are placed in track, all kinks must be removed as this would be difficult to achieve once the rail has been installed and would be a source of lateral acceleration of the train bogies.

Jointed rail in curves with a length shorter than 36 metres must be bent true to the curve radius to avoid the rail ends from kicking straight causing severe lateral acceleration of the train bogies which in turn may cause discomfort to passengers, shifting of loads, damage and wear to rails and wheels and combined with other track defects, may lead to a derailment. Refer to Figure 48 for a classic example.

Rail is bent using equipment designed for this purpose (Figure 49) which can be either hand or hydraulically operated.

2.14. Rail Lubrication

2.14.1. Objectives

The objectives of curve lubrication are to:

- reduce the friction between the rail and the wheel (CHAPTER 4 paragraph 4.5);
- reduce rolling contact fatigue of the rail gauge corner and the running surface due to lateral acceleration of the wheelset;
- reduce wear (see paragraph 2.18.2);
- reduce energy usage; and
- to improve vehicle steering.

Rail lubrication is normally required for all curves up to a radius of 1,200 metres. Lubricators should also be placed on curves where the offset on a 10 m chord is 10 to 15 mm, or just before the point on the rail where there are signs of contact between wheel flanges and the rail.

The lubricant separates the two surfaces. The applied load of the objects is carried by pressure generated within the fluid due to the frictional viscous resistance to motion of the lubricating fluid between the surfaces. It is important to select appropriate grease approved for the purpose together with a proven lubricator.

2.14.2. Wayside Lubricators

Various models are available that are capable of holding different volumes of grease. Rail lubricators must be filled to ensure that they never run dry and maintained to ensure their functionality. Lubricant must be deposited approximately 20 mm below the running surface of the rail crown.

The units are generally installed in the tangent track before a curve or series of curves.

For mechanised maintenance rail lubricators must be removed and replaced by qualified personnel.

Figure 48: Typical Example of Rails not Bent in Curves

Figure 49: Rail Bender

2.14.3. Mobile Lubrication

Lubrication equipment may have been installed during the initial construction of a line, but is often neglected and very seldom will one find a working unit in many developing countries. The theft of the grease is often part of the problem. Where these problems are encountered it is recommended that a road/rail vehicle be used with equipment to pump grease onto the rail in curves. It is also particularly cost effective for low traffic lines.

It is usually done using a light delivery vehicle specially fitted with the lubrication equipment and road/rail steel and rubber wheels. The vehicle often doubles as an inspection and light emergency repair vehicle. The vehicle lubricates the track at intervals determined by the depletion rate of the grease applied to the line. The applicator system can be operated by the driver of the vehicle, although automatic systems based on GPS data are also available.

The lubrication equipment applies a thin bead of grease (about 0,4 mm) under high pressure on an intermittent basis to the side of the rail. The equipment consists of a grease reservoir, a unit to pressurise the grease, a grease pump and supply lines culminating in an application nozzle sitting in the shade of the wheel flange so as not to be damaged by obstructions like crossing noses or axle counters on the track. Grease application rates are pre-set for a trip and can usually not be adjusted from the cab while the vehicle is running. The vehicle is usually equipped with two cameras, each aimed at one nozzle and two monitors in the cab, showing the driver how the grease is being applied.

Figure 50: Two Types of Rail Lubricators

Figure 51: Road/Rail Rail Lubricating Machine (Transnet)

2.15. Closure Rails

It is often required to install a short section of rail after cutting out a rail break or rail defects. Poor maintenance practice to install a very short section (Figure 52) is common and probably due to ignorance, a lack of training and unavailability of maintenance standards. The shortest length of closure that should be installed is 4,2 metres (South African standards).

2.16. Check Rails and Guard Rails

Check/guard rails are provided on the low leg of curves to prevent excessive wear of the high leg but they increase the curve resistance. Curves of 150 metre radius or less in main or running lines must be check railed. In yards they must be provided where excessive side wear occurs on the high leg or where other conditions call for their provision e.g. where trains tend to derail. On bridges guard rails are installed to keep derailed rolling stock from falling off the bridge, striking the structure or piling up in a tunnel.

Check/guard rails may have to be removed to permit mechanised tamping and replaced again depending on conditions and tamping unit design.

The area between the check rail and the running rail is called the flangeway. This area must be kept clear of ballast.

2.17. Introduction to Rail Defects

Rail defects develop for many reasons such as rolling contact fatigue, dynamic loading, external impacts due to, for example, damaged wheels and ballast imprints which may cause a stress raiser from where cracks can develop.

It is not always possible to establish or categorise rail defects by visual inspection alone and even if the cause and type of defect can be established by visual or laboratory investigation, it is no simple task to group types together since there is a subjective element to it. Spalling for example which is visible as a surface defect may rather be as a result of a subsurface crack that developed due to a manufacturing defect.

Grouping and categorising of rail defects consistently according to a coding system is very important for statistical and rail maintenance management purposes.

In the absence of a system the classification of rail defects proposed by the International Union of Railways (UIC) publication UIC 712 can be used. It will still however require the opinion of an expert in this field.

Rail defects in this manual is however broadly divided between rail wear (paragraph 2.18), rail cracks (paragraph 2.19), damaged rail (paragraph 2.20) and broken rail (paragraph 2.21). These can be defined as follow:

Figure 52:
Poor Maintenance Practice

- Rail Wear – Wear takes place as a consequence of the relative motion between the wheel and the rail and involves the loss of material from either or both.
- **Cracked Rail** – A crack can be defined as a gap in the rail material, visible or not, and has the potential to rail fracture if the gap grows in length. Rail cracks can be caused by thermal loading or mechanical loading.
- **Rail Damage** – In the context of this manual rail damage refers to any rail defect that that cannot be classified as wear or a crack and is mostly related to dynamic loading.
- **Broken Rail** – Rail is said to be broken/fractured if it has separated in two or more pieces, or a piece of the rail becomes detached, causing a gap of more than 50 mm in length and more than 10 mm in depth in the running surface.

Figure 53 illustrates the location of some of the typical defects that will be discussed in the following paragraphs.

2.18. Rail Wear

Wear (loss of material) takes place as a consequence of the relative motion between the wheel and the rail (see Fundamentals of Rail and Wheel Interaction in CHAPTER 4). The following wear mechanisms on the rail can be identified:

- Adhesive Wear – This type of wear takes place as a result of wheel burns where extreme high heat is created.
- Surface fatigue – This is the most common form of wear between wheel and rail due to relative slip and creep forces. The fatigued material will eventually lift off the rail surface.
- Traffic Loading – In this context corrugations are listed as a form of wear due to traffic loading.

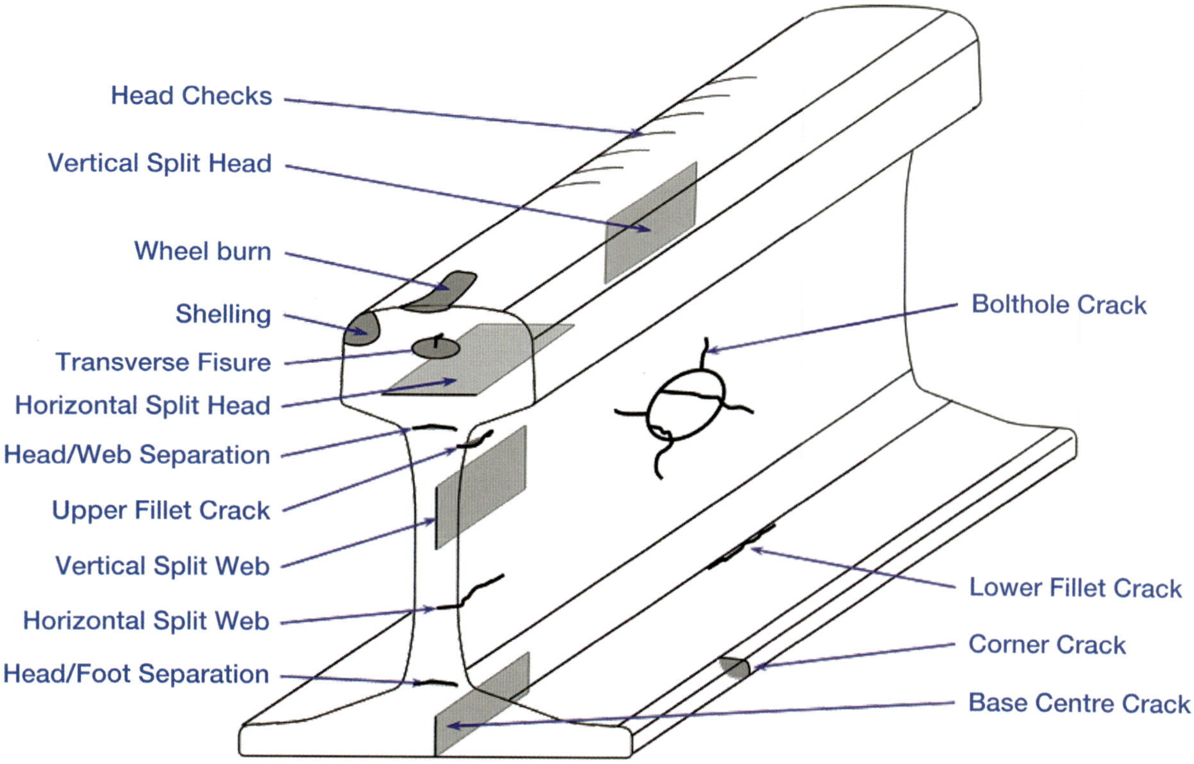

Figure 53: Typical Rail Defects

2.18.1. *Crown Wear/Rail Head Wear/ Vertical Wear*

Rail wear is measured using a rail wear gauge (such as the simple gauge illustrated in Figure 54) or one of various electronic measuring devices available. Crown wear is a perpendicular measurement in the centre of the rail running surface (Figure 55).

Once the maximum wear (according to the standard for the specific railway, rail type and rail profile) has been recorded, the rail must be replaced. Crown wear due to rail/wheel interaction is generally minimal on well maintained and profiled rails and wheels. However, rolling contact fatigue due to the rail/wheel interaction must be removed during rail grinding causing an artificial wear of the crown. Grinding is however essential to get the maximum life from the rail.

Crown wear is one of the parameters that can be measured using an infrastructure measuring vehicle equipped with rail scanning equipment. Refer to paragraph 2.2.

2.18.2. *Side Wear/Lateral Wear/Bevelling*

On tangent (straight) track and flat curves the flange of the rail wheel will not make contact with the rail unless defects of the track or bogie cause a lateral force on the wheel set.

However, when the train enters a tight curve, centrifugal forces will push the flange against the gauge side of the high rail (outside rail) if gauge widening has not been implemented in short radii curves, check rails are not installed and where the rail/wheel contact band is not well defined. If the rail in the curve is not lubricated (see paragraph 2.14) either, the wheel flange contact with the rail will cause damage to the gauge side of the rail crown (see Figure 57). In the presence of sharp wheel flanges as well, this area will be a high derailment risk area.

Side Wear is measured 14 mm below the crown of the rail using the same rail wear gauge illustrated in Figure 54. Once the rail crown has exceeded the maximum wear for the specific rail profile, the rail must be replaced.

It is preferred maintenance practice, however, to prevent the wear through gauge widening, grinding of the rail and the use of high quality lubricants but in terms of corrective maintenance, the rail should rather be transposed before the wear exceeds the usefulness of the rail.

The side worn high leg rail can sometimes be transposed with the low leg rail but in the event of severe side wear of the high leg, metal flow of the crown on the low leg is often pertinent (see Figure 75), in which case it is better to transpose the curved rail with the tangent track before or after the curve. The bending of the rails must, however, be considered.

Side wear can also be measured using an infrastructure measuring vehicle equipped with rail scanning equipment. Refer to CHAPTER 9.

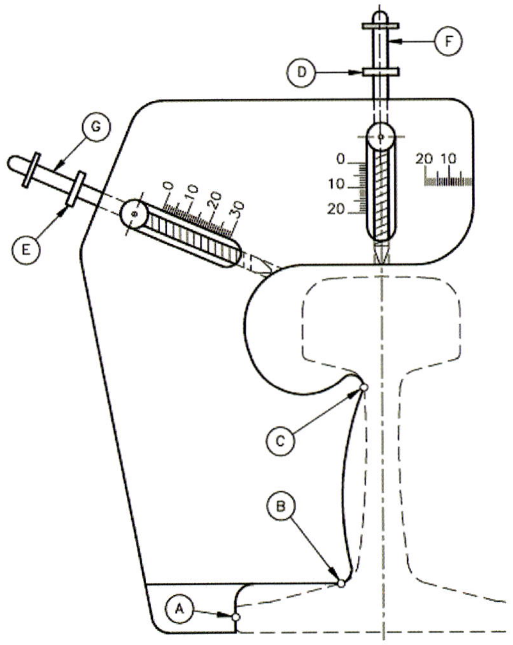

Figure 54: Rail Wear Gauge

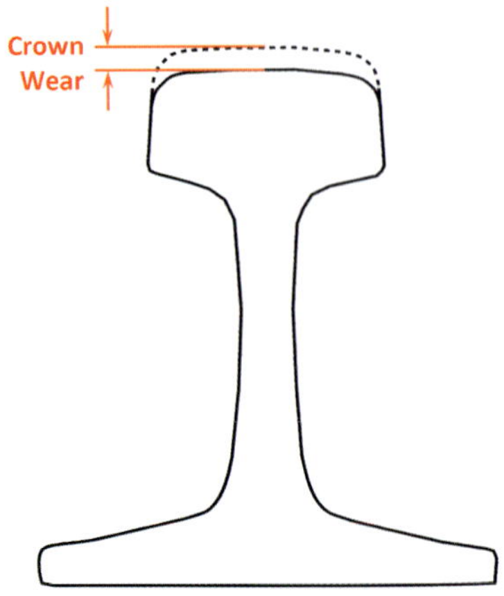

Figure 55: Crown Wear

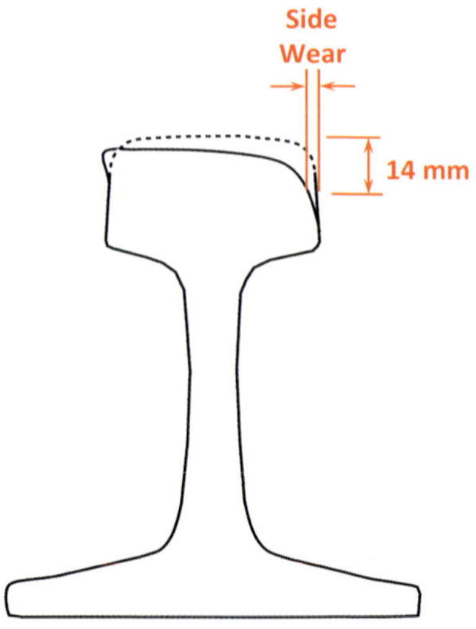

Figure 56: Side Wear

2.18.3. Rail Corrugations

Rail corrugations is a form wear but can also involve plastic flow in severe cases. Corrugations can be of a short-pitch (30 to 90 mm) or long-pitch (80 to 300 mm) in wavelength. Shorter pitch corrugations are more common on light axle loading.

Figure 57: Severe Side Wear on Curves

Rail corrugations are most common in areas where the ballast is severely fouled and therefore exhibit little elasticity. These short wave marks on the rail crown (Figure 58) are the source of noise and increased stress on all the track components due to dynamic loading.

Corrugations are also related to high wheel/rail contact stresses, poor wheel and rail profiles, differential wheel diameters, high bogie suspension stiffness etc. Long-pitch corrugations generally occur on the low leg of curves. Al of these potential causes must be inspected to find the root cause/s.

Rail corrugations can be repaired by re-profiling the rail crown in the longitudinal level with either a reciprocating or rotating rail profile grinding machine but in severe cases of plastic flow rail milling may be required (see CHAPTER 16). The condition of the rails pads must also be inspected and replaced if they have exceeded their life or rail pads of lessor stiffness must be used.

However, unless the ballast is cleaned to reinstate the elasticity of the ballast bed, the corrugations will return very quickly.

Figure 58: Short-Pitch Corrugations

2.19. Rail Cracks

2.19.1. Surface Cracks

There are several types of surface defects which are related to excessive shear stresses at the rail and wheel interface as a result of rolling contact fatigue (RCF). RCF is removed during preventive rail grinding. However, if grinding does not take place or the grinding programme falls short of maintenance practice, small cracks start developing in the work-hardened crown which can grow into the crown over time, eventually causing a rail break with potentially grave consequences as was witnessed with the Hatfield high speed train accident (England) on 17 October 2000 during which 4 people died and 70 were injured.

Another major problem with untreated surface crack defects is that they prevent the ultrasonic rail defect detection signal to detect larger and deeper defects which are more dangerous for potential rail breaks.

There are several causes for RCF and surface cracks related to shear stresses including:

- acceleration and braking of the wheels, i.e. traction and creep forces;
- different radii of wheels on the same axle which will cause slippage of the one or both wheels and also a torsional force on the axle;
- wheelset steering forces in sharp curves;
- vehicle hunting;
- high wheel/rail stresses due to poor rail profile, hollow wheels or poorly profiled wheels, small radii wheels increasing contact stress;
- etc.

The following surface cracks can be identified:

2.19.1.1. Head Checks

High contact stresses occur on the high rail gauge corner on curves causing head checks when the material ductility is exceeded and the material ruptures.

Head checks (Figure 59) occur at about 2 to 20 mm spaced intervals along the gauge corner of the rail crown and can grow 2 to 5 mm in depth. Light head checking will be considered crack lengths less than 10 mm and heavy checking can be longer than 30 mm in length. Once head checks have reached 20 mm and longer in length the probability of downward propagation has already developed with a significant breakage risk. If cracks have reached a depth of 25–30 mm there may be an immediate breakage risk.

Head checks can be seen at a variety of angles and develop into a variety of shapes (straight, elongated s-shaped, V-shaped, etc). If not rectified, they start flaking out (spalling). Areas where the surface has spalled are also categorised as "severe" (Railtrack PLC).

Head checks are more common on high rails in sharper curves where the wheels tend to hunt. On flat curves with a high level of cant deficiency, the head checks will tend to move towards the centre of the crown.

2.19.1.2. Running Surface Checking

This RCF rail surface defect is snakeskin-like pattern on the rail running surface (Figure 60). If rail grinding is not initiated in time to remove the cracks and the RCF, the cracks will "connect" and start breaking out in the form of spalls and flakes that can be 10 to 15 mm wide and up to 3 mm deep.

2.19.2. Internal Rail Cracks

2.19.2.1. Shelling/Spalling

Shelling and spalling (Figure 61) develops at a depth 2 to 8 mm below the gauge corner or contact patch and is visible as dark spots on the surface. It is associated with delamination of the rail head at the depth of the maximum shear stress (see CHAPTER 4, paragraph 4). In advance stages cracks will develop on a horizontal or longitudinal plane (Figure 62) and then either; spall out into a shell causing increasing

Figure 59: Advanced Head Checks and Spalling/Flaking

Figure 60: Snakeskin-like Pattern of Running Surface Checking

dynamic loading of the rail with passing traffic; or turn down and form transverse defects which may result in a rail break. Either way the rail deterioration will grow exponential ending with a catastrophic failure. Shelling and spalling is mostly associated with traffic loading, non-conformal wheel and rail profiles, high adhesion conditions or increased stresses at false flanges.

2.19.2.2. Squat Defects

Squats appear as dark spots (Figure 63) on the surface of the rail crown due to a depression of the rail surface which are caused by subsurface cracking 3 to 5 mm below the rail surface. Squats appear similar to wheel burns. The subsurface cracking can grow transversely down into the head resulting in a rail break.

Squats can also initiate from head checks spreading from the gauge corner towards the running surface on the crown.

Squats are often found at both Thermite and flash butt welds and at areas of corrugations.

2.19.2.3. Transverse Fatigue Cracking, transverse Fissure or Tache Ovale

Transverse cracking (Figure 64) initiate near the centre of the rail head at a depth of approximately 15 to 20 mm and grow transversely towards the crown. They resemble
a kidney in shape.

Transverse cracks develop during the rail manufacturing process or during welding due to the presence of high levels of hydrogen in the rail steel or non-metallic inclusions.

Slow cooling of the rail during manufacture will allow most of the hydrogen to diffuse out of the steel. However, if the cooling is too fast, the hydrogen atoms will combine in inclusions which cause an increase in the internal pressure thereby greatly increasing the risk of crack development in a transverse plane.

Figure 61: Spalling (Source: UIC 712)

Figure 62: Shelling with Longitudinal Cracking (Source: UIC 712)

Figure 63: Squat Defect Visible as Dark Spot

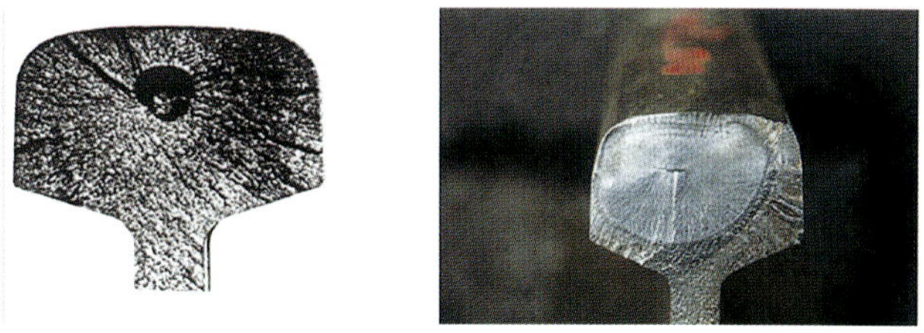

Figure 64: Transverse Fatigue Crack (Source: UIC 712)

2.19.2.4. Vertical Split Head or Longitudinal Vertical Crack

Vertical split head defects are separations along the vertical plane of the rail crown (Figure 65), which tend to split the rail head in two parts and can propagate down into the web. The crack can be very long down the length of the rail potentially weakening a large section of the rail. They normally form near the rail centre line and when they are still small will not be visible without ultrasonic testing.

At an advanced stage of a vertical split it can be visible with the naked eye as widening of the rail head, a depression of the running surface, a line of rust running along the length of the rail on the web/head fillet intersection or a black line running on the rail running surface.

Failure could mean breaking out of halve the crown over some length of rail and if this happens on the gauge side, a derailment would very likely.

Vertical split heads originate from the manufacturing process on older rails produced by means of ingots as opposed to more modern continuous casting processes.

High impact loads, hollow wheels and false flange wheels which applied high loads off the centre of the crown will accelerate crack growth.

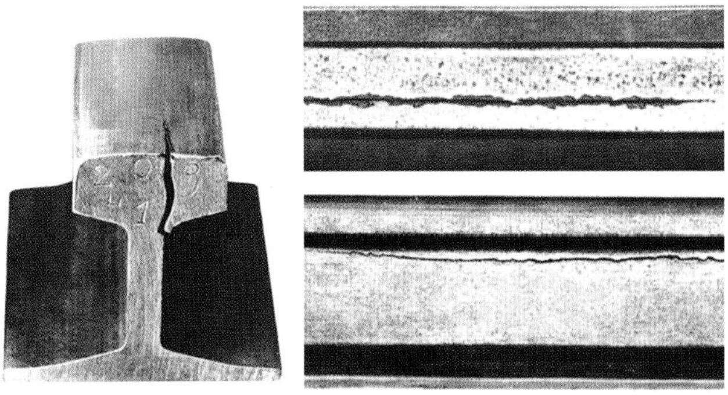

Figure 65: Vertical Split Head (Source: UIC 712)

Figure 66: Horizontal Split Head (Source: UIC 712)

2.19.2.5. Horizontal Split Head or Longitudinal Horizontal Crack

Horizontal split head defects are horizontal separations in the rail crown, which tend to split the rail head along the horizontal plane, at a depth of 10 to 20 mm from the running surface growing horizontally (parallel to the running surface), both across and along the rail head. It can be visible as crushing of the rail end at joints, widening of the rail surface and as the cracks grows, it may be visible running along the length of the rail crown. It may also result in a depression of the running surface which is visible as a dark patch in contrast with the shiny running surface.

This type of defect can be of considerable in length of 50 to 100 mm and cannot be seen when the crack is still small unless ultrasonic testing is carried out.

Failure could mean breaking off of the top halve of the crown over some distance with clear consequences.

Horizontal split head defects are also mostly found in older rails as a manufacturing defect. Clusters of non-metallic inclusions and abnormal vertical service stresses can be the cause of this defect.

2.19.2.6. Head/Web Crack

This is a horizontal crack that develops parallel with the rail/web fillet radius which may curve up or down and lead to rail head separation from the web. Typical causes are wheel flats, poor joints and other rail surface defects that cause high dynamic loads, inclusions in the rail and high residual rail stress (stress induced during manufacturing of the rail).

Figure 67: Crack in the Head/Web Fillet Radius (Source: UIC 712)

2.19.2.7. Vertical Split Web (Piping)

Horizontal split webs tend to separate the web and are also known as piping. These cracks run longitudinally along the length of web and may cause the rail to fracture. It can be observed as swelling of the web

Figure 68: Vertical Split Web (Piping; Source: UIC 712)

2.19.2.8. Fish-Bolt Hole Crack

Cracking that develops at fish-bolt holes and other holes such as rail-bonds. It is also known as star-cracking due to its appearance that radiate from the bolt-hole at 45 degree angles.

These cracks may run towards the head or foot of the rail and is related to poor joint maintenance or where the minimum distance of approximately 100 mm to the rail end has not been observed when the rail was cut to remove the joint for welding the rail into continuously welded rail. The development of these cracks depends on the quality of the drilling and finishing of the hole. The crack will be initiated by traffic loads exacerbated by dynamic loading from rail surface defects.

Figure 69: Fish-Bolt Hole Crack (Source: UIC 712)

2.19.2.9. Web/Foot Crack

This crack develops in the web/foot fillet radius which may grow down and lead to separation between the rail web and foot. It may also grow up towards bolt holes at rail ends. Typical causes are high vertical or lateral loads and high residual rail stress. It may also develop at rail ends. It develops parallel to the web-foot fillet and may curve upwards or downwards causing a fracture.

Figure 70: Web/Foot Separation (Source: UIC 712)

2.20. Rail Damage

The following rail damage is not related to cracking or wear but rather as a result of impact loads or defects during the manufacturing process of the rail:

2.20.1. Long Groove and Line Defects

These are longitudinal manufacturing defects which depths do not exceed a few millimetres.

With a long groove defect a strip of metal of up to several meters in length breaks away from the running surface. A line defect only becomes visible soon after the rail is laid.

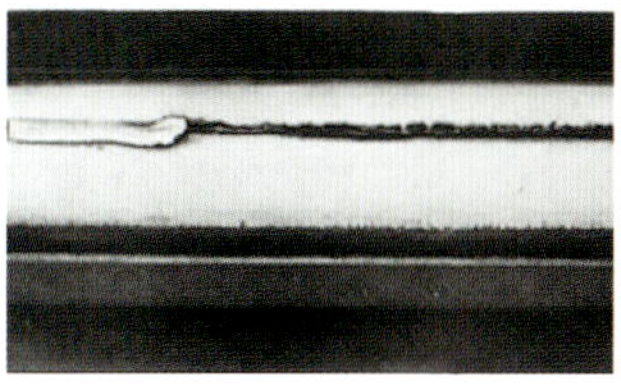

Figure 71: Long Groove Defect (Source: UIC 712)

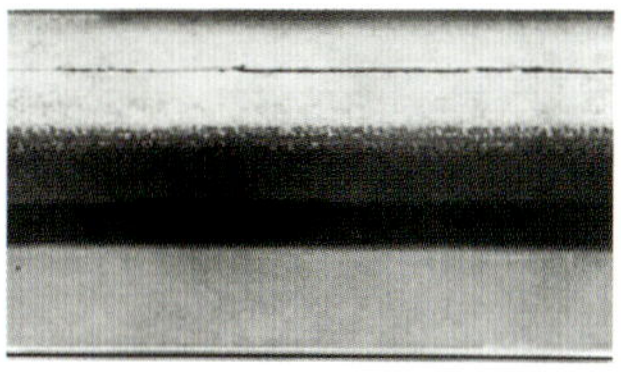

Figure 72: Line Defect (Source: UIC 712)

Figure 73: Battered Rail End

Figure 74: Dipped Joint

2.20.2. Battered Rail Ends and Dipped Joints

The dynamic wheel loading due to the discontinuity at rail joints causes battered rail ends in the absence of effective joint maintenance. Typical examples of battered rail ends can be seen in Figure 73. Battered rail ends must be cropped (cut) taking notice of the minimum space between the rail end and the first fishplate hole should the rail be welded together.

The dynamic loading will also tend to bend the rail ends causing a dipped joint (Figure 74), also called a hogged rail. Dipped joints can be bent out using equipment designed for the purpose.

2.20.3. Metal Flow

Metal flow (plastic flow) is common where the rail profile has deteriorated and the crown of the rail becomes flat. The rail wheel contact area becomes vague, especially if wheel profiles are also hollow. The wheel tends to hunt across the rail profile causing metal flow to develop on the rail crown. If this is not addressed, the metal flow can start breaking off (Figure 75) and even cause a crack to develop into the crown.

If the rail profile has been allowed to deteriorate to this point, rail grinding would not be capable of removing the metal flow due to the grinding stones not being able to reach at that angle. Generally this would require the rail to be replaced.

Where economics allow, a rail planning machine can be used which, in the case of the SBM 140 used in South Africa, is equipped with a burr cutting wheel designed for this purpose (Figure 76). See also CHAPTER 16.

Figure 75: Metal Flow

2.20.4. Wheel Burn

Wheel burns are caused by continuous slipping (spinning) of the locomotive wheels on the rail when they exceed their adhesion limit. The result is either gouging of the rail surface or the rapid increase in the temperature on the rail surface and the fast cooling that follows which may form martensite (a hard and brittle phase). The martensite will tend to crack and break out of the rail crown or form a transverse defect.

Wheel burns must be treated with the same urgency as a rail break. It causes severe dynamic forces to be transferred to the sleepers, ballast and formation and will cause premature failure of these components.

Wheel burns less than 1 mm deep can be repaired by grinding, using equipment such as shown in Figure 35. If the damage is up to 6 mm deep, the wheel burn mark must be properly grinded out, checked for cracks, welded up according to maintenance practice standards and the rail profile reinstated with grinding. Damage deeper than 6 mm must be cut out and a closure rail of minimum 4,2 metres inserted.

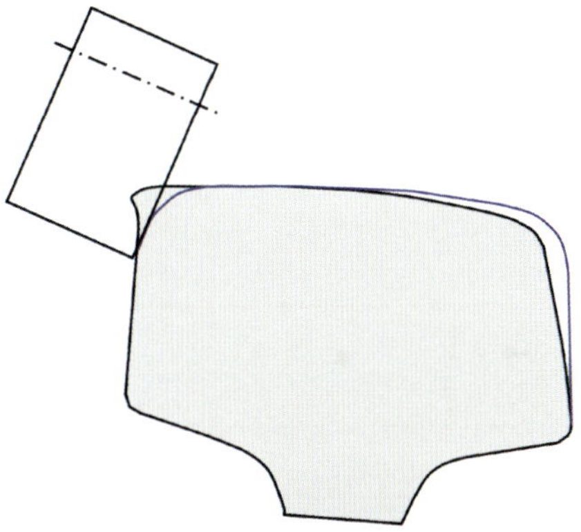

Figure 76: Burr Cutting Wheel on the SBM140 Rail Planing Machine used in South Africa

2.20.5. Crushing

Crushing of the rail head is caused by traffic loading which appears as a depression of the running surface and a widening of the tread. In an advance stage it may appear like metal flow with a lip protruding over the head. It may be as a result of cracks below the rail running surface.

Figure 77: Wheel Spin Damage

2.21. Rail Breaks

Rail breaks are very dangerous and known causes of derailments. Rail breaks can have many causes such as internal inclusions and cracks, rail surface cracks, old fatigued rails, poorly made exothermic or even flash butt welds, rail stress (see paragraph 2.12), etc.

If a rail break occurred due to a tensile stress, the track must be destressed first for at least 80 sleepers on either side of the break before a closure rail is inserted and welded.

Figure 78: Crushing of the Rail Head

2.22. Prevention and Control of Rail Defects and Rail Breaks

The prevention of external rail defects will depend on the cause of the defects which may require one or more of the following actions after proper investigation:

- Prevention starts with inspection which includes periodic visual as well as ultrasonic (see Figure 79 and CHAPTER 10) and or eddy current examination to detect external and internal rail defects in the early stages of development in order to control it.
- X-ray testing of Thermite welds after installation.
- Installation of rail steel grades more suitable for the application such as head hardened rails which have a higher resistance against plastic flow and higher shear stress limits preventing surface cracks.
- Regular geometry correction by tamping will prevent dynamic loading and improve vehicle dynamics.
- Maintaining ballast fouling to within standards by ballast cleaning and maintaining the cross-sectional profile with ballast regulating machines will retain the elasticity of the ballast bed and reduce shock loads on the rails.
- Regular rail profiling such as grinding will remove the RCF layer preventing cracks by improving the wheel and rail interaction and remove surface cracks that have already developed.
- Regular wheel maintenance will also improve the rail wheel interface with less lateral movement, preventing flange contact with the gauge corner, preventing false flanges exerting high loads on the extremities of the crown, etc.
- Rail lubrication where required such as at sharp curves, turnouts etc. will reduce the rail surface stress. However, if checking is not controlled, grease can actually increase crack propagation.
- Where especially internal rail flaws have been detected, such defects exceeding the limits set must be removed and a closure inserted. Should it not be possible to remove the defect immediately, the area of the defect may have to be secured using clamps such as the Robel rail clamp in Figure 80 until the defect has been removed. If high speed Robel clamps are not available and g-clamps with joggle plates are used, speed restrictions have to be imposed.
- Internal rail defects must be removed as a matter of urgency to avoid potential rail breaks. It is important to follow the correct procedure for the installation of a closure rail.

Figure 79: Ultrasonic Rail Flaw Detection Vehicle

Figure 80: Robel Rail Clamp

3. SLEEPERS

3.1. Functions

The sleepers (also referred to as ties in the US) have the following functions:

- receive the load from the rails and distributes it over the supporting ballast;
- provide resiliency and absorption of some impact though concrete sleepers require pads between the rail base and the sleeper to provide a cushioning effect;
- hold the fastening system to fasten the rail to the sleeper; and
- maintain the track gauge.

3.2. Life Expectancy

Concrete sleepers in particular do not require maintenance. However, the condition of the rest of the track structure has a deciding effect on the life expectancy of a sleeper.

On well-maintained lines they may have a life expectancy of up to 50 years but are very susceptible to damage where:

- rails have surface defects that cause impact loads (see Figure 81), more so on concrete sleepers though;
- ballast are very fouled causing return shock waves, breaking sleepers; and
- the fines in the ballast bed got hard after having been saturated causing centre binding (Figure 82).

External damage is also caused to sleepers due to poor maintenance of rolling stock. Hotboxes (overheating of the bearings on axles), may cause the axle to burn off and derail. One derailed axle may not be detected by the train driver until many sleepers have been damaged (Figure 83).

3.3. Sleeper Types and Designs

Mixing of different types of sleepers or fastenings in the same sub-section, must be avoided. The following are the more popular sleeper types in use.

3.3.1. Steel Sleepers

Steel sleepers are more common on old lines and lines with a light structure such as rails of 30 to 40 kg/m and shallow ballast. Steel sleepers are often used for specialised plant locations or areas not favourable to the use of other sleepers types such as tunnels with limited clearances.

Shovel packing of steel sleepers are more problematic and tamping of sleepers are often very slow, of differing quality due to a hard sausage-like core forming in the hollow of the sleeper and the rounded outer ends of the sleepers that are buried in the ballast resisting alignment. The ballast bed at the end of steel sleepers must therefore be opened up to expose the sleeper ends and the ballast must be loosened before re-alignment is attempted.

Figure 81: Damaged Sleeper due to Impact Loads caused by Rail Surface Defects and Hard Ballast

Figure 82: Centre Binding Causing Sleepers to Break in the Centre due to Fouled Ballast

Steel sleepers are also not cheap anymore and are also susceptible to damage such as corrosion (Figure 85) and even bending due to wheel derailments.

Steel sleepers do not damp impact loads very well and will therefore cause crushing of ballast much quicker than timber for example.

On electrified lines insulating pads must be used and keeping the ballast clean to reduce conductivity is of greater importance than with other sleeper types.

Steel sleepers are not a common choice for new or renewed lines today.

3.3.2. Timber Sleepers

Timber sleepers used to be very popular due to the availability of hardwood. Softwood sleepers and laminated sleepers are also in use but have a limited application. However, environmental pressure and timber becoming less available made this type of sleeper less popular for use with new construction or track renewal today.

Despite the greater flexibility and resiliency of timber, it is susceptible to rotting and breaking which diminishes its ability to hold gauge (Figure 86 and Figure 87). They can also be affected by pests or weeds. Timber sleepers should therefore be pressure-treated with preservatives to protect them from insect and fungal attack.

Figure 83: Damaged Sleepers due to Wheel Derailments

Figure 84: Steel Sleepers

Figure 85: Corrosion of Steel Sleepers

Timber is still very common on some lines around the world especially older lines and under older turnouts due to the advantage that the position of the base plates can be drilled in the right place without restricting a sleeper to fit into a specific position in the turnout. A wide range of sleeper drilling machines are available (such as seen in Figure 88) that simplify this process.

Reusing of used timber sleepers are not advisable since experience shows that the part of the sleeper that was previously buried in the ballast deteriorates rapidly once exposed to the air. If second-hand sleepers have to be used it should be inserted the originally orientation.

3.3.3. Monolithic Pre-Stressed Concrete Sleepers

These sleepers are the most popular sleepers today, especially on high speed and heavy axle load lines. Due to their design, weight and great surface area, maximum ballast contact can be ensured with reduced ballast pressure and high resistance to lateral and longitudinal displacement. Rail inclination is also casted in the concrete and is therefore guaranteed.

Concrete sleepers are manufactured with the fastening system attachment casted in the concrete which provides greater reliability. This also has a disadvantage since when concrete sleepers are used in turnouts, the sleeper must be manufactured to fit in a specific position in the turnout, complicating the assembly process. However, for higher axle loading and speeds, these turnouts are far superior in terms of stability, lower maintenance requirements, longer life and lower lifecycle cost.

Figure 86: Timber Sleepered Track in Poor Condition

Figure 87: Rotting of Timber Sleepers

Figure 88: Sleeper Drilling Machines

Figure 89: Track on Monolithic Concrete Sleepers

Due to their weight concrete sleepers can be laid on curves with a radius of 250 to 400 metres (depending on the railway standard) while still maintaining continuous welded rail.

Another consideration for turnouts on concrete sleepers is its weight which is a benefit but require turnout tamping machines with long wheelbases and other specialised equipment to tamp effectively. Refer to CHAPTER 11 for detail on specialised turnout tamping machines.

Concrete sleepers are available for a wide range of axle loading and fastening systems. Different designs are also available for specialised use in tunnels and stations, replacement sleepers for timber sleepers, checkrail sleepers, etc. The cross-sectional shape and weight is dependent on the application of the sleeper.

Concrete sleepers are less resistant to bending and will crack where centre binding is experienced. They are, however, more resilient to wheel derailments. Though wheel derailments will probably require replacement of the sleepers, they will not cause the gauge to narrow or complete loss of gauge easily.

Concrete sleepers are also less resilient to absorb impact loads requiring a specialised pad between the base of the rail and the sleeper to cushion and absorb the load. See paragraph 4 below.

3.3.4. Twin-Block Concrete Sleepers

These sleepers can be cheaper and simpler to manufacture and are popular in some developing countries but also still in use in European countries such as France, Belgium and Southern America. Twin-block sleepers consist of two non-prestressed reinforced concrete blocks connected by a rigid steel beam in order to keep the track gauge (Figure 90).

They have a higher resistance to lateral displacement than timber sleepers due to their weight and greater surfaces on the sides of the block. Since the area between the blocks does not have a seat area it also prevent centre-binding. However, they are very susceptible to bending due to wheel derailments or corrosion of the steel beams causing narrow gauge (Figure 91).

3.3.5. Composite Material Sleepers

These sleepers are made of waste material such as ground up rubber tires, plastic containers, etc. Polymers are added to these materials to produce a sleeper that meets the required criteria. Composite material sleepers are not yet widely used.

Figure 90: Twin Block Sleeper

Figure 91: Narrow Gauge Caused by Sleeper Damage

3.3.6. Other Sleeper Developments

Research is still on-going in sleeper design to reduce the load transmitted to the ballast, increase stability and reduce track maintenance requirements. See typical examples in Figure 92.

3.4. Under Sleeper Pads

Related to sleeper development is the growing use of under sleeper pads which is an elastic layer between the concrete sleeper and the ballast. Test results have been very positive as it increases the contact area between the sleeper and ballast, reduces impact loading between the sleeper and ballast and have shown to extend tamping cycles. Testing over 10 years has also shown that the ballast thickness can be reduced and it reduces structure-borne noise.

3.5. Sleeper Spacing

The narrower the sleeper spacing the less the loading on the ballast and the formation will be. However, more sleepers and therefore a greater capital investment would be required. On heavy haul and a heavy track structure, sleeper spacing is approximately 600 to 650 mm. The economic benefit of 600 mm should, however, be carefully considered against the additional capital cost.

On branch lines (secondary lines) and general freight lines a sleeper spacing of approximately 700 mm is adequate. However, if very light rail profiles are used, rail fatigue must be considered and a narrower sleeper spacing should be considered.

Of more importance, however, is to ensure that the sleeper spacing is equal and that the sleepers are squarely positioned. This will ensure that the production of tamping machines are not negatively affected, especially if two, and even more so, if three or four sleeper tamping machines are used.

3.6. Sleeper Replacement

A wide range of mechanised sleeper replacement and track renewal systems are available (CHAPTER 20) which produces unequalled production and precision. However, in the light of labour-intensive versus mechanised maintenance and construction methods in developing countries and small or low capacity railways, the use of these machines would be very difficult to motivate.

A combination of mechanised and labour-intensive methods might be more appropriate. Refer to CHAPTER 20 for typical semi-mechanised track renewal systems, using a combination of various machines and labour to achieve comparable results in quality, albeit at a lower production rate.

Figure 92: On-Going Sleeper Development

4. FASTENING SYSTEMS INCLUDING INSULATORS, BASE PLATES AND RAIL PADS

4.1. Functions

Fastening systems are the connection between the sleepers and the rail and their function is to retain the rail at the correct gauge against the sleepers, to resist the longitudinal movement of the rail (rail-creep), contain vertical movement of the rail and to prevent rail-roll-over. Insulators and rail pads insulate the rail from ground where signalling and overhead traction is used whereas rail pads have the additional functions to resist rail-creep and provide resilience.

4.2. Components of a Fastening System

There are many different types of fastening systems. Figure 93 provides a typical layout of the components of a fastening system (Pandrol e-clip illustrated). The following is a more detailed description of these components:

4.2.1. Clips

Clips may have a working life in excess of 20 years under normal traffic conditions. However, due to corrosion and other elements, even these must be replaced from time to time to prevent unwanted movement of the rails, especially on lines near the coast and those facing the sea. An average service life of 12 to 20 years can be expected.

The clips can generally be divided between rigid (such as spikes and coach screws) and elastic (spring loaded clips such as Pandrol and Vossloh clips). Elastic clips will allow and contain vertical upward movement (the lift-off wave in front of the wheel) or compressive forces on the pads or timber sleepers due to the spring loading in the clip whereas rigid fastenings will tend to lose their firmness on the rail due to loosening of the screw by dynamic loading, compression of the pads or timber, etc. Rigid fastenings are less likely to be found on new lines today since they are not suitable for modern track for higher speeds, axle loading, heavier structures or continuous welded rail. The following are some of the more popular elastic clips used:

4.2.1.1. Pandrol e-Clip

The Pandrol e-clip (Figure 94) is extensively used worldwide. The e-clip is very easy and quick to install or remove though the process is difficult to mechanise.

The downside of the "quick installation/removal" is that they get stolen in some countries for scrap metal, requiring railway authorities to weld them in place, defeating the objective.

4.2.1.2. Pandrol Fastclip

The Pandrol Fastclip is probably one of the most popular clips around the world today. This is mainly due to its ease of installation and removal and the fact that the process can be fully mechanised.

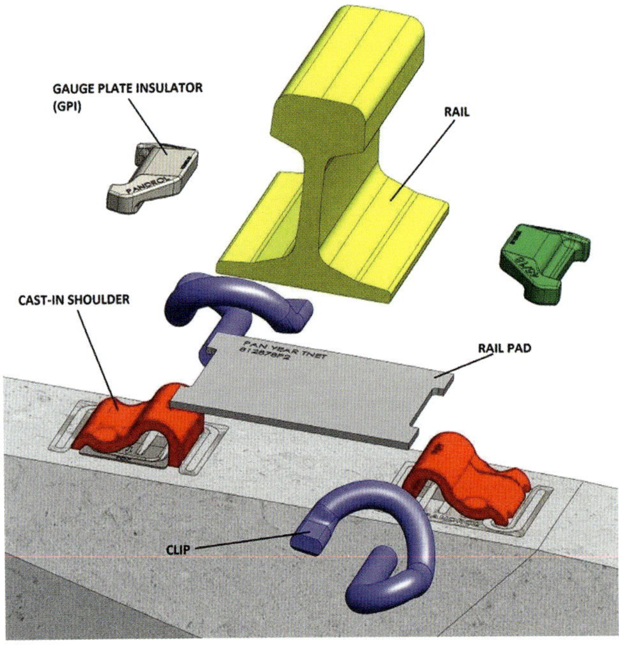

Figure 93: Components of a Fastening System (Pandrol e-Clip) (with Courtesy Pandrol)

Figure 94: Pandrol e-Clip (with Courtesy Pandrol)

4.2.1.3. Vossloh Fastenings
A variety of Vossloh fastening configurations are available for concrete, wood or steel sleepers.

4.2.2. Insulators
Insulators are placed between the clips and the rail on electrified and signalled lines. These insulators also have a finite life and must be replaced at the same time as the clips.

4.2.3. Base Plates
Base plates provide a uniform bearing surface for the rail onto the sleeper and prevent the rail from cutting into the sleeper.

Base plates are generally shaped in such a way that the correct track gauge is maintained. In the area of the rail foot they present an inclination corresponding to the required rail inclination where such are not designed into the sleepers such as in concrete sleepers.

4.2.4. Rail Pads
Rail pads are made of different materials depending on the application and are located between the rail and the sleepers or rail and base plate. Rail pads have various functions, including providing resilience for the rail and sleeper system, damping of wheel induced vibrations, providing electrical insulation for the track signal circuits, protecting the sleeper and/or rail against wear etc.

Rail pads have a maximum life of around 700 million gross tons for the highly resilient Hytrel pads under heavy haul conditions. However, the lifespan of High Density Polyethylene (HDPE) pads may be as longs as the life of the sleepers.

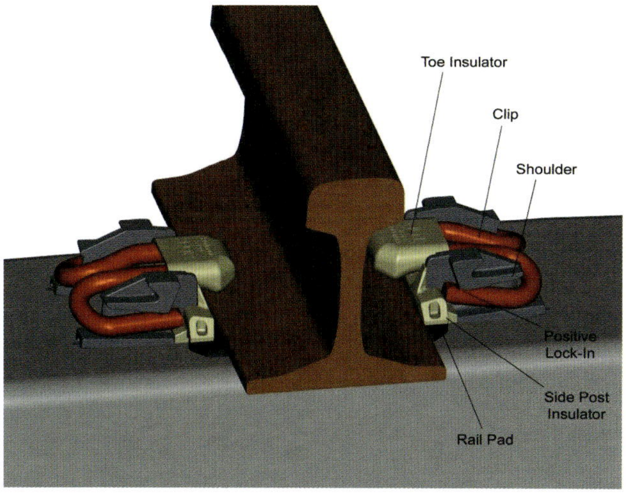

Figure 95: Pandrol Fastclip (with Courtesy Pandrol)

Figure 96: Vossloh Fastening

Figure 97: Studded Rail Pad

Figure 98: e-Clip Installation and Removal Tool

4.3. Attaching and Removing Common Elastic Fastening Systems

The Pandrol e-clips and Vossloh fastening systems cannot be fully mechanised and some hand labour would be required, whereas the installation and removal of the Fastclip on the other hand can be fully mechanised using small machinery such as the one shown in Figure 99. Various sizes of machines are available. Track renewal and sleeper replacement machines can be supplied with automated Fastclip extraction and installation equipment as part of the machine's functions.

The installation and removal of coach screws and Vossloh fastening systems can also be simplified with machinery such as those shown in Figure 100 to Figure 102.

Figure 99: Fastclip Installation and Removal Machine

Figure 101: Impact Wrench

Figure 100: Universal Power Wrench

Figure 102: Precision Power Wrench

5. BALLAST

Ballasted track has remained virtually unchanged for centuries and is still the most cost-efficient and maintainable track design. The ballast has various very important functions (paragraph 5.1). However, to fulfil these functions the ballast material must conform to specified characteristics (paragraph 5.2) and the ballast bed has to conform to a specified profile (paragraph 5.3).

Too little or sometimes no ballast in the track (Figure 103) and or severely fouled ballast may have disastrous consequences if not rectified. This is often due to theft of ballast (as in the case of Figure 103 taken in an African country) but generally due to a lack of proper ballast maintenance such as ballast screening using a ballast cleaning machine. Poor ballast maintenance affects all the other track material as well, which results in a rapid deterioration of the entire track structure. Well-maintained ballast, such as in Figure 104, will guarantee the most cost effective track with the longest possible life.

Figure 103: No Ballast on a Running Line (This Line was Still Open for Traffic)

Figure 104: Good Ballast Profile

5.1. Functions

5.1.1. Resisting Forces Applied to the Track

The ballast resists vertical, lateral and longitudinal forces applied to the sleeper in order to retain the track in its required position. Adequate ballast in the cribs resists longitudinal forces, especially on steep grades for heavy axle loads, whereas adequate ballast at the sleeper shoulders resists lateral forces from traffic or temperature kick-outs.

5.1.2. Elasticity

The depth of the ballast provides the elasticity of the track and reduces pressures from the sleeper-bearing area to acceptable stress levels for the underlying formation. See Figure 105. The pressure induced by passing traffic is distributed over a wider area the deeper down the ballast it goes, referred to as the pressure bulb for its actual bulb-like shape. The greater the area of pressure distribution, the smaller the point pressure per square area.

This is required due to the lower bearing capacity of the formation than the ballast. The bearing capacity refers to the ability of the soil to sustain pressure without shearing.

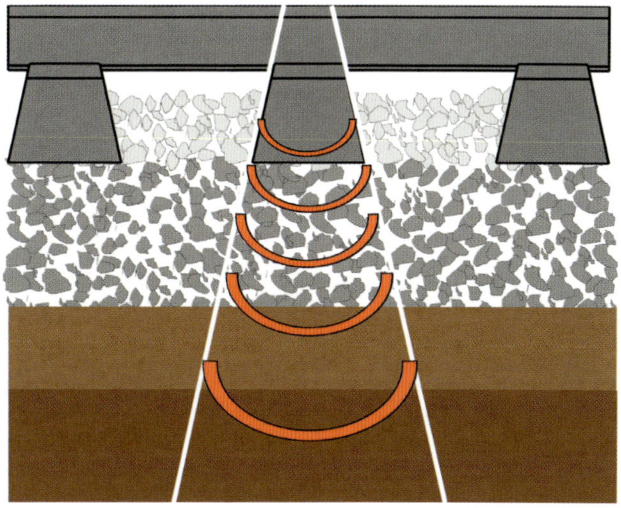

Figure 105: Resiliency Provided by Ballast Bed Thickness

5.1.3. Drainage

The ballast bed provides large voids between the ballast stones for the movement of particles through the ballast bed and to allow effective drainage of water entering the ballast bed due to precipitation.

Effective drainage means the ability to drain water out of the ballast bed to prevent saturation of the formation which will lead to a reduction in the load-carrying capacity and life expectancy of the formation.

5.1.4. Correction of Track Geometry

The ballast allows correction of the vertical and horizontal alignment defects of the track (track geometry) through a process of lifting, levelling, aligning and tamping. Refer to CHAPTER 11.

Geometry correction is required due to the track progressively moving vertically and laterally under repeated loading from traffic, as a result of for example, dynamic loading by flat wheels, rail surface defects, etc. This deviation from the desired geometry is generally irregular, therefore passenger ride quality decreases and dynamic loads increase. This causes rapidly increasing geometry deterioration.

5.2. Ballast Bed Material

The ability of the ballast bed to perform its function is also controlled by the characteristics of the ballast material, which should be specified in a standard specification, to control the supply of stone from the quarry and to assist in ballast condition assessment.

5.2.1. Ballast Stone Specification

Good ballast stone must be specified to have have the following characteristics:

5.2.1.1. Cubic and Angular

A good ballast stone should have as many sides as possible. This will ensure that the ballast stones can interlock and remain in that stable position. Particle angularity increases the shear strength of the ballast bed. It therefore follows that rounded and flat (flaky) stones will not comply with the specification.

Round stones are usually smooth and because they do not have flat surfaces they cannot interlock and are therefore not stable. Rounded ballast decreases the shear strength of the ballast. The stones will slip and change their orientation the moment a force is applied to them.

Flaky stones on the other hand have too large a surface which reduces the void spaces. This will cause a very inelastic ballast bed which will cause damage to the rest of the structure.

Figure 106:
Poor Track Geometry

5.2.1.2. Abrasion and Wear Resistant

Availability and economic considerations have been the prime factors considered in the selection of ballast material. Therefore a wide variety of materials have been used for ballast.

However, good ballast stone must be hard and wear resistant to resist breaking down into fine particles due to crushing from impact loads by traffic and causing fouling of the ballast bed. Hard rock (granite, basalt, diabase etc.) is most suitable as the compressive strength of soft rocks (limestone, dolomite, sedimentary rocks etc.) is inferior.

5.2.1.3. Broadly Graded and Free of Fine Material

Broadly graded refers to a mixture of small and larger stones which will ensure effective interlocking between the stones, increasing the shear strength of the ballast bed. However, to ensure effective tamping, the maximum area of contact between sleepers and ballast, and to ensure the stability of the track, it follows that there should be a maximum ballast stone size. Furthermore, to ensure effective drainage of the ballast bed, the void spaces between the ballast stones must be large enough and clear of fine material that will retain moisture, implying that a minimum ballast stone size will also be specified. A ballast specification will therefore specify this mixture of small and large stone in a ballast grading curve. See Figure 107 for the ballast grading curve adopted by Transnet in South Africa for the general freight lines. The grading curve for heavy haul lines is however different in that the minimum and maximum stone sizes are larger.

On the horizontal axis of the ballast grading curve, standard sieve sizes in use are given. A sieve of, for example, 19 mm can be described as a wire mesh with square holes of 19 mm by 19 mm. It will therefore only allow stones that can fit through a hole of 19 mm by 19 mm.

The vertical axis shows the percentage of a representative ballast sample by mass that may pass through a given sieve. Mass is easy to measure and therefore allows a simple method of ballast grading measurements as illustrated in Figure 108.

For example, from the ballast grading curve it is evident that if you have a 100 kg ballast sample that complies with specification, at the large end of the scale all 100 kg (100%) must pass through the 63 mm sieve and at the low end of the scale, a maximum of 0.25 kg (0.25%) may pass the smallest sieve, being 13.2 mm to comply with specification.

The amount of fine material in the ballast will determine how fouled the ballast is i.e. the mass of fine material as a percentage of the sample smaller than that specified. A representative ballast bed according to the blue line in the ballast grading curve would therefore represent a ballast bed that is 0% fouled.

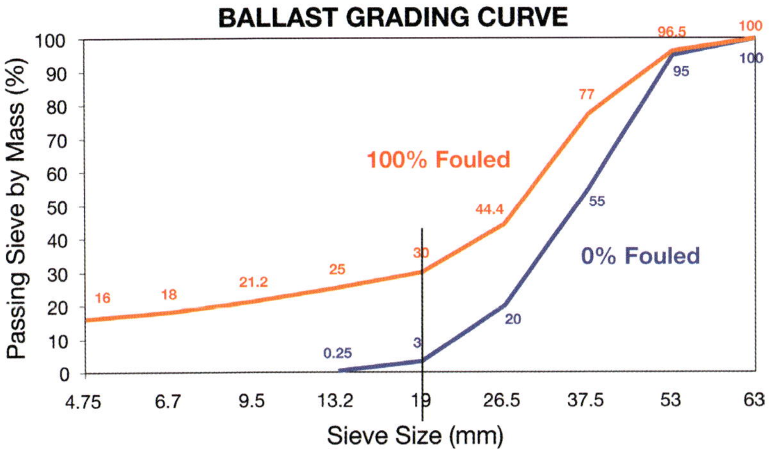

Figure 107 : Ballast Grading Curve (Transnet for Non-Heavy Haul Lines)

Figure 108: Simple Method of Determining Ballast Fouling

5.2.2. Deviations From the Desired Ballast Material Characteristics

The ballast layer is consistently subjected to contamination or fouling by fine material. There are several reasons or sources for this contamination, the extent of which should be measured and once it exceeds a maximum level, ballast cleaning (screening) is required to ensure the long term integrity and life of the entire track structure.

5.2.2.1. Causes of Ballast Fouling

- **Operational loads** – Before passing over a sleeper, the wheels rolling over the track make the sleeper ahead lift off slightly. The moment the wheel passes on the sleeper, the sleeper rebounds abruptly on the ballast, which causes ballast splintering at the contact points, re-arrangement of the stones and friction processes leading to a rise in the portion of the fines.

- **Internal degradation of the ballast bed caused by ballast fracture and abrasion –**
  - due to flat spots on wheels of rolling stock (dynamic loading);
  - rail surface defects such as locomotive wheel skid marks;
  - tamping; and
  - chemical weathering.

- **External infiltration of alien fines –**
  - from the surface (windblown or washed in during heavy rain); and
  - dropped from trains such as coal dust.

- **Infiltration from underlying granular layers –**
  - due to a shallow ballast layer which inadequately distributes transmitted wheel loads onto the formation causing over stressing and rising of fines into the ballast bed;
  - softening of the subsoil as a result of an insufficient formation cross-fall trapping water or fouled ballast retaining moisture eventually causing saturation of the formation as well; and
  - subballast particle migration from inadequate gradation.

5.2.2.2. Measuring the Extent of Ballast Fouling

The extent of ballast fouling is a theoretical value generally expressed as a percentage of fines of a sample by mass. On the ballast grading curve in Figure 107, the red line represents a theoretical 100% fouling. It therefore follows that it is possible for the ballast to be more than 100% fouled.

Various methods for establishing the extent of ballast fouling have been developed. They include lab tests, visual measurements and a simple sieve test, such as the method shown in Figure 108, whereby a 19 mm sieve is used and the fines that pass the sieve are weighed and expressed as a percentage of the total sample. With the simple sieve test, if the weighed fines are 30% of the sample weight, the ballast is 100% fouled (Figure 107).

In practice the finer material will vibrate down to the bottom of the ballast bed where it will start to fill the void spaces between the ballast stones and gradually build up to the bottom of the sleeper. The height to which the ballast bed is filled with fines is also a direct measurement of the degree of fouling of the ballast bed.

Figure 109 shows a ballast bed which is approximately 10 to 20% fouled. If a template is made which is cut to the design depth of the ballast bed, the fouling will also be visible with the fines filling the bottom 10 to 20% of the template.

Figure 109: 10 to 20% Fouling

Figure 110: 100 to 120% Fouled

The ballast bed in Figure 110 is 100–120% fouled. The whole ballast bed from the formation to the bottom of the sleeper consists of almost only fines. This ballast bed will have no or very little elasticity and no draining ability. Of note is the clean ballast in the upper levels and in the cribs which may give the appearance of a clean ballast bed, if the ballast bed is not opened from the side and the ballast directly beneath the sleeper inspected. It is therefore important that a representative ballast sample for fouling analysis be taken correctly according to laid down specifications.

Producing a representative sample is however very difficult as upper layers of the ballast bed will normally have larger stones and if these fall into the sample, a distorted outcome will result. A recommended method that can be used is sampling via the box test which is described in the European Rail Research Institute (ERRI) report D184 (Unified Assessment Criteria for Ballast Quality and Methods for Assessing the Ballast Condition in the Track), document RP2 (Assessment of Ballast Condition in the Track).

5.2.3. Maintenance of Ballast Bed Material

The point at which ballast must be screened to reinstate the specified characteristics is called the critical fouling point. Ballast screening involves the removal of the fine material from the ballast bed and replenishing it with new ballast.

The critical fouling point (as described by Esveld, see references) is at that point at which it can be shown that the fines in the ballast are likely to cause either an unacceptable loss of geometry retention, an inability to tamp effectively, or, due to water retention by the fines, are causing damage to the ballast and formation.

Very often fouled ballast goes hand in hand with other deviations from the required design criteria of ballast, such as too little ballast depth, rounded and smooth stones etc. The result is that damage will be caused and geometry retention lost long before the ballast is 100% fouled. The critical fouling point is therefore a unique value for each unique line condition and varies between 70 and 85%.

When the fouling has reached these measures ballast screening should be a high priority. This process can only be carried out effectively and economically with the use of mechanised ballast cleaning machines (Figure 111 showing the widely used RM74/76). Using hand labour for this purpose has proven to be ineffective since the ballast underneath the sleepers is often neglected and an uneven formation is created by the hand tools used. This unevenness creates areas where water collects leading to saturation of the formation.

Refer to CHAPTER 14 for detailed description of the ballast cleaning process using ballast cleaning machines.

5.2.4. The Effect of Fouled Ballast

The ballast material progressively loses its function as it starts deviating from its design criteria. This loss in function has various detrimental effects on the track structure, maintenance costs and track life if not attended to timely. These include:

5.2.4.1. Prevents Effective Drainage

As the storage capacity (void space) starts to fill with fines, the drainage capacity of the ballast decreases. The fines will hold the moisture which will lead to a wet ballast bed.

Figure 111: RM 74 Ballast Cleaning Machine

Experiments done by Prof Selig showed that the ability of the ballast to drain rain water decreases dramatically with an increase in fouling. His experiments showed that clean ballast can drain rain water at a rate of 150 mm per hour. It was also interesting to note that if the ballast shoulders are cleaned but the ballast under the sleepers are highly fouled, a common practice, the ballast bed can only drain rain water at 35 mm per hour.

If the entire ballast bed is uniformly fouled to ± 75% by volume (fouling index of 30), it would only be able to drain 1,5 mm per hour of rainfall. With a hard spell of rain the ballast bed will therefore remain wet for a very long time.

5.2.4.2. Reduces the Life Expectancy of the Track

Fouled ballast also drastically reduces the life of the track. This can clearly be illustrated by a nomogram (Figure 112) prepared by Dr Mülke in his thesis "Storm Water Drainage of Rail Track".

This nomogram is used to design a new track. The scale on the left shows the number of axle passes the track is designed for. The second scale shows the expected stress on the formation. This figure is known and is dependent on the axle load for which the track is designed. A straight line (the blue line) from the required traffic through the known formation stress will provide a weighted modulus.

This line can then be extended through the track design profile which would be one with clean ballast and a 4% cross fall to provide the required soil characteristics of the formation. The unit of measure for the soil is the California Bearing Ratio.

The California Bearing Ratio (CBR) is a penetration test for the load-bearing capacity of subgrade soils. The pressure required to penetrate a soil sample with a plunger of standard area is expressed as a ratio of the measured pressure required to achieve an equal penetration on a standard crushed rock material. The higher the load bearing capacity of a soil, the higher the CBR rating will be. Moist sand has a CBR of approximately 10 whereas high quality crushed rock has a CBR over 80. The standard material for this test is crushed California limestone which has a value of 100.

This nomogram can now be used in reverse to illustrate the effect of fouled ballast on the life expectancy of the track. For example; with a good quality formation and minimum required cross fall (which is usually not the case) but the ballast is fouled, the line will give a new weighted modulus. Extend the line through the known formation stress (which would also be higher when the ballast is fouled) and it will show the expected life of the track under these conditions. The fouled ballast will reduce the track life from 40 million axle passes to only 17 million axle passes.

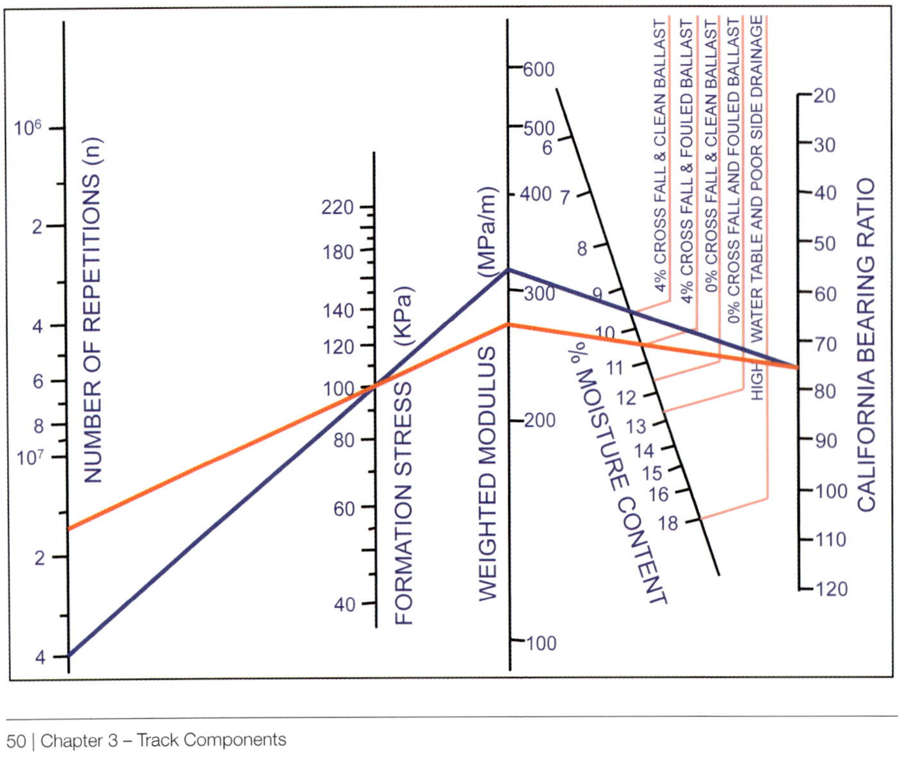

Figure 112: Nomogram used to Determine Track Life

5.2.4.3. Reduces The Elasticity Of The Track

Fines in the void spaces will reduce the elasticity of the ballast. In addition, if the fines in the ballast bed become wet and are allowed to dry again, the ballast bed can be cemented into a rock hard layer.

The resultant lack of elasticity will cause reflected shock waves which is likely to cause rail corrugations and traffic noise. These shock waves will also cause damage to sleepers, fastenings and rolling stock.

Figure 113 illustrates the effect of centre binding due to the cementing of the ballast. The weight of traffic will force the sleeper to bend over a hard centre. The concrete will crack in the middle due to the stress and expose the reinforcing which will corrode and weaken the steel, as well as cause further cracking of the sleeper due to the expanding corroded steel. These sleepers will not be able to hold the gauge for very long and could cause a derailment.

5.2.4.4. Poor Durability after Maintenance Inputs

The lack of void spaces due to ballast fouling will restrict the effective rearrangement of the ballast stones during tamping. The track memory is therefore not removed. The result is that the track will revert to its original geometry in a very short time. This would require the tamping machine or gang to repeatedly return to the same spot with the resultant effect on train operations.

Even if the ballast can be sufficiently re-arranged to remove the track memory, the fines clinging to the ballast surface will still act as a lubricant, thus reducing the friction between the ballast stones, resulting in poor durability of tamping input.

Where labour gangs are used to tamp by hand with pickaxes and hand held vibratory tampers, the repeated tamping will damage the sleepers. See also Figure 224. These sleepers will soon have to be replaced since they will have very little resistance against forces. Replacing these sleepers would require a full occupation, severely affecting train operations.

5.2.4.5. Reduces the Production Rate of Ballast Cleaning

If the fouling in the ballast is allowed to increase to beyond specified levels, mechanised ballast cleaning becomes increasingly ineffective and more expensive as the moisture levels in the ballast increases due to moisture retention by the fines. Even in dryer regions, the ballast bed will tend to become very compacted and hard which will also reduce the production rate of the machine. Any perceived savings for extending the ballast cleaning cycle will be consumed many times over due to lower production rate of the ballast cleaning machine.

In Figure 114 a typical example of ballast with very high moisture content is shown which will clog the screens and may even bring the process to a complete halt.

Not only does it become expensive to ballast clean at such low production rates, but the effectiveness of the process also decreases dramatically. This would mean that the ballast cleaning cycle will reduce, resulting in another track possession (occupation) for the ballast cleaner to return only a few years later.

5.2.4.6. Leads To Formation Failure

If the ballast bed remains wet due to the absorbed moisture by the fines, the formation will also become

Figure 113: Centre Binding of Sleepers

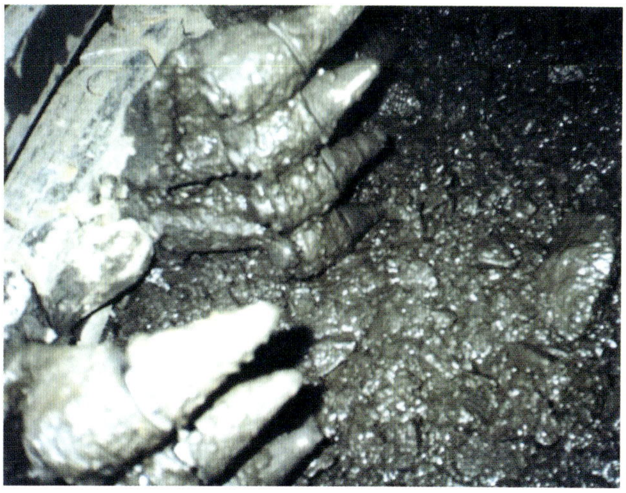

Figure 114: Ballast Cleaner Cutting Chain in Wet Ballast

saturated which will lead to failure. When the formation fails, slurry from the subgrade together with the fines inside the ballast will start pumping through the ballast and will be visible on the surface of the ballast bed. This is often referred to as sleeper pumping, mudhole or wet spot.

In the photos of Figure 115 the mud pumping has reached an advanced stage. Total rehabilitation of the formation now becomes necessary which would require a speed restriction until formation rehabilitation can be carried out.

Formation rehabilitation is a very extensive and expensive process during which the line will remain closed for an extended period. It therefore pays to avoid this scenario altogether. Once the extent of failure has reached this level, mechanised formation rehabilitation is probably no longer possible due to the depth of the damaged formation.

5.2.4.7. Speed Restrictions

Fouled ballast with the results that have been discussed above will cause line speed restrictions resulting in:

- Late deliveries or arrival of goods or passengers which damages the image of the Railways. Lost customers to the roads are difficult to recover. This is an indirect cost which is very difficult to measure.
- Lower rates – Customers use slow deliveries as a bargaining tool for lower freight rates, further reducing the income of the Railways.
- Opportunity cost of volumes – Slower moving volumes have an opportunity cost of higher volumes (more train slots) on busy lines.

5.3. The Desired Ballast Bed Profile

5.3.1. Ballast Bed Profile Specification

For the ballast to effectively perform the aforementioned functions, the ballast bed profile should conform to the following (refer to Figure 116):

- The sleeper should rest inside the ballast with the cribs filled with ballast. A shoulder of approximately 250 to 300 mm (depending on track alignment such as curve radius) is required to resist the lateral movement of the track.

Figure 115 : Mud Pumping

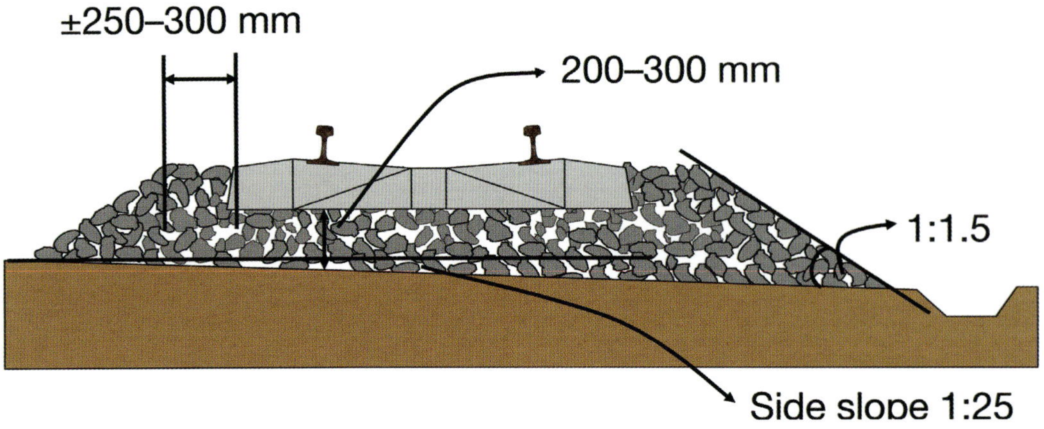

Figure 116: Specified Ballast Bed Profile

- The ballast depth should be kept at between 200 to 300 mm under the sleeper (measured at the lower edge of the sleeper in superelevated curves) depending on the type of line and axle loading. Together with the ballast at the shoulder the ballast volume will therefore be between 1,200 and 1,600 m^3 per kilometre for cape gauge lines (1067 mm).
- The slope of the ballast should be at approximately 34° (1:1.5). This is the natural fall angle of the ballast stones.
- The formation should have a cross fall towards the drain at a slope of 1:25. This is required to direct the water, which drains through the ballast, towards the drain and away from the formation.

5.3.2. Deviations from the Desired Ballast Bed Profile

5.3.2.1. Too Little Ballast

The most common of ballast profile defects is too little ballast under the sleeper, at the sleeper shoulder and/or in the cribs. Refer back to Figure 103. This is due to various factors including:

- Neglect or a lack of ballast profile maintenance. A common cause is the unavailability of track maintenance machinery or inability to maintain the machinery in a working order.
- People and animals walking across the track spread the ballast over a wide area exposing the sleeper shoulders.
- Theft of ballast is a big problem in developing countries due to the inherent value of the aggregate for the building of dwellings.
- The crushing of the ballast by traffic, excessive dynamic loads caused by rail profile irregularities, flat wheels, etc. which causes ballast stones to settle closer together reducing the void spaces and a reduction in ballast bed volume.
- The tamping process, which if not followed by ballast regulating, sometimes lifts the sleeper out of the ballast resulting in no or too little ballast in the cribs and at the sleeper shoulders.

5.3.2.2. Excess Ballast

If the ballast cleaning cycle is too long or the tamping cycle is too frequent the sleepers are continually lifted out of the ballast and new ballast is then added to compensate. This does, however, not have severe track stability implications except for the uneconomical use of ballast and clearance problems to the overhead electrical structure and other fixed structures such as bridges and tunnels.

5.3.3. Maintenance of the Ballast Bed Profile

Both too little and excessive ballast is rectified through a process of mechanised ballast cleaning (CHAPTER 14) and ballast regulating (CHAPTER 13) as part of a ballast management programme.

Ballast management would include the offloading of ballast once it is clear that the ballast volume has been reduced to less than that specified. Ballast wagons with bottom discharging doors (Figure 117) are generally used for this purpose and are often in short supply, though it is a very important piece of equipment for effective track maintenance.

Often, however, the ballast volume is adequate but the ballast has spread over a wide area leaving the cribs empty and sleeper shoulders bare. The ballast bed then needs re-profiling using one of a vast range of ballast regulating machines (Figure 118) available for ballast management which includes machines capable of picking up excess ballast in some parts of the track and offloaded elsewhere where it is required.

5.3.4. Effects of a Poor Ballast Profile

Inadequate support at the sleeper head by the ballast shoulder will weaken the track's lateral resistance to displacement. On continuous welded rails this can result in temperature kick-outs.

If the fines in the ballast are saturated, the ballast will have a decreased resistance to shear deformation due to the clayey or silt particles and water present at the ballast contact points. The fines will act as a lubricating agent that leads to a decrease in track support and geometry.

Figure 117: Ballast Transportation Wagons

Figure 118: Ballast Regulating Machine

6. FORMATION

The next track component is the subballast and subgrade levels (refer back to Figure 12) which are collectively referred to as the formation.

6.1. Functions

6.1.1.Subballast

The subballast is a placed soil that consists of a broadly-graded naturally occurring or processed sand-gravel mixture. The functions of the subballast are to prevent upward migration of fine material emanating from the subgrade and to allow drainage of water away from the subgrade. In many instances the subballast has been omitted during the construction phases of many lines, which became the cause of many track defects later on in the life of the track.

6.1.2. Subgrade

The subgrade is the platform on which the track structure is constructed. Its main function is to provide a stable platform for the overlying layers. Sometimes soil can be brought in if the local soil is considered to be inadequate in bearing capacity or to raise the platform to the required elevation for the rest of the track structure.

6.2. Formation Failure

The formation cannot be maintained. Its functional life expectancy and lifecycle cost is dependent on the initial quality of construction and material used and the level of maintenance to the other track components as described above. The lack of ballast maintenance is the biggest cause of formation failure (see paragraph 5.2.4.6 above). The presence of water and the presence of clay are two of the main causes of formation failure. Following is an explanation of the causes and prevention of both.

6.2.1. The Presence of Water

Three sources of water entering the track can be identified being precipitation (rain), surface flow and subsurface seepage. To look at the three sources individually, a typical example of a track in a cutting is illustrated in Figure 119.

6.2.1.1. Precipitation

The ballast bed covers and protects the formation. As mentioned in paragraph 5, one of the functions of the ballast bed is to provide immediate drainage of rain water, which is only possible if the fine material in the ballast is limited to enable the drainage of water as fast as it enters or at least faster than the absorption rate of the formation.

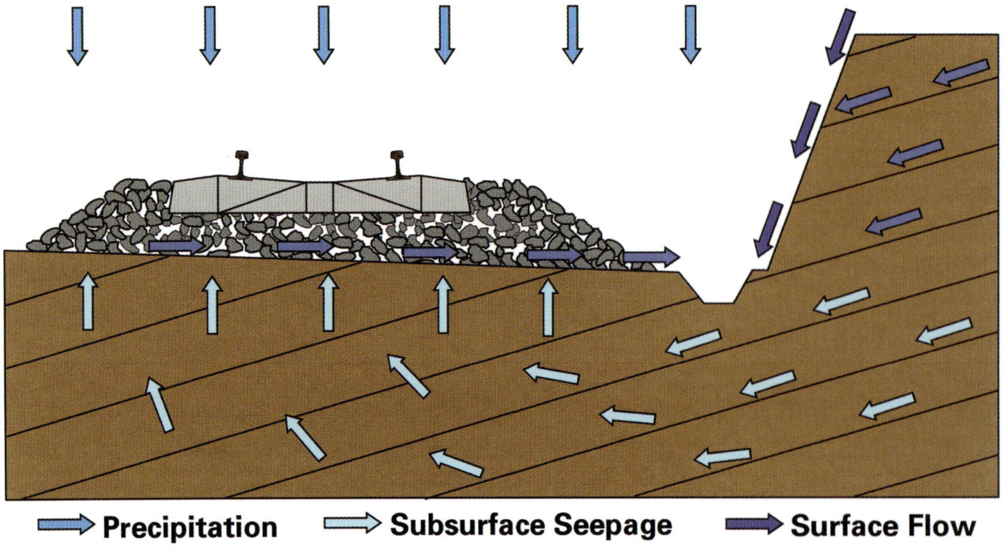

Figure 119: The Sources of Water Entering the Track

As the fouling of the ballast increases, the ballast bed will retain the moisture for longer and if this period is longer than the absorption rate of the formation, the formation will become saturated as well.

This will cause failure of the formation due to its reduced shear strength as well as an increase in formation stress. When the formation fails, slurry from the subgrade together with the fines inside the ballast will start pumping through ballast and will be visible on the surface of the ballast bed (refer to paragraph 5.2.4.6 again). The whole track structure will start sinking into the mud slurry with the obvious danger of a derailment.

To prevent the ballast from retaining moisture and saturating the formation, the ballast must be screened using mechanised ballast cleaning machines.

6.2.1.2. Surface Flow

A cutting is likely to cut through layers of sediment (one example is illustrated in Figure 119). These layers are channels for transporting water. Where the sediment layer exits on the side of the cutting, water being transported in the sediment will seep through, running down the cutting towards the track.

Rain will also collect on the surface of the cutting and start flowing down the slope towards the track.

This water must be prevented from entering the subgrade. A suitable edge drain can be a means of carrying away the water. Edge drains must be well maintained but is unfortunately very often one of the most neglected of track components. Refer to paragraph 9 which covers side edge drains.

6.2.1.3. Subsurface Seepage

The water running along the sediment layers that crosses underneath the track will experience a sudden drop in pressure due to the cutting. The water will tend to seep upward into the subgrade due to hydrostatic pressure.

Here a normal side edge drain may not be adequate. To prevent this seepage, the drain must be cut deep enough or until an impenetrable high density layer is entered Figure 120. The flow of water through the drain must in all cases be greater than the absorption rate of the subballast.

Various other scenarios of subsurface seepage exist which can be remedied by various design subsurface drains.

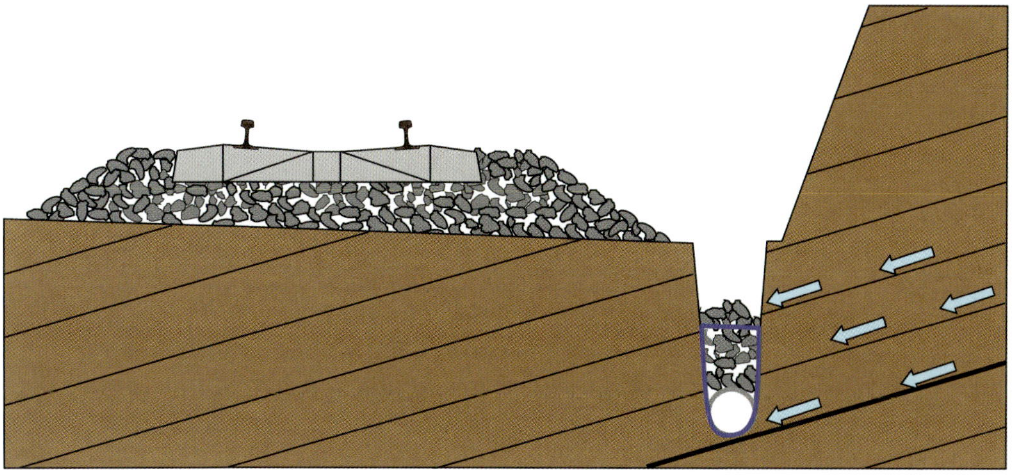

Figure 120: Subsurface Drain Preventing Subsurface Seepage

Figure 121: Magnified Clay Particles

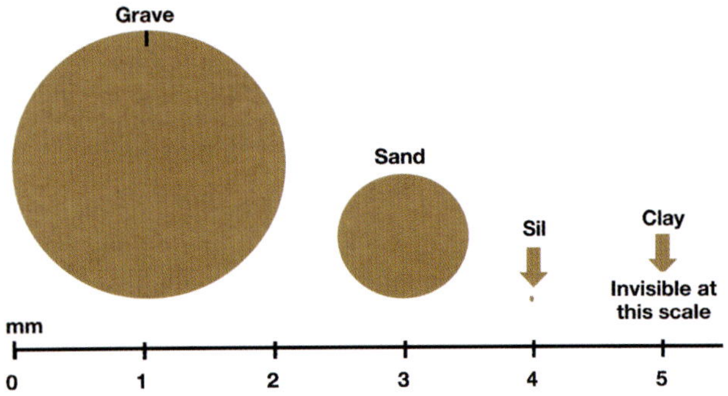

Figure 122: Relative Soil Particle Sizes (Taken from www.pottersfriend.co.za)

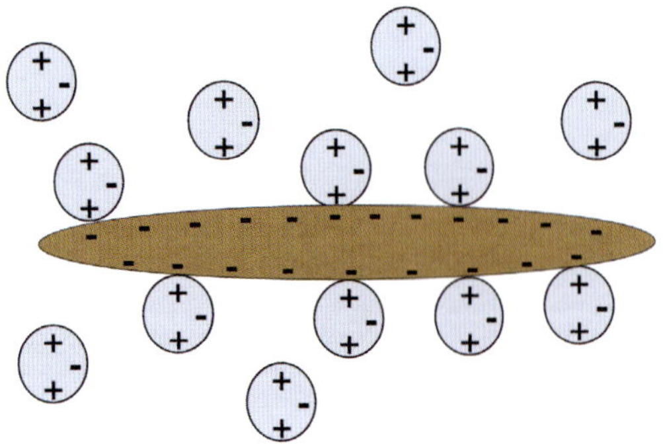

Figure 123: Clay Particles Attracts Water Molecules

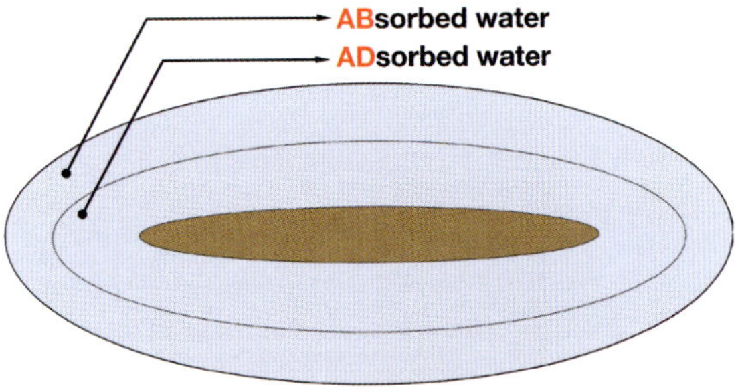

Figure 124: Adsorbed and Absorbed Water Attached to a Clay Particle

Figure 125: Clayey Slurry Pumping Through the Ballast

6.2.2. The Presence of Clay

Clays are formed by chemical decomposition of rock and soil in the presence of water, higher temperature and pressure. The result is plate or tube-shaped particles with a very high surface area to volume ratio (see Figure 121). Clay particles are very small (2 micrometres) and compact into a very dense impenetrable matrix.

The clay particles have unsatisfied negatively-charged surfaces that attract positively-charged molecules (Figure 123). Water molecules are positively charged and if available a film of water will attach itself to the clay particle through an electrochemical process to satisfy the negatively-charged clay particle.

This film of water is called adsorbed water (Figure 124) and can be 80% greater in volume than the clay particle. Adsorbed water can only be removed or the very strong electrochemical bond between the water layer and the clay particle only broken under the following conditions:

- A temperature exceeding 600°C
- A pressure exceeding 550 MPa
- Ion exchange, that is, replacing the water particle with other stronger positively charged ions

A second layer of water will also attach itself to the first layer which is called absorbed water. Absorbed water can however be atmospherically dried due to its weaker electrochemical charge.

Even if the source of water to the substructure can be stopped and the absorbed water be dried, the adsorbed water in the clay will still be present. Due to the swelling of the clay, the capillary space between clay particles will be drastically reduced, restricting the permeability of the soil, with the result that water cannot flow freely to drain the soil. With a repeated supply of water, the soil will become saturated, reach its flow limit and form slurry (Figure 125).

Once this stage has been reached, the only practical solution would be to remove as much of the clay as possible which requires complete rehabilitation of the formation.

6.2.3. The Presence of Water and Clay Combine to Form a Mudhole

The clay slurry visible on the ballast bed, which is the evidence of a mudhole, could either have been formed in the ballast or in the subgrade. The two causes of a mudhole can be treated in different ways.

6.2.3.1. Formed in the Ballast

The ballast is exposed to air, the sun, high vibration and pressure from passing trains. If the ballast is also fouled, water will be retained by the fines due to the low drainage ability of the ballast bed (Figure 127). The ideal conditions have been created for rapid decomposition of the fines in the ballast bed.

As the fines become saturated, clayey slurry will in due course become visible on top of the ballast. In Figure 126 a secondary effect of a mudhole can be seen; as the absorbed water dries out, the ballast bed becomes very hard and solid. Centre binding will cause the breakage of concrete sleepers.

This type of mudhole can be treated by cleaning the ballast, but if left untreated, the trapped water will increase the rate of decomposition and clay forming of the subgrade.

6.2.3.2. Clay Formed in the Subgrade

Whether drainage of water is slowed down due to fouled ballast, poorly maintained side drains, badly designed side slopes, subsurface seepage or the like, progressive settlement of the subgrade can eventually reverse the cross fall and can cause depressions which trap water beneath the track.

The trapped water saturates the ballast, subballast and subgrade soil which decreases the carrying capacity of the materials. Ballast migrates down into the subballast and the subgrade due to repeated pressure on the ballast. Recurring slacks after tamping is one of the first indications of formation failure (Figure 128).

The ballast is now unable to fulfil its function to reduce the pressure transferred, due to water not being compressible. The result is that higher pressures of 600 kPa can now be applied to the subsoil, whose carrying capacity has reduced to as little as 1 kPa due to the loss of bearing capacity of the substructure.

The weakened subgrade will yield under this higher load and the result is pumping of subballast and soil grains into the ballast. A mudhole has been formed.

This type of mudhole cannot be treated by cleaning the ballast only. The only practical long term solution would be to open the substructure, remove as much of the clayey soil as possible and reconstruct the subgrade. Refer to CHAPTER 21.

Figure 126: Centre Binding Due to Hard Ballast after Absorbed Water Dried Out

Figure 127: Saturated Ballast Bed

Figure 128: Recurring Slacks is a Sign of Formation Failure and is Often also Evidenced by White Spots in the Ballast Bed

6.3. The Effects of a Failed Formation

A failed formation and mudhole have many primary and secondary consequences including:

6.3.1. Recurring slacks due to subgrade failure:

It is not economically viable to send a tamping machine to tamp short sections of recurring slacks. Teams with pick axes and ballast forks will therefore tamp these sections at regular intervals which has an opportunity cost of other preventive infrastructure maintenance. As a secondary consequence, regular hand tamping damages the ballast, increasing the rate of substructure failure and also causes extensive damage to sleepers (refer to Figure 129 below) which will later have to be prematurely replaced.

6.3.2. Speed orders and train delays:

Due to the danger of a derailment at areas of formation failure, speed orders (temporary speed restrictions) are imposed. This causes train delays, late deliveries and loss of customers.

6.3.3. Low ballast cleaning production rate:

Where mudholes are encountered as part of the section to be screened, production rates in these areas are particularly low. Refer back to Figure 114 which shows a ballast cleaner cutting chain in a very wet ballast bed. This clayey soil will clog the screens, reducing or even stopping screening production. As a secondary consequence, the opportunity cost of not screening other sections of track is mudholes in the making.

6.3.4. Reduction in track life and increase in lifecycle cost:

Premature rehabilitation of the track will significantly increase the lifecycle cost which will increase fares of travel and transportation of goods. As a secondary consequence, an uncompetitive transport mode will lose customers to cheaper and more convenient modes of transport.

Prevention is clearly better than any cure. It pays to monitor ballast fouling and to screen the ballast before water retention will cause damage to the formation. The same applies to edge drain design and cleaning. Formation rehabilitation is a very expensive and disruptive process and must be avoided at all cost. Refer to CHAPTER 21 for formation rehabilitation processes and machinery.

Figure 129: Damaged Sleepers due to Manual Tamping

7. TURNOUTS, SLIPS AND CROSSINGS

7.1. Functions

Turnouts, slips and crossings (collectively called "sets") are essential to form networks, i.e. for the lines to diverge and converge. The performance and line speed of a railway is heavily influenced by the number and type of sets used. Sets must therefore be designed to achieve a dynamically favourable layout of the line.

Sets demand special attention with regards to selection, transport, installation and maintenance. They represent a considerable portion of the fixed assets of the permanent way installations of a railway administration. Switches and plain track must be suitable for safe and smooth running at the highest permissible speed. At the same time it is required, that the cost of maintenance is kept as low as possible and a long service life is attained.

Due to their design, switches are subject to higher stress than tangent tracks. Deficiencies in level occur in the area of the frog because of the effect of continual impacts. Lateral impacts, caused by the wheels being forced between rail and check rail are the reason for the occurrence of deficiencies in line. A correct geometric position is an absolute necessity, not only for switches located on main through tracks and negotiated at maximum line speeds, but also for the remaining switches and crossings, in order to assure operational safety and a long service life.

The good negotiability and the service life of switches depend, therefore, to a high degree on their condition. To achieve a good geometric position and a long service life of a switch, it is imperative that all the switch parts are connected securely to each other and that the fastenings are well tightened. If the defects caused by stress on the switches are not remedied in due time, this leads to rapid wear of individual switch components and unacceptable deviations in level and line.

They are weak links in the track structure and often the source of derailments which can generally either be caused by excessive speed (over design speed) or a lack of maintenance. Speed is not under the control of the Infrastructure Maintenance Department but the maintenance of sets certainly is.

7.2. Types of Sets

Turnouts enable vehicles to pass from one track to another without interrupting their run. Crossings are the intersection of two tracks.

The illustrations and information contained herein is predominantly based on turnouts from voestalpine (VAE) as a leading manufacturer of turnouts. Some of the Illustrations and information was supplied by and used courtesy of VAE South Africa.

7.2.1. Diamond Crossing

A diamond crossing allows two lines to cross one another without the possibility to cross over to the other. Refer to Figure 130.

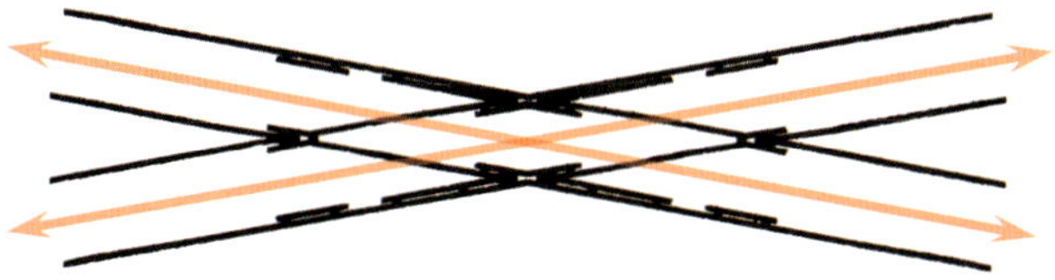

Figure 130: Diamond Crossing

7.2.2. Single Slip

A single slip is essentially a diamond crossing with a slip to allow traffic to cross over from both lines in a curve to the other (red line) as well as one line remaining in the curve (blue line). Refer to Figure 131.

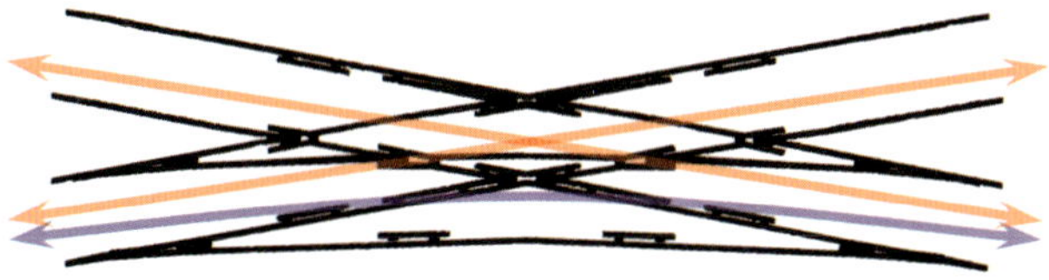

Figure 131: Single Slip

7.2.3. Double Slip

A double slip is essentially a diamond crossing with slips to allow traffic to cross over from both lines in a curve to the other (red line) as well as remaining in the curve (blue line). Refer to Figure 132.

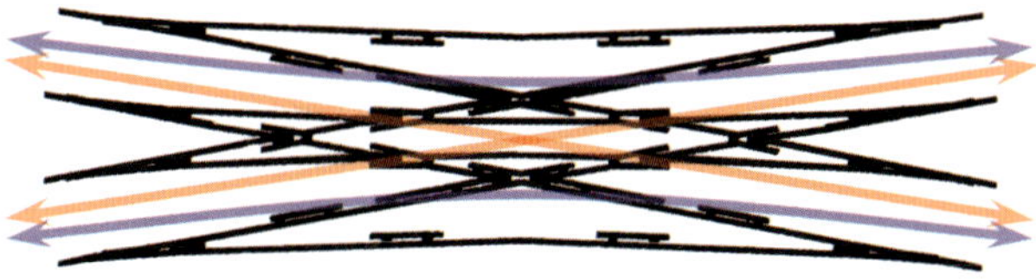

Figure 132: Double Slip

7.2.4. Scissors Crossing

A scissors crossing is generally found on a double main line which allows traffic to either carry straight on (blue line) or allows traffic from both lines to cross over from one to the other (red line). Refer to Figure 133.

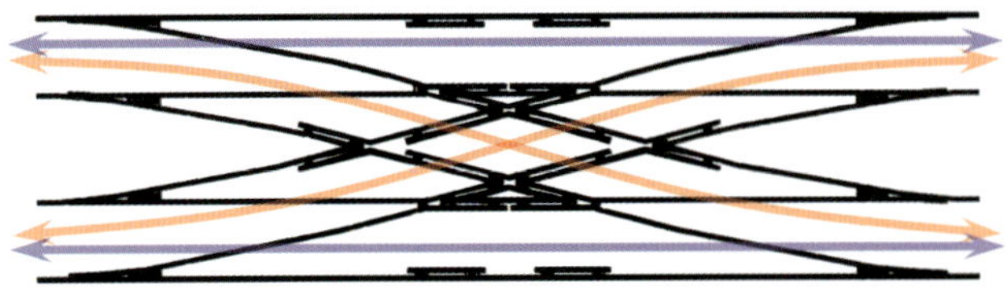

Figure 133: Scissors Crossing

7.2.5. Straight Single Turnout

A turnout allows a main line (generally the line that carries straight on) to cross into a secondary line to the left or the right. Figure 134 illustrates a straight single turnout to the left.

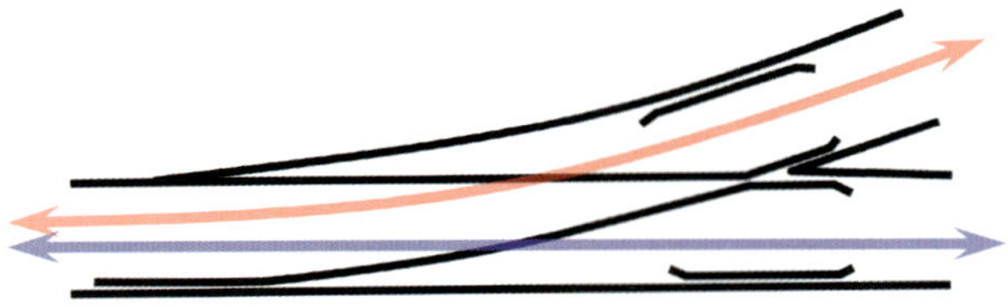

Figure 134: Straight Single Turnout

7.2.6. Tandem Turnout

In tandem turnouts, two other tracks diverge from the straight track section (main track). It can either be to opposite sides as illustrated in Figure 135 or can even be to the same side. These will normally not be found on a mainline due to the limited allowable speed and is more common in shunting yards where space is very limited.

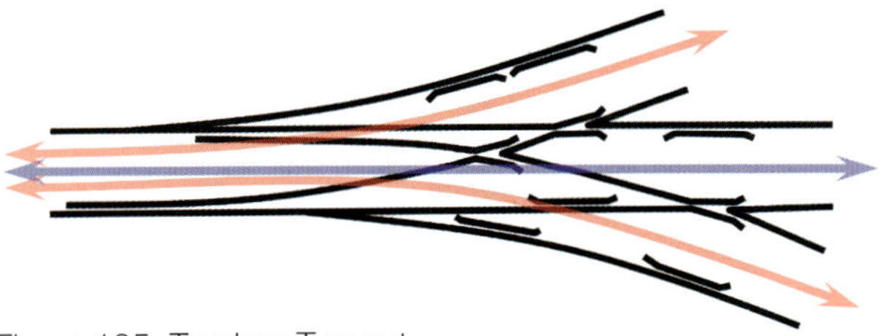

Figure 135: Tandem Turnout

7.2.7. Compound Curved Turnout

These turnouts have a continuously curved progression that can be installed within a curve. They are also referred to as similar flexure turnouts where the curve centres of the main and the branch tracks are on the same side.

To avoid the manufacture of special structures for turnouts with two curved branches, curved turnouts are produced by bending straight single turnouts.

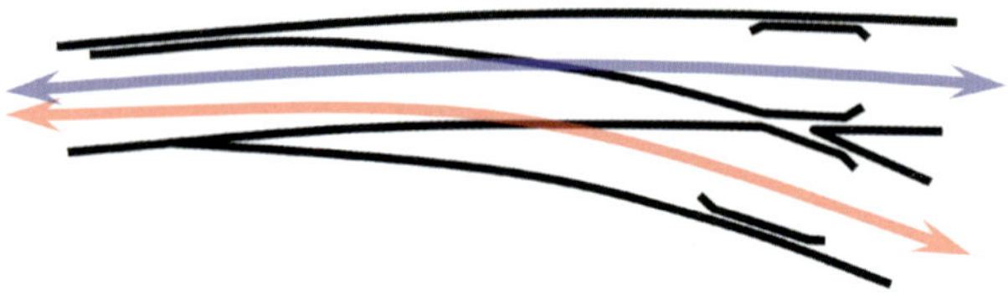

Figure 136: Compound Curved Turnout

7.2.8. Equal / Unequal Split Curved Turnout

These are also referred to as contrary flexure turnouts which have the curve centres of the main and the branch tracks on opposite sides. These are common at the end of double track where the two track converge into one track.

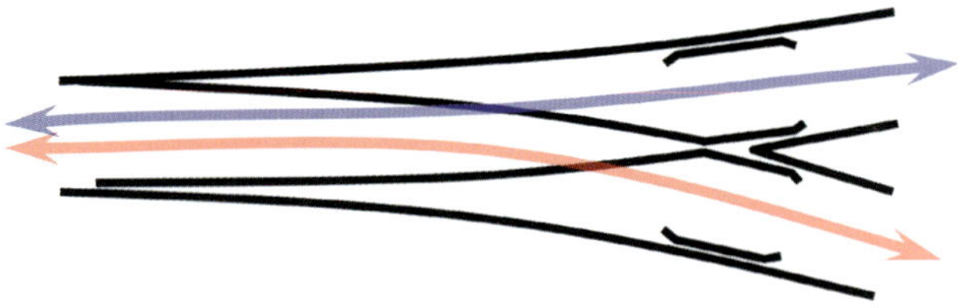

Figure 137: Equal/Unequal Split Turnout

7.3. Identifying a Turnout and Turnout Components

For the remainder of this section, the focus will be on turnouts as this is the most common of sets in the track infrastructure. Refer to Figure 138 for a typical example of a turnout identifying the more important components of a turnout.

7.3.1. Right Hand or Left Hand Turnouts

Turnouts are identified by the hand (left hand or right hand turning turnout). The entry of a turnout is from the switchblade side and the exit is at the crossing. Facing the switchblade at the entry of the turnout, if the turnout curves left from the straight section it is referred to as a left hand turnout and vice versa. The turnout in Figure 139 is therefore a right hand turnout.

Similarly, facing the switchblade at the entry of the turnout, components on the left hand of the turnout are referred to as left hand components and vice versa. The side on which the component lies comes first in the identification of a component and then the hand of the turnout.

7.3.2. The Angle of the Turnout (Take-Off Angle)

This is one of the most important turnout identification characteristics. Selecting the angle of the turnout during the design phase is generally determined by the speed that is required across the turnout and is expressed as a ratio. Costs and space may also influence the selection criteria. The most common turnout angles (α in Figure 140) in developing countries are 1:6, 1:7, 1:8, 1:9, 1:12 and in South Africa 1:20 turnouts are also very common. The turnout angle in Figure 140 is a 1:12. On high speed lines in Europe the turnout angle can be a lot flatter to allow for higher speeds through the turnout.

The greater the ratio (smaller turnout angle α), the faster the permissible speed is across the turnout due to the lower centrifugal forces. However, the greater the ratio, the longer, heavier and more expensive the turnout becomes.

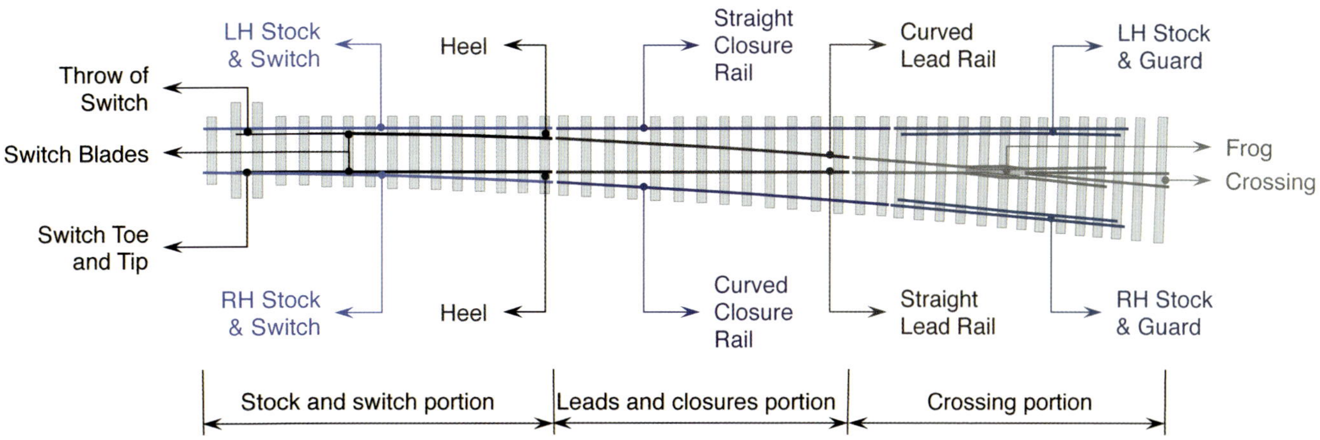

Figure 138: Turnout Components

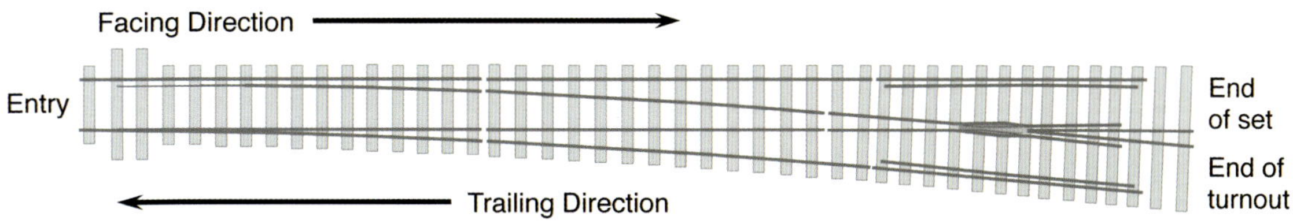

Figure 139: Direction of a Turnout

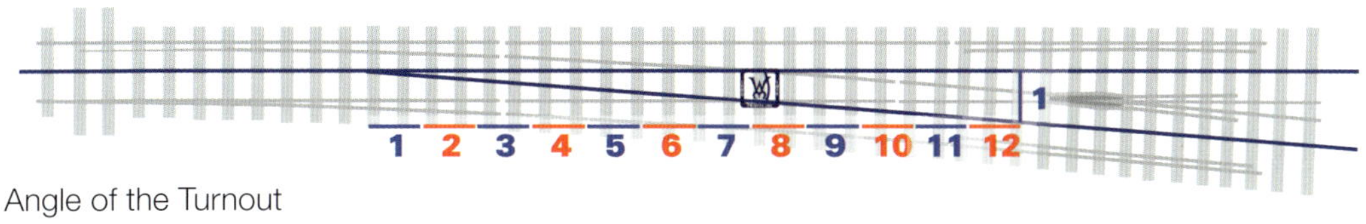

Figure 140: Angle of the Turnout

B – D Distance

1:6 = 178 mm
1:7 = 155 mm
1:8 = 137 mm
1:9 = 124 mm
1:12 = 96 mm
1:20 has a swing nose crossing

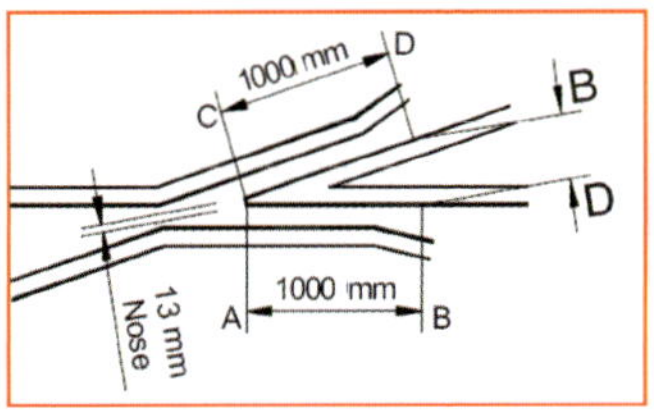

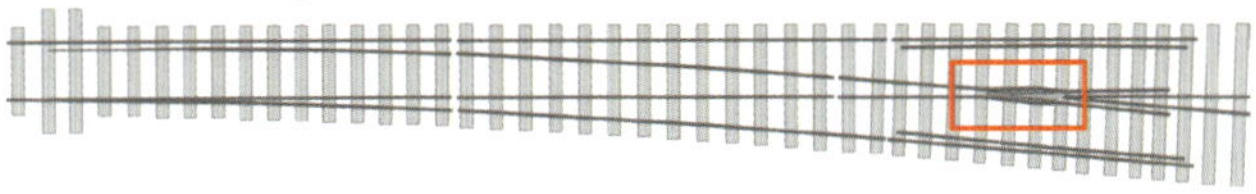

Figure 141: Measuring the Turnout Angle

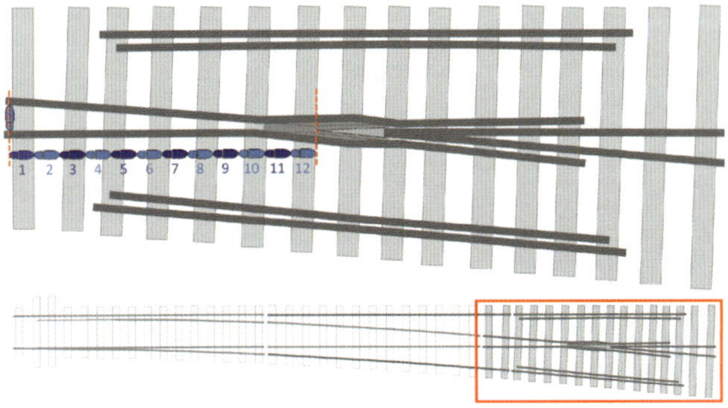

Figure 142: Using Footsteps to Determine the Turnout Angle

Figure 143: Turnout on Concrete Sleepers

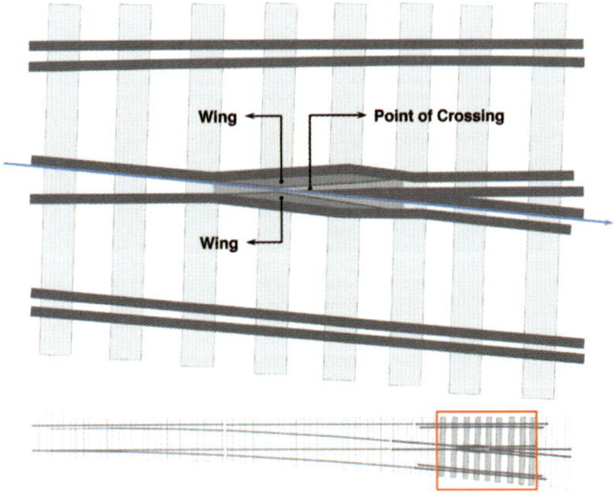

Figure 144: Path of the Wheel Flange and Transfer from Wing to the Point of the Crossing

Very often the type and angle of the turnout is no longer available and may have to be determined. There are two ways of determining the angle:

- Measured: From the 13 mm nose of the crossing measure one meter back along both running edges of the crossing (A–B and C–D in Figure 141). Measure the distance between B–D and refer to the table below. Bear in mind any side wear or mushrooming of the rail or running edge.
- Using Footsteps: Refer to Figure 142 for a graphic example of this simple method to determine the turnout angle. The person making the measurement must place his foot between the curved and straight lead rails at the point where the front and back of the shoe touches the rail. From this point step heel to toe and count the number of shoe lengths to the point of the crossing. The example in Figure 142 is therefore a right hand 1:12 turnout.

7.3.3. Turnouts on Timber or Concrete Sleepers

Older turnouts were all fitted with timber sleepers though most new turnouts are fitted with concrete sleepers. Timber sleepers have the advantage that they can be drilled on site and are lighter, whereas concrete sleepers are made for a specific position in the turnout.

Experience in South Africa has shown that the new generation VAE turnouts on concrete sleepers are more stable and require less tamping than older generation turnouts on timber sleepers. They are, however, much heavier and require specialised tamping machines and third rail lifting to tamp them effectively. Refer to CHAPTER 11.

7.3.4. Type of Crossing

One of the most critical points of a turnout is the crossing. The transfer area from the wing to the point of the crossing is exposed to impacts and forces from the wheels, as they transfer from one to the other across a gap.

Refer to the blue line in Figure 144 which represents the path of the wheel flange as it jumps the gap from the wing to the point of the crossing. Various designs and material specifications for the crossing have therefore been developed to withstand the impact, especially on lines carrying high axle loading and or high speeds. The crossing's wear capabilities are only as good as the steel composition from which it is made. The following types of crossing will now be discussed; the rail manufactured crossing, the cast manganese crossing and the swing nose (moveable point) crossing.

7.3.4.1. Rail-Manufactured Crossing

These crossings are manufactured from rail sections and distance blocks welded to the point to maintain its position. The wing rails are also made from rails that are machined and bolted to the point using distance blocks. See Figure 145 and Figure 146.

The advantages of this type of crossing are:

- Due to the bolted design of the rail-manufactured crossing they perfectly adapt to the elastic behaviour of the track.
- They are economical to manufacture even in small numbers due to no tools and patterns required.
- The crossing can be bent to a curve.
- The crossing can be fish-plated or Thermite welded to the track.

The disadvantages of this type of crossing are:

- The transfer area on a rail-manufactured crossing is not manufactured to suit the wheel profile, which creates accelerated wear.
- The crossing is held together by bolts, nuts and blocks which tend to vibrate loose over time.

Figure 145: Rail-Manufactured Crossing

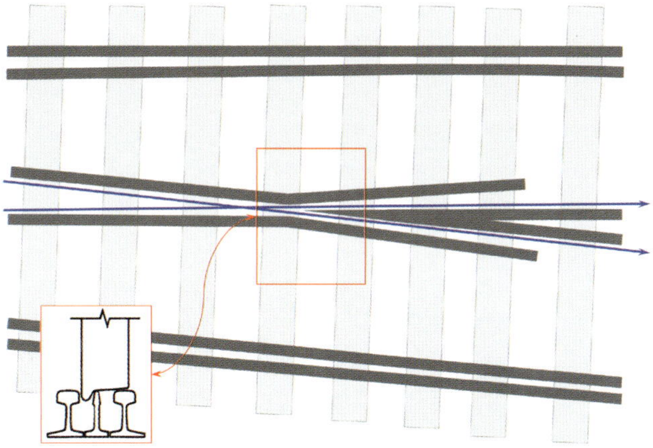

Figure 146: Crossing

7.3.4.2. Cast Manganese Crossing

To improve the wear characteristics of crossings, the crossing is cast of a manganese steel compound consisting of approximately 11 to 14% manganese.

The first manganese crossing had the complete crossing casted in one manganese steel block, referred to as the monoblock manganese crossing (Figure 147).

The advantages of this type of crossing are:

- It eliminates the need for and the disadvantages of bolts.
- The crossing was cast to suit the wheel profile of a 1:20 angle, which reduced the impacts and forces applied throughout the transfer area.

The disadvantages of this type of crossing are:

- The large monoblock manganese casting is very expensive compared to that of a rail-manufactured crossing.
- Manganese steel and rail steel cannot be welded together, which means that the joint between the casting and the rails at both ends of the crossing must be undertaken by fishplates.

Figure 147: Manganese Monoblock Crossing

- The two different metals, having different compositions and therefore a different wear rate, lead to a mismatched joint and loosening of bolts.

VAE has solved some of these disadvantages by creating the Centro manganese crossing which has four closure sections (see the blue rail sections in Figure 148) made from standard rails that are welded to the manganese crossing using a special patented welding process. This means that the manganese crossing can now be Thermite welded to the track.

To reduce the cost of the monoblock manganese crossing the point of the crossing and the overrunning area of the wing rails are cast in one piece (see the green insert in Figure 149). This is referred to the railbound manganese crossing. The supporting surfaces and fitting areas as well as the complete running edges and surfaces are milled. The closure rails are machined at the head and foot in the front area and welded together.

7.3.4.3. Swing Nose Crossing

Despite improvements made in crossing design with the monoblock crossing, it still has a transfer area across a gap from wing to nose. Although this transfer area has been machined to the wheel profile and hardened, this is still a high wearing area due to differently worn wheels which cause impact and additional forces through this area.

This lead to the development of the swing nose crossing which nearly acts like tangent track. In Figure 150 it can be seen that the through line is uninterrupted. The gap for the flange of the wheel is closed creating a continuously supported transfer area. As a result there is hardly any more impact or forces exerted on the transfer area than that of a tangent track.

This type of crossing is mostly used in high speed, high volume and high axle loading lines; this is mainly due to the cost of such a crossing.

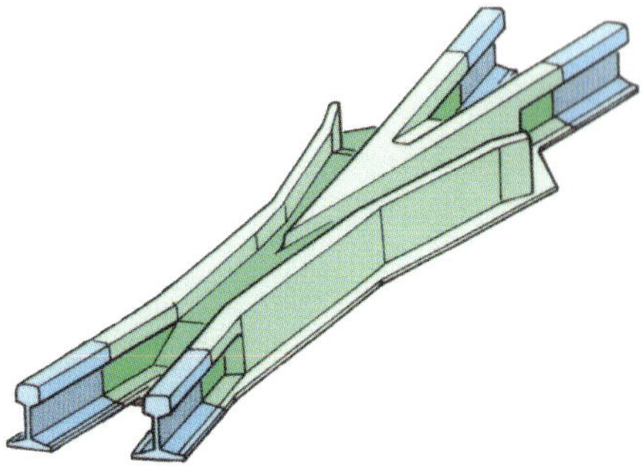

Figure 148: Centro Monoblock Manganese Crossing (Courtesy VAE)

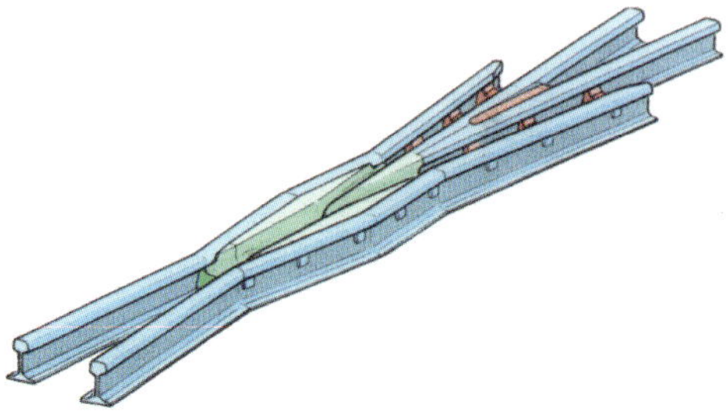

Figure 149: Railbound Manganese Crossing (Courtesy VAE)

Figure 150: Turnout with Swing Nose Crossing

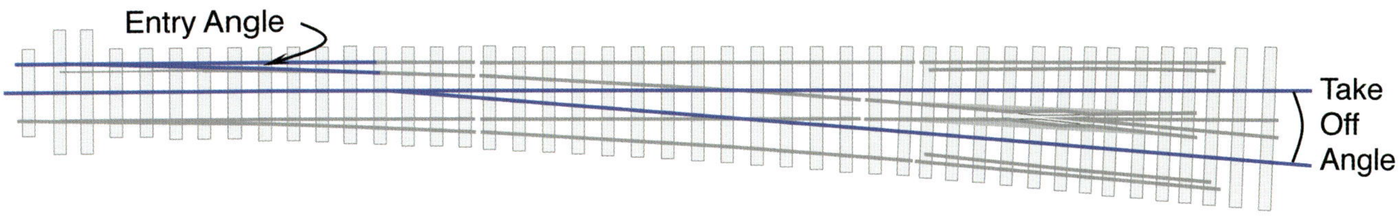

Figure 151: Entry Angle and Take-Off Angle

The advantages of this type of crossing are:

- Increased running comfort (no impacts and noise at the open fishplate joints or transfer area).
- No check rails required.
- Reduction of wheel wear and increase in service life.
- It provides a hardened manganese transfer area.
- The crossing can be exothermically welded into the track.

The disadvantages of this type of crossing are:

- It is expensive compared to even monoblock crossings.
- The crossing needs an additional switching mechanism, one for the crossing and one for the switch blade.
- Turnouts with swing nose crossings are very long.

7.3.5. Switch Blade

Together with the crossing, the switch blade is the fastest wearing component of a turnout. Turnout design has therefore gone through an evolutionary process over many years to reduce the wear on the switch blade, to increase traffic speed through the turnout, to lower impact loads and also to lower the lateral acceleration experienced by the wheel set, which causes discomfort to passengers and may cause cargo to shift. This objective was achieved by various design changes over the years, including reducing the entry and take off angles of the turnout and by curving the switch blade.

There are various types of switch blades including loose-heel switch blades, flexible switch blades with switch blade plates, spring switch blades and flexible switch blades with switch blade plates.

7.3.5.1. Entry and Take-Off Angles

A smaller entry angle, also referred to as the "attack angle" (Figure 151), creates a more gradual change in direction for the wheel set. Smaller entry angles are achieved by extending the length of the turnout and by curving the switch blade (paragraph 7.3.5.2).

A longer turnout length will also result in a smaller take-off angle. Reducing these two angles will reduce the lateral acceleration experienced by the wheel set, which means less wear and higher speeds through the turnout. It is for this reason that turnout angles of 1:12 are common on main lines and 1:20 turnouts common on heavy haul and high speed lines.

7.3.5.2. Switch Blade Curvature

- Straight switch blade – The first switch blades were straight rail sections which met the stock rail at an angle. This sharp angle caused the wheel set to experience a very sudden change in direction, resulting in rapid wear of the lead rail and rail wheels, high lateral acceleration and discomfort for passengers. These turnouts could therefore only be crossed at low speed. Refer to Figure 152. There are still many straight switch blade turnouts in use in developing countries.

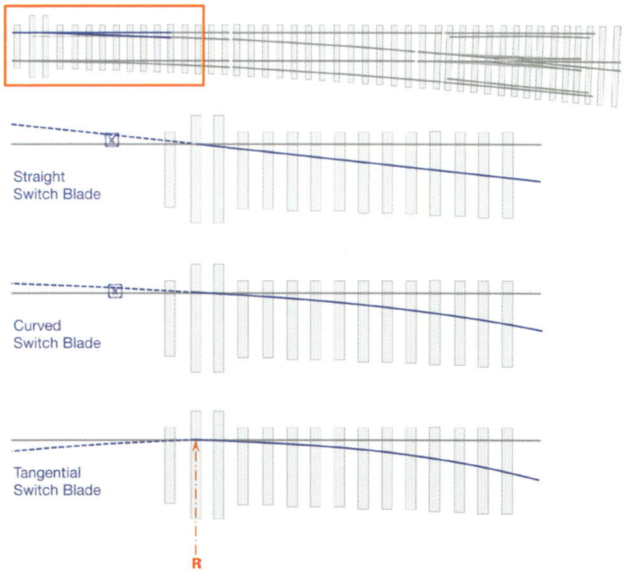

Figure 152: The Evolution of Switch Blade Designs

- Semi-curved switch blade – It was soon realised that if the switch blade is curved, a more gradual transition will be experienced by the wheel set with much lower lateral accelerations and wear. The semi-curved switch blade is curved up to where the machining starts on the blade or the blade meets the stock rail. From that point onward the blade is straight and machined to blend in gradually with the stock rail.
- Fully-curved switch blade – The fully-curved switch blade further reduces the entry angle. Curved switch blades are referred to as Secant switch blade design and this is the most common type in developing countries.
- Tangential switch blade – The demand for higher axle loads and higher speeds through turnouts resulted in the development of the tangential switch blade design. The switch blade follows a curve with a fixed radius (Figure 152) resulting in an almost 0° entry angle. The stock rail crown of tangential turnouts is also under cut to allow the tip of the switch blade to seat flush with stock rail, protecting the switch blade tip.

7.3.6. Turnout Mechanisms (Points Machines)

Turnout mechanisms are required to move the switch-blade (and moveable crossing in swing nose turnouts). These mechanisms can either be electrical or hydraulic (Figure 153) with remote operating or manually (hand) operated mechanisms (Figure 154).

Specialised hollow steel sleepers have been developed within which the rodding and other turnout mechanism components are housed to protect it from outside effects and damage. The hollow sleeper is bent from one piece with welded on parts to increase the stiffness and holes drilled in for rail fastening plates and cradles for the drive and detection unit. The sleeper replaces two standard concrete sleepers.

Electricity supply is generally not reliable or available in some countries, resulting in manually-operating turnout mechanisms still being the most popular system.

There are various types of manually-operated turnout mechanisms available on the market. These must be maintained through greasing at regular intervals.

Figure 153: VAE Electrical/Hydraulic Turnout Mechanism for Remote Operating

Figure 154: Manually-Operated Turnout Mechanism (Tumbler)

The turnout mechanism deserves a lot of attention since, in the case of remote switching, the turnout can be far from a depot and if the turnout does not complete the stroke, it can cause a derailment or at least cause delays to traffic.

7.3.7. Switch Blade Roller Slide Chairs

The forces required to move switch blades are primarily the result of friction between the switch blade and the contact surface of the slide baseplates. Switch moving forces are frequently the cause of turnout failures often resulting in train delays.

Towards the end of the 1990's Schwihag, in order to reduce this type of failure, developed a switch roller system which could be integrated into a slide baseplate or cradle system (Figure 155).

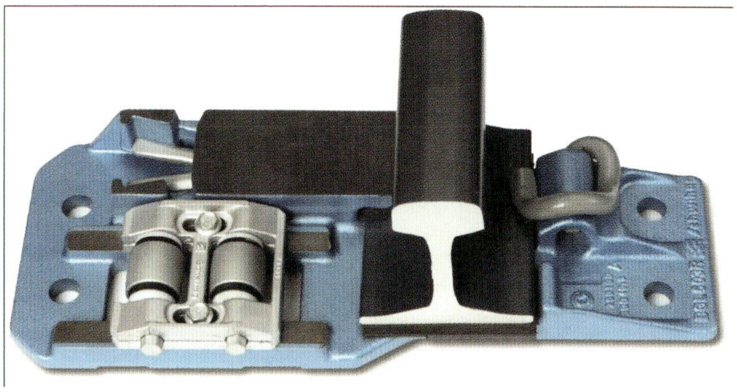

Figure 155: Switch Roller System Integrated into Slide Baseplate

To maximise the performance of this roller system an eccentric axle was developed ensuring the roller could be set to suit both the new switch assembly and any subsequent conditions through the life of this switch assembly.

The adjustable, maintenance free roller system has a known performance and a life well in excess of that of the switchblade. This enables planned maintenance to be managed beyond the life of the incumbent turnout assembly.

This roller system from Schwihag has gained acceptance worldwide, from high-performance metro systems, local transport networks through to high-speed lines and heavy freight transport with axle loads over 30 tons.

Figure 156: Switch Blade Assembly

It has become a uniformly accepted method of reducing friction in turnout assemblies and can be applied to all track gauges, switch and stock rail sections.

7.3.8. The Check Rail

In switches, check rails are provided on the running rail opposite the fixed crossing. They lead the wheelset over the gap in the throat of the crossing in order to prevent the wheel flange from striking the crossing nose. Once again there are various design check rails.

7.4. Transportation of Turnouts

Turnouts must preferably be pre-assembled in a workshop to a jig or template for perfect alignment. This, however, causes complications for the transportation of the turnout to the installation site since the long sleepers of a turnout generally foul the adjacent line and the distance to wayside structures such as bridges, tunnels and electrical overhead structures, where applicable.

To transport the pre-assembled turnouts, turnout transportation wagons were developed (Figure 157) which hydraulically tilts the sections with long sleepers to fit within the moving vehicle structure gauge. Flat wagons are also used to transport the panels with normal sleepers. Refer to CHAPTER 18, paragraph 3.

7.5. Installation of Turnouts

Turnouts on timber sleepers are still relatively simple to install by hand since the holes can be drilled in the sleepers where required. However, despite this, it is still recommended that a turnout be assembled on a template, even if it is next to the track. Modern turnouts on concrete sleepers are more complex since every concrete sleeper is made to fit in a specific position in the track.

Whether the turnout is assembled in a factory or next to the track on a template, specialised equipment is required to carry the turnout into the installation gap. Turnout installation machines (Figure 158) are available to carry the turnout in the minimum number of sections into the installation gap with the least amount of deflection on the turnout to ensure its integrity and long life. Refer to CHAPTER 18, paragraph 4.

Figure 157: WTW Tilting Turnout Transportation Wagons

Figure 158: Turnout Installation Machine

7.6. Maintenance of Turnouts

7.6.1. Component Replacement

Switch blades and crossings have a finite life and needs replacement once the wear has reached a determined level according to laid down standards. Through proper maintenance of the turnout, the life of the components can be significantly extended.

7.6.2. Lubrication

To reduce the wear on switch blades, switch lubricators must be installed or grease or graphite must be applied by hand.

Switch blades and slide chairs must be black-leaded over the first 1,25 metres. Lubricating oil may not be used though this practice has been witnessed in developing countries (Figure 159).

7.6.3. Welding and Grinding

Maintenance welding and grinding must take place in accordance with specifications before wear exceeds set standards.

Burrs which interfere with the fit of switch blades against stock rails or with the operation of locking bars, must be removed by grinding according to specification to prevent breakouts.

Specialised railbound and road/rail turnout grinding machines are available but not always affordable. Small turnout grinding machines (see Figure 161) are affordable and should be applied at regular intervals to ensure the long life of this expensive component.

Figure 159: Poor Maintenance Practice of Used Oil to Lubricate Turnout

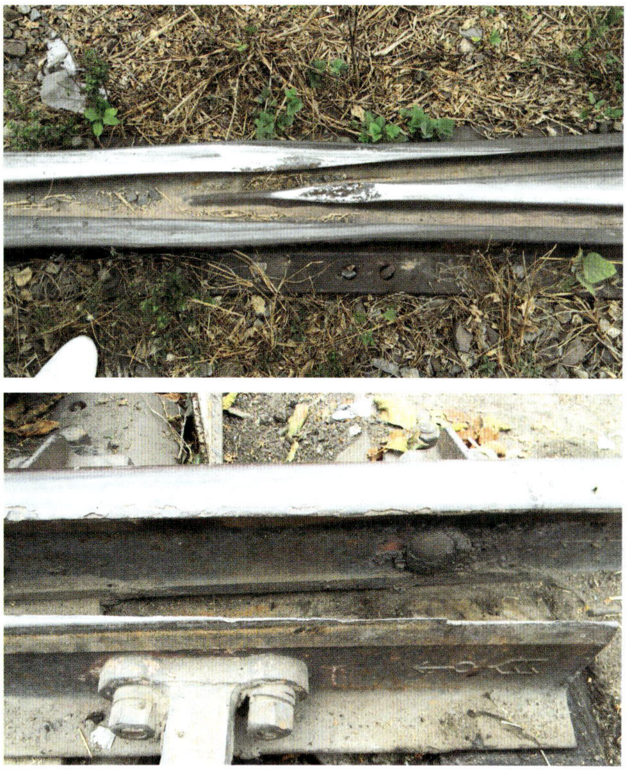

Figure 160: Damaged Crossing and Switch Blade due to Lack of Grinding

Figure 161: Switch Grinding and Deburring Machine

Figure 162: Portable Milling Machine

Once the wear on the crossing has reached the advanced level as seen in Figure 160, grinding can no longer repair the crossing. The crossing must be ground and built up by welding. Though grinding has been the norm for many years, Robel Bahnbau-maschinen offers a small portable milling machine (Figure 162) as an alternative which has various advantages above grinding. Milling provides perfect 90° evenly treated surfaces with very low heat applied during the process and less chance of the welding chipping off.

Just as a note with regards to Figure 160; manganese crossings require specialised welding processes, skills and equipment and are difficult to apply in track. It further supports affordable incremental preventative maintenance due to the very high costs of corrective maintenance.

7.6.4. Tamping

Due to the dynamic loading in most types of turnouts, they settle with defects in the vertical and horizontal plane; generally requiring more frequent tamping than the adjacent track.

Due to the restrictions created by the switchblade, crossing and turnout rails (the highlighted area in Figure 163), turnouts pose a problem for normal tamping machines to grip the rail for lifting and space for the tines to enter the ballast around these restricted areas. A tine on the one side of the tamping unit will hit an obstruction as illustrated in Figure 164.

Turnouts must then either be omitted or specialised tamping machines designed for tamping turnouts must be used. These machines may be equipped with either tilting tines (Figure 164), split units (Figure 165) or a combination of the two technologies. Various other features are also required for effectively tamping turnouts such as slewing tamping unit frames to reach the turnout rail while standing on the straight section, specialised lifting units to grip the rail in restricted areas, a long wheel base to lift the heavy turnout without stressing the rail and third rail lifting to lift the long sleepers with turning out rails without over-stressing the fastenings. Refer to CHAPTER 11.

A weak point of sleeper supports are the areas of the inside lock and the centre lock. If the inside or the centre locking devices are located between two sleepers, these two sleepers can only be tamped mechanically on one side. Tamping these two sleepers on the side of the drive and closure rods is scarcely possible even with handheld vibratory tampers (Figure 166). As the sleeper crib is not ballasted, good compacting becomes impossible. This leads to instability in this particularly critical area of the set of switches.

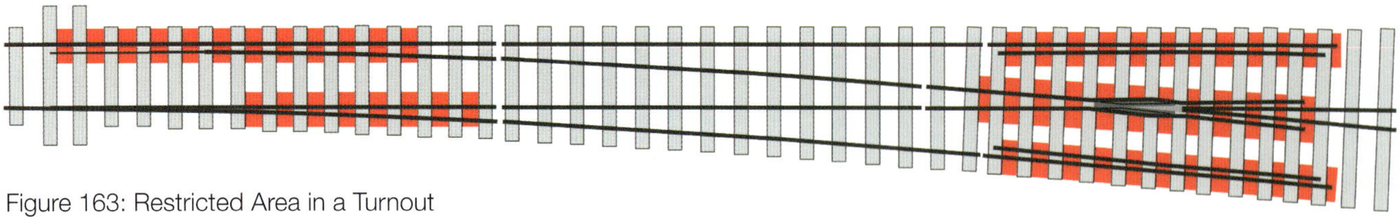

Figure 163: Restricted Area in a Turnout

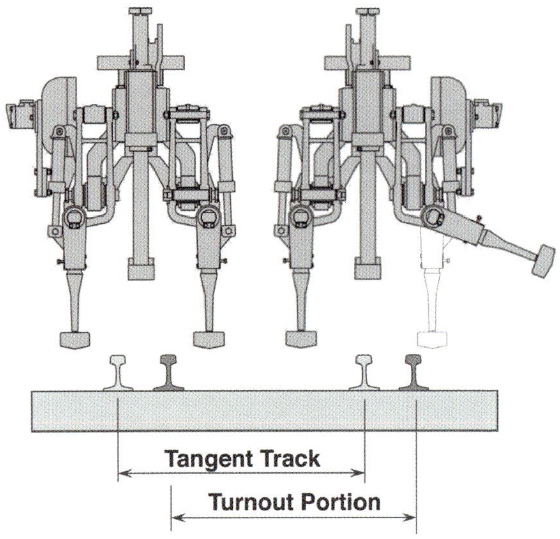

Figure 164: Tilting Tine Units

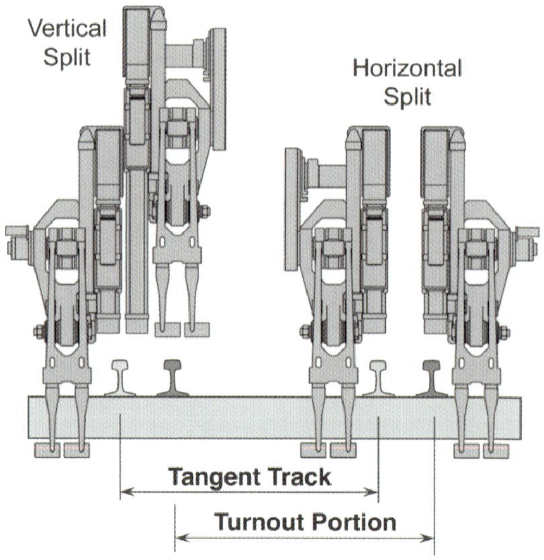

Figure 165: Split Units

hollow sleeper which contains all the necessary rods for shifting of the switch. See also Figure 153. With this design, there are no more rods between the sleepers, which mean all sleepers can be tamped on both sides using a tamping machine of good design and quality.

7.6.5. Ballast Cleaning

Turnouts, especially those on high speed lines and modern designs on concrete sleepers are very heavy, long and wide. Therefore, ballast cleaning under turnouts has always been a challenge. Ballast cleaning of this most expensive and critical track component has therefore been mostly neglected.

In South Africa ballast cleaning of turnouts has been done to a limited extent using the RM74 ballast cleaning machine adapted for turnout screening (see Figure 167) using a wide cutterbar with adapted chutes. This is, however, only possible on turnouts but not on scissor crossings, diamond crossings or slips, or even turnouts on multiple lines such as busy metro lines.

Now continuous ballast screening in turnouts can be achieved with the development of the Plasser & Theurer URM 700 universal ballast cleaning machine which allows screening any turnout, crossing or slip with infinitely variable cutting widths without obstructing the adjacent line, even on the multiple lines so often found on metro lines.

The specially designed sword shaped cutterbar can be inserted and removed without dismantling the turnout or excavating an entry hole for the excavating chain (Figure 169).

Refer to CHAPTER 14 paragraph 3 for more information on turnout screening.

Figure 166: Tamping Around Switch Rods

Figure 167: Screening of a 1:20 Turnout Using the RM74 Ballast Cleaning Machine

Figure 168: Stabilising of the Cutterbar at the Other End

Figure 169: Sword Shaped Cutterbar with Shoulder Excavating Units Visible Behind

8. LEVEL CROSSINGS

Level crossings have a very limited application in this book since it is not a maintenance activity that can be mechanised. Also, it is not the intention here to consider the safety or signalling of level crossings. However, level crossings are maintained by the track maintenance department and therefore deserve mention.

Poorly constructed and maintained level crossings are a danger to road traffic and rail traffic alike (Figure 170). Level crossings must be properly constructed using concrete blocks (Figure 171) designed for the purpose. Cracked or broken blocks must be replaced.

The ballast just before and after the level crossing must also be profiled by hand when the blocks are removed for tamping since the ballast regulating machines can generally not reach in the restricted area.

Figure 170: Poorly Constructed and Maintained Level Crossing

Figure 171: Well-Constructed Level Crossing on the Sishen to Saldanha Iron Ore Line South Africa

9. DRAINS

The surface drain (table drain) is the track component which is often neglected the most. The drain intercepts water entering the track substructure and removes water draining out of the track as explained in paragraph 6.2.1 and Figure 119. Drains are either constructed of concrete (Figure 173) or may be a cleaned area alongside the track (Figure 174). The fall of the drain is important to consider to direct water out of the cutting into a runoff.

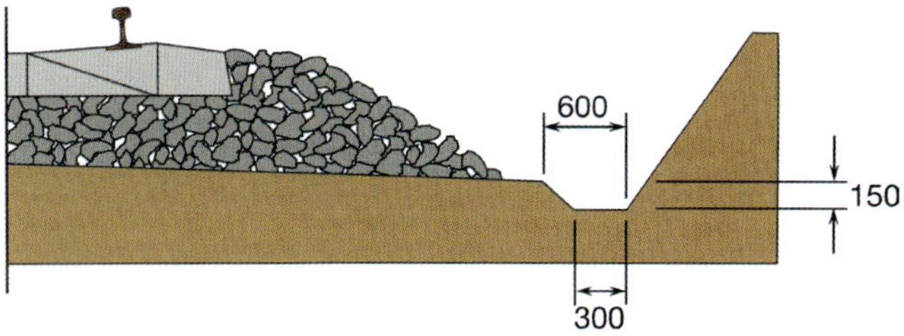

Figure 172: Minimum Table Drain Dimensions (Transnet Specification)

Drain cleaning can either be done by hand labour or using mechanised machines. The greatest advantage of mechanised methods is the removal of the material that was excavated. This could be problematic in long cuttings with labour methods.

Subsurface drains may be constructed when:

- the groundwater level needs to be lowered (depth <1.5 m below top of rail);
- the soil surrounding the track structure needs to be drained;
- surface and seepage water needs to be captured and drained off; and
- to drain off the water from water-carrying layers; or
- when there is a danger that open ditches will be filled up by deposits.

Figure 173: Concrete Drain

Figure 174: Cleared Drain In a Cutting. However, Lack of Weed Control will Prevent Drain to Direct Water out of the Ballast

Figure 175: Before and After Drain Cleaning

Chapter 4
Fundamentals of Rail and Wheel Interaction

1. INTRODUCTION

The contact area between steel wheels running on steel rails is a tiny patch which provides for a very low rolling resistance. It also allows for very heavy axle loading and low energy requirements compared to any other form of land transportation.

This system is however only efficient if the wheel and rail interaction is properly managed for which a good understanding of the interaction between rail and wheel is essential. Figure 176 illustrates some of the terminology relating to the wheel and rail interface that will be used in this chapter. See also CHAPTER 3, paragraph 2 on rails.

One of the best literatures on this subject is the International Heavy Haul Association's "Guidelines to Best Practice for Heavy Haul Railway Operations – Management of the Wheel and Rail Interface".

2. PRINCIPLES OF STEERING

Railway rolling stock wheelsets comprises two wheels attached to a common axle without relative movement being possible between the wheels. The rail wheels are shaped in the form of a complex cone with nonlinear changing radii between the large and small radius of the conical wheel.

Steering of a train around a curve is made possible by the radius difference on the two wheel treads on the same axle and where and how the wheel makes contact with the rail, i.e. rail/wheel interaction.

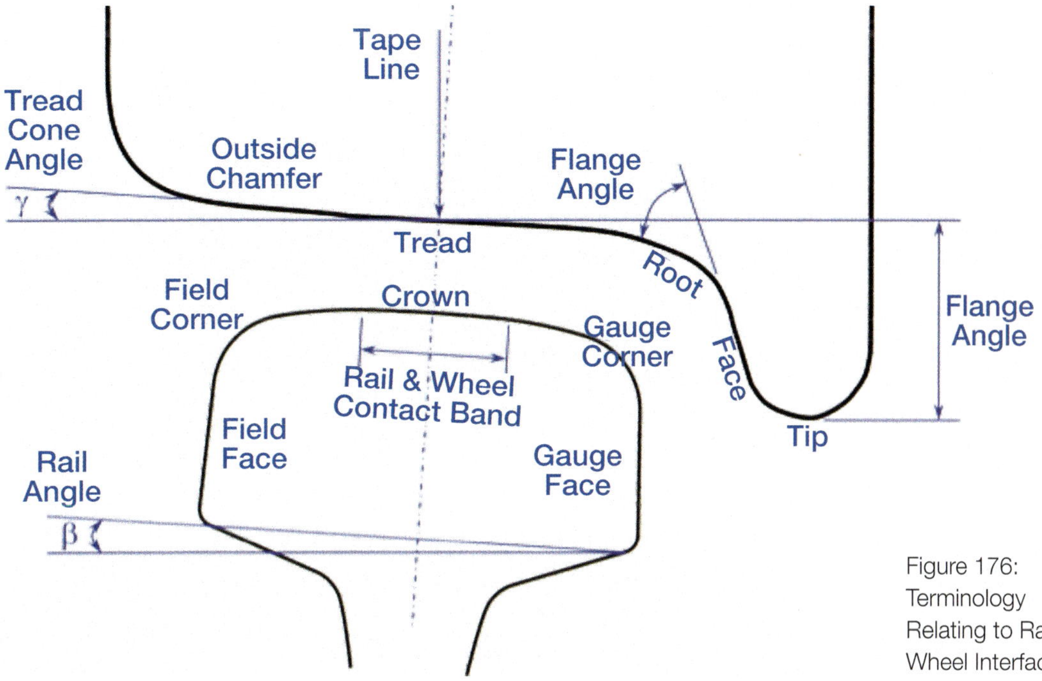

Figure 176: Terminology Relating to Rail/ Wheel Interface

For the sake of explaining the concept, take the wheel tread as being cut to a linear angle, Gamma (γ) (Figure 177). The circumference of the wheel is measured at the centre of the cone at the "tape line". Rail wheels are flanged to avoid derailment of the wheel under severe conditions but play no role in guiding the wheel on tangent (straight) track. On sharp curves the flange may however play a role in guiding the wheel and could be beneficial from a wear point of view.

The rail crown has an equally complex shape consisting of various radii across the crown specifically designed to match the conicity of the wheel. Refer to CHAPTER 3, paragraph 2.2.

The two rails are attached to a sleeper at a specified gauge, i.e. the distance between the two rails. To perfectly position the wheel's ideal running radius in the centre of the tread to the centre of the rail crown, the two rails are positioned at an angle β (Beta) on the sleeper. This also assists in controlling the bending stresses on the rail web as well as rollover of the rail. This angle is in the order of approximately 20 degrees.

The sleeper is canted (superelevated) at an angle α (Alpha) around curves to counter centrifugal forces, unloading of the inner rail and rollover forces on the train (Figure 178). The angle of the superelevation depends on the radius of the curve and the track section speed.

Each of the above parameters, wheel profile, rail profile, rail angle, curve radius and superelevation has a significant influence on the steering ability of the wheelset, comfort of passengers and wear and stress of the rail and wheel. Maintenance of these track parameters are all addressed in this manual.

The conical wheel profile provides a "rolling radius difference" between the flange side and the field side of the cone. On tangent track the two cones will self-centre so that the both wheels will be running on the same radii of the wheels on the centre line of the rail crown (Figure 177).

On curved track the outer wheel has a longer distance to travel than the inner wheel. When the unconstrained wheelset enters the curve, the centrifugal forces together with the superelevation (cant) of the sleeper and the conicity of the wheels will move the wheelset laterally until the radii of the wheels achieve a state of "pure rolling" where the "distance" the wheel travels on that specific radius matches the distance of the inner versus the outer rail around the curve without wheel slippage. The bigger radius of the outer wheel will travel a longer distance than a smaller radius of the inner wheel for the same revolutions of the common fixed axle (Figure 178).

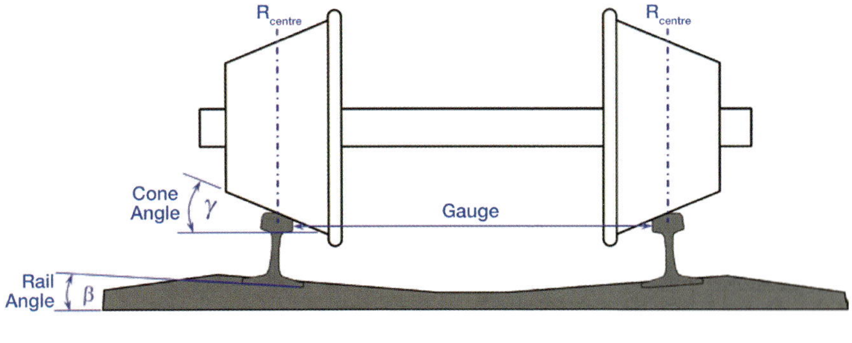

Figure 177: Self-Centring of Rail Wheels on Tangent Track

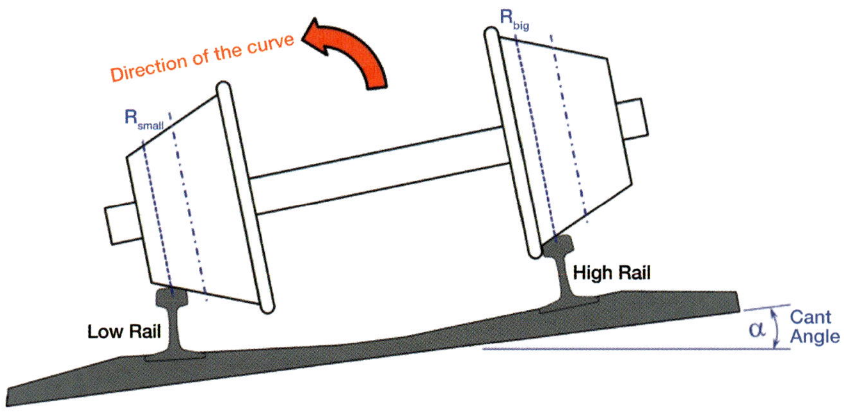

Figure 178: State of Pure Rolling on Curves

3. RAIL/WHEEL CONTACT

The above explanation of steering and self-centring of wheelsets assume wheels of the same diameters and wheels and rails with a profile within design specification. However, deviations from the above will cause forces acting upon the wheelset resulting in accelerated wear of both rails and wheels and material flow (plastic deformation). Understanding rail/wheel contact will contribute towards better rail and wheel wear management and achievement of the objectives of maintenance, i.e. longer component life and lower lifecycle costs.

There are many influences on the position of the rail and wheel contact patch that may have a significant effect on the service life of both rails and wheels and the steering ability of the wheelset which then also affects the ride comfort to passengers or in severe cases may even cause a derailment where the deviation from the required contact area causes the wheel flange to climb the rail, causes roll-over of the rail etc.

3.1. Ideal Rail/Wheel Contact

Ideally the contact patch between the rail and the wheel should be in centre of the rail and centre of the wheel tread as illustrated in Figure 179 on tangent (straight) track. This would be the case on new rails and wheels and where a very good maintenance regime is followed regarding wheel profiles and grinding of the rails. It also implies that the rail and wheel profile must be designed to be compatible which requires expert knowledge of contact dynamics considering all the influencing parameters of the specific track alignment, rolling stock type, axle loading, etc. Rail grinding often changes the original profile of the rail to optimise rail/wheel contact depending on the position of the rail in the track such as in curves. Grinding profile templates must therefore be specified by a person having considered contact dynamics.

On shallow curves (800 to 2000 metre radius) the contact patch will move slightly towards a larger radius on the wheel on the high rail (Figure 180) with the lateral movement of the wheel as it finds a state of pure rolling as described above.

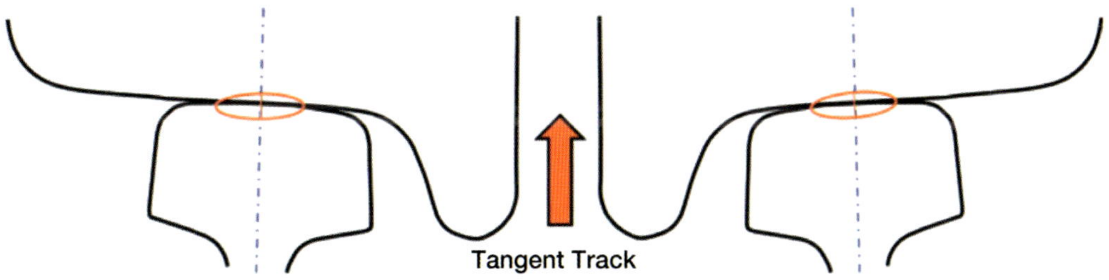

Figure 179: Ideal Rail/ Wheel Contact on Tangent Track

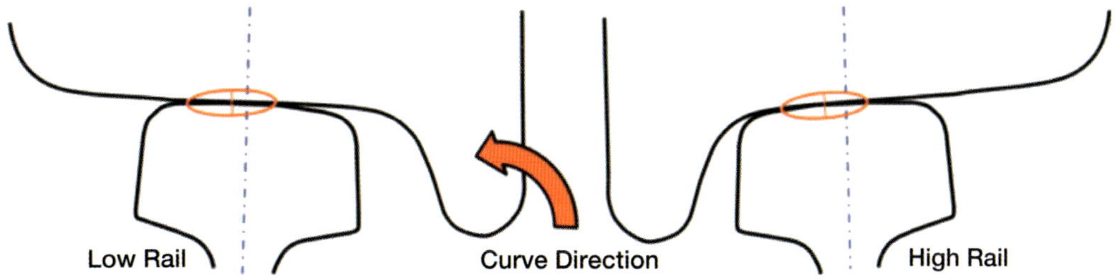

Figure 180: Ideal Rail/ Wheel Contact on Shallow Curves

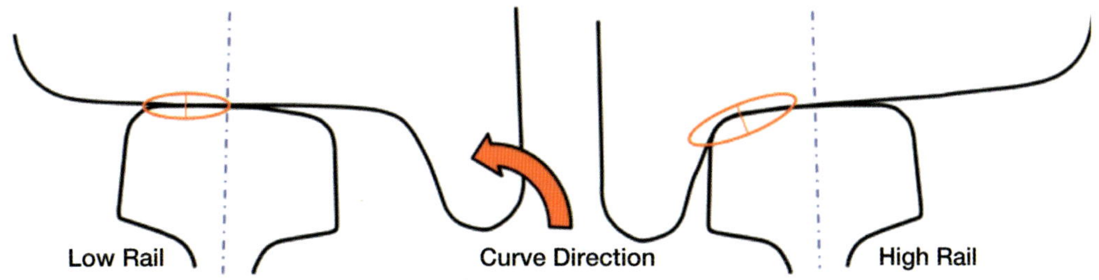

Figure 181: Ideal Rail/ Wheel Contact on Tight Curves

As the radius increases the contact patch on the high rail will move further towards the flange until a conformal contact is achieved on very tight curves of small radii (Figure 181).

3.2. Poor Rail/Wheel Contact due to Poor Rail Profiles

There are many variations of poor rail profiles as a result of no or poor rail profile management regimes caused by the forces on the wheelset and poor wheel profile management. A poor rail profile will negatively affect steering and accelerate wheel and rail wear such as in Figure 182 where a flat rail crown can be seen with no defined contact band. Another typical example (Figure 183) is side wear on the high rail of tight curves with probable cant deficiency, where gauge widening is not practiced, curves are not lubricated and grinding is not carried out. This is a severe case indicating no rail/wheel interaction management whatsoever. This side wear implies wheel flange contact with the gauge face of the high rail resulting also in wheel defects as can be seen by the sharp flange in Figure 184.

Figure 185 illustrates a typical scenario with side wear on the high rail but with wheels that are still within specification. A similar scenario may result with over-grinding of the gauge corner. On the low rail the wheel/rail contact may still be within specification but on the high rail the state of pure rolling cannot be achieved due to a very undesirable two-point contact.

The two point contact will cause longitudinal creep forces in opposite directions (tread versus flange) resulting in accelerated gauge wear, rolling contact fatigue and plastic flow of both the high rail and the wheel, especially the wheel flange. Furthermore, the flange contact on the high rail will cause a lateral force on the rail resulting in higher loads and fatigue of the fastening systems and potential rail roll-over. The net result may be gauge widening as a result of the side wear and together with a thinner flange against a side worn rail may cause the low rail wheel to derail. In the severe cases illustrated by Figure 183 the flange may actually climb the high rail causing a derailment.

Figure 182: Flat Rail Crown with no Clear Contact Band

Figure 183: Severe Side Wear on the Gauge Side of the High Rail

Figure 184: Sharp Flange and Hollow Wear

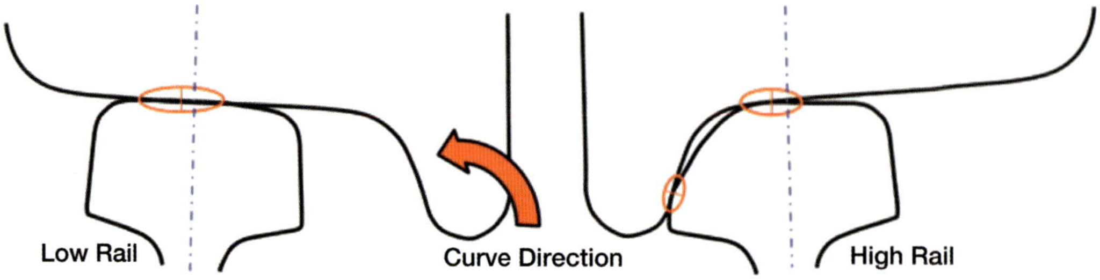

Figure 185: Poor Rail/ Wheel Contact Due to Poor Rail Profile

Figure 186: Hollow Wear

3.3. Poor Rail/Wheel Contact due to Hollow Wheels

Hollow wheel wear develops on the wheel tread and is specifically a problem on very straight long lines with very flat curves using dedicated rolling stock typically found on heavy haul lines. If wheel profiles are not properly managed, the wheel pattern can wear hollow as illustrated in Figure 186 and Figure 187.

Well managed hollow wear is normally limited to between 2 and 4 mm maximum.

The hollow wheel will cause a very large contact area between the rail and the wheel. On the field side of the wheel tread a false flange develops (Figure 187).

Since there is no distinctive wheel radius across the wheel treads, pure rolling cannot be achieved in curves. Due to centrifugal forces in the curve and the resulting lateral shift of the wheelset, the hollow wheel will cause very high contact stresses on a very small area of the rail gauge corner on the high rail resulting in severe rolling contact fatigue and even plastic flow of both the wheel flange and gauge corner (Figure 188). The false flange will similarly cause very high contact stresses on the field side of the low rail promoting plastic flow of the field corner.

In turnouts the hollow wheel false flange may make contact with the stock rail as illustrated in Figure 189 causing high contact stresses and a lateral force on the stock rail placing strain on the fastening and potential rollover of the stock rail.

If the wheels have asymmetrical wear, either in terms of diameter or hollow wear, the difference in radius of the wheels will cause skew bogie movement and rapid wear and contact fatigue of the rail and wheel with very poor steering.

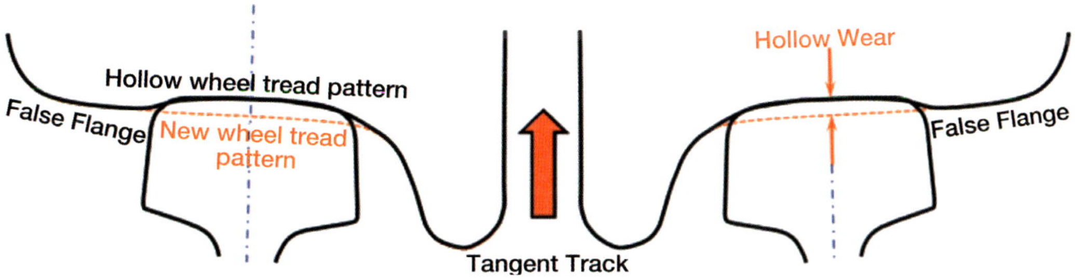

Figure 187: Large Contact Area due to Hollow Wheels

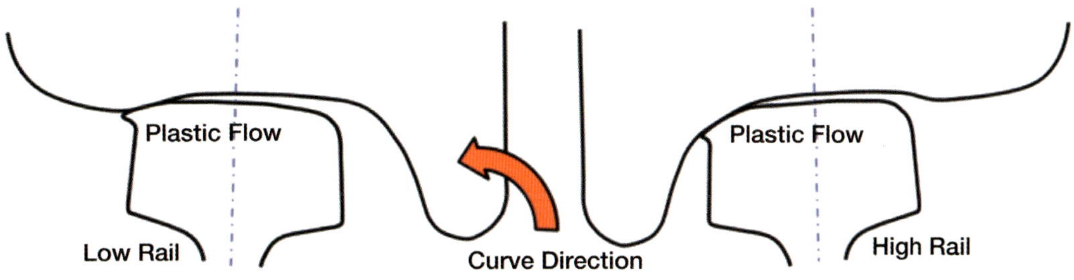

Figure 188: Plastic Flow due to Hollow Wheels

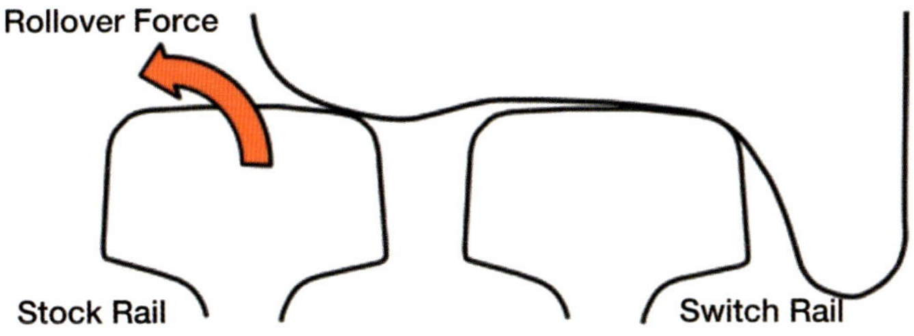

Figure 189: High Lateral Forces on the Stock Rail due to Hollow Wheels

4. CONTACT MECHANICS

4.1. Contact Patch

The contact patch between the rail and the wheel is about the size of a coin, depending on the wheel diameter and rail profile, with an elliptical shape according to the Hertzian contact theory. On a typical heavy haul 30 ton axle loading wagon this tiny contact patch carries 15 ton! However, because of equal shear stresses on the surface, deformation of the rail and wheel is minimal resulting in the wheel and rail to sustain very high loads at a very low rolling resistance. These stresses act in the three principle directions (x, y, z) which decay at different rates under the rail surface beneath the contact patch. The unbalance between the stresses creates a shear stress which may cause failure such as shelling or spalling.

4.2. Forces Acting on the Contact Patch

- **Normal Contact Forces** – The wheel loading due to the weight of the rolling stock and the pay load applies a vertical static load on the rail. Due to the conical shape of the wheel, the gravitational force as a result of the axle loading is not only vertical but also at an angle resulting in an additional lateral force on the contact patch. When the wheelset moves off the central position on the rail as it enters a curve for example, the lateral force will become unequal between the two rails due to the nonlinear conical shape of the wheel resulting in a net force acting on the wheelset. Due to the equal shear stresses on the surface the rail and wheel can sustain this load.
- **Dynamic Contact Forces** – The static normal contact forces will increase exponentially with rail and wheel surface defects and track geometric defects as a result of dynamic loading with rolling motion. The dynamic loading will increase as the speed increases. Dynamic loading can exceed the normal contact forces by a factor of 5 and is very damaging to the rail and track alike and a major cause of accelerated track deterioration.
- **Adhesion due to Traction** – Normal traction forces will result in adhesive wear (friction) of the contacting surfaces. High traction forces can cause the shear yield stress of either the wheel or the rail to be exceeded causing abrasive wear, plastic flow and/or subsurface or surface crack formation.
- Slip due to loss of Traction – Forces that cause deviations from pure rolling will result in relative slippage between the rail and wheel in the longitudinal as well as lateral direction. This slippage is called creepage which arises if the longitudinal distance travelled by the rotating wheel is longer or shorter than the circumferential length (one rotation) of the wheel typically during braking or acceleration. Lateral creepage is associated with the wheelset entering or exiting a curve and the lateral movement of the wheel relative to the rail to find radius on the wheel that will provide pure rolling.

4.3. Forces Across the Contact Patch

Forces as a result of traction across the contact patch leads to the initiation of rolling contact fatigue in the wheel and/or rail. Surface fatigue is generally influenced by:

- the ratio between creep forces and the normal contact force; and
- the ratio between the material yield strength and the maximum contact stress.

Some of the factors influencing the forces across the wheel/rail contact patch are:

- **Bogie Suspension Design** – The development of self-steering bogies dramatically reduced lateral creep forces. On the other hand, conventional bogies with misaligned wheelsets due to poor maintenance practice will experience increased creep forces.
- **Wheel and Rail Profile Design** – Wheel profile design to match the rail profile design as well as the maintenance thereof with tight tolerances will have a significant effect on the contact stresses and the dynamic behaviour of the rail vehicle as described in paragraph 3 above.
- **Track Geometry** – Track geometry defects such as horizontal alignment defects causes a rapid lateral acceleration of the wheelset across the rail. The lateral acceleration will increase with an increase in train speed.
- **Track Topology** – The alignment of the track with regards to sharp curve radii will cause greater lateral movement of the wheel across the rail.

4.4. Forces on the Wheel and Rail (Not on the Contact Patch)

Compressive and tensile forces are transmitted between the rolling wheel and the rail due to the deformation of the rail and wheel material as illustrated by Figure 190. As can be seen, the rail and wheel are in oppositely deformed states both in front of and behind the wheel. To change from one state to the other state of deformation, there will be relative slip between the rail and the wheel causing fatigue and wear.

4.5. Friction

Friction is a force in the opposite direction to the direction of motion that one object encounters when moving over or against another object.

Two objects in relative motion will experience friction between the objects which arises from a combination of inter-surface adhesion, moisture, temperature, surface roughness, surface deformation and surface contamination. The frictional force is directly proportional to the applied load (weight) but independent of the area of contact. The complexity of these interactions means that friction cannot be expressed as a notional value but rather an empirical coefficient of friction (CoF) Mu (μ).

This can best be explained by Figure 191 with object 1 being stationary on object 2. The pulling force $mg\sin\Theta$ will tend to pull object 1 down the slope. If the object is however stationary it means that the frictional force μF_n is equal to pulling force $mg\sin\Theta$ where F_n is the force created by the load of the object. In other words, the frictional force between two stationary objects is equal to the force that would cause motion of the object.

Lubricating the area between the surfaces in contact will reduce the coefficient of friction (see CHAPTER 3 paragraph 2.14).

Friction falls within the study of tribology which is the science of interacting surfaces in relative motion. Tribology also includes the study of lubrication and wear.

The CoF can be measured with a tribometer which measures sliding conditions. Though not a representative measurement for rail/wheel friction, it provides a basis to manage from. Using a tribometer the Transportation Technology Centre Inc and National Research Council of Canada have established the following best practice guidelines for rail/wheel friction:

- Gauge face on curves: less than 0.2μ
- Gauge corner on curves: less than 0.25μ
- Top of rail (curves and tangent track): 0.3 to 0.4μ
- Maximum difference between top of any two rails (curves and tangent track): less than 0.1μ

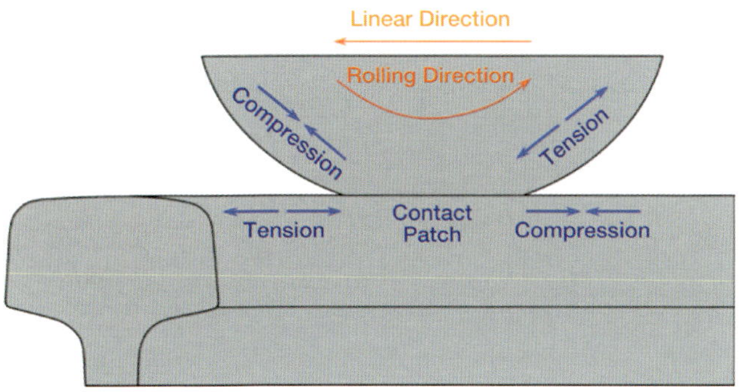

Figure 190: Forces Transmitted between the Rolling Wheel and the Rail

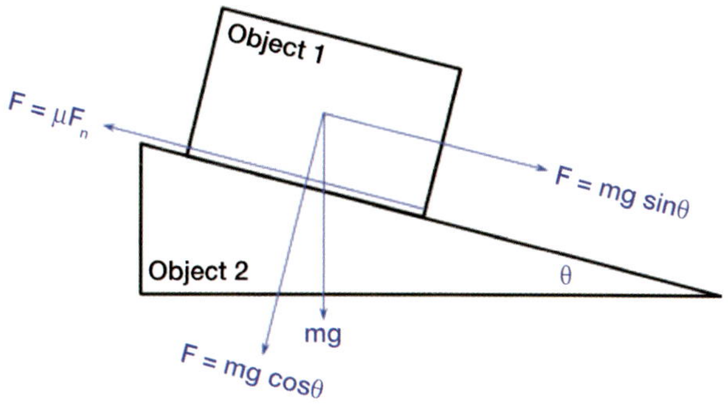

Figure 191: Frictional Force

Chapter 5
Track Deterioration

1. INTRODUCTION

All the track components are subjected to deterioration either through the forces exerted on the components and/or wear and/or contamination. It is the objective of the maintenance programme to slow down the deterioration and for this an understanding of the causes of deterioration is required.

2. FORCES EXERTED ON THE TRACK

One of the most important functions of track maintenance is to reduce the forces exerted on the track as these forces will rapidly reduce track life. The forces discussed here has a net effect on the track structure and should not be confused with the micro forces as discussed in CHAPTER 4 on the fundamentals of rail and wheel interaction. Forces on the track act from 3 main directions onto the track:

2.1. Vertical Forces

Vertical forces originate from static loads (wheel loading as a result of rolling stock weight) and dynamic loads caused by track geometry faults, rail running surface faults (corrugations, wheel spin marks, poor joints, cupping on welds, etc.), rolling stock defects such as flat spots on wheels, etc.

With the rail supported intermittently by sleepers the wheel loading is a fatigue load on the rail due to rail deflection. This fatigue load can be considerable if the rail profile is light and the sleeper spacing is large. Ideally the fatigue loading should be limited so as to give the rail a longer fatigue life than a wear life. Increasing rail fatigue will be manifested in increasing number of rail breaks.

In curved track the wheel loads will also vary by as much as 20 to 30% between the high and low rails depending on the track superelevation and operating speeds.

The vertical forces on the rail have to be taken up by the individual components of the system of the superstructure. The very high surface pressure between wheel and rail is gradually reduced down to the subsoil which is the weakest component of the track structure (see Figure 192 below and also Figure 105). In this example a wheel load of approximately 12,5 ton (axle load of 25 ton) is used.

Especially dynamic forces are very damaging to the entire track structure causing over stressing of the ballast and formation and breakage of concrete turnouts. It causes pulverisation of the ballast resulting in fouling of the ballast bed, reduction in ballast depth, inelasticity of the ballast bed, retention of moisture; all of which will very quickly result in formation failure ending the track life prematurely and increasing lifecycle cost.

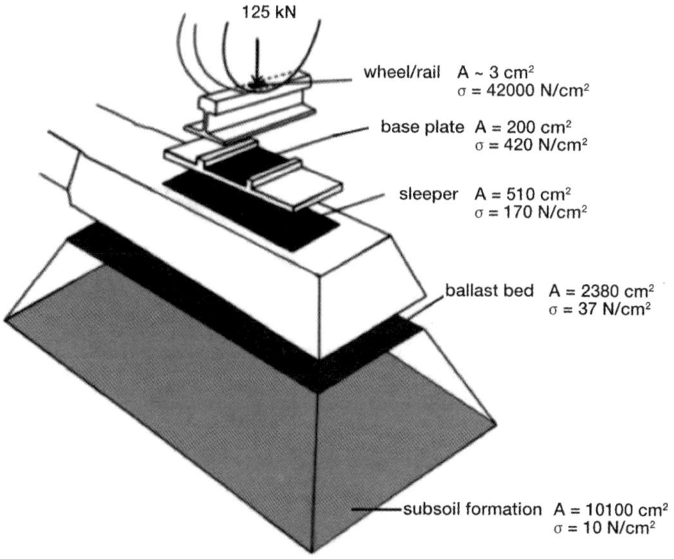

Figure 192: Dissipation of Vertical Forces

2.2. Longitudinal Forces

- **Longitudinal forces due to rail/wheel interaction** – Creep forces at the wheel/rail contact patch in the longitudinal direction (see CHAPTER 4) are generated by the tractive force of the locomotive. These forces will be transferred to the sleeper causing longitudinal forces on the entire super structure.

- **Longitudinal forces due to rail stress** are either compressive or tensile of nature –
  - Compressive forces are caused by temperature increases (the resultant elongation of steel) and heavy axle loads on downgrades as a result of braking and/or rolling action of the wheels (causing compressive forces at the one end and tensile stresses at the other). Compressive forces may result in kick-outs (Figure 193) and closed gaps in fish-plated joints such as at turnouts.
  - Tensile stresses develop at low temperatures and at the top of downgrades as explained above. Tensile stresses my result in a rail break (Figure 194).

If the longitudinal forces are greater than the frictional forces created by the fastenings, rail creep can occur relative between the rail and sleeper causing skew sleepers and accelerated wear of rail pads. Longitudinal forces must be compensated for by regular rail stress measurements and destressing the rail by cutting the rail, rolling the stress out and re-welding the rail during the neutral temperature range.

Rail creep can also be prevented by replacing fatigued fastening systems or worn rail pads or replacing them with higher creep resistant components if creep persists or prevent creep altogether by the installation of rail anchors.

Adequate ballast in the cribs and at the ballast shoulder will also resist kick-outs and skewing of sleepers.

Refer to CHAPTER 3, paragraph 2.12 for a more in depth look at rail temperature, rail stress, rail destressing and neutral temperature destressing range.

2.3. Lateral Forces

Horizontal forces transverse to the track (lateral forces) can be caused by centrifugal forces of trains negotiating a curve, wind acting on the side of the train, a result of compressive stresses in the rails due to temperature increases etc.

The centrifugal forces caused by a train negotiating a curve is compensated for by cant (see CHAPTER 8 paragraph 3.1.3), adequate ballast shoulder at the sleeper head (see CHAPTER 3 paragraph 5.1.1) and making the curve radius as large as possible (see CHAPTER 8 paragraph 2.2.1.1). The curve radius is determined during the design phase based on various factors such as topography and manmade obstacles but the first two are maintenance issues.

Figure 193: Temperature Kick-Out due to Compressive Forces in the Rail

Figure 194: Rail Break due to Tensile Stresses in the Rail

Dynamic lateral forces can also occur as a result of wagon/locomotive hunting from the one rail to the other causing damage to the gauge face of the rail and high dynamic lateral forces (Figure 195). There are various reasons why this hunting phenomenon may start including rail profile irregularities (poor welds which are staggered on the two rails i.e. not across from one another), track geometry defects (such as twist faults), dynamic behaviour of the rail wagons (a combination of the centre of gravity of the wagons, wheel base, hollow wheels, etc.), misaligned of self-steering bogies on locomotives etc. Most of these are maintenance issues that can be avoided.

2.4. Bending Forces on the Rail

Bending of the rails can occur because of a range of forces, including:

- The rail will bend vertically between the two sleepers as the wheel sets pass over that area. The longer the sleeper spacing the more the vertical deflection. This leads to tensile longitudinal stresses in the rail foot and the fishing surface. See vertical forces causing fatigue in paragraph 2.1 above.
- Lateral wheel loads will cause the rail head to move laterally relative to the foot causing tensile vertical stresses in the rail web and tensile longitudinal stresses on the fishing surface located on the field side.
- Vertical loads applied to the side of the rail head off the centre line by wheel false flanges, hollow wheels etc. leads to a bending stress on the rail web and tensile longitudinal stresses on the fishing surface.

Bending stresses on the rail may result in crack propagation and rail fracture.

3. WEAR OF COMPONENTS

The rail is the most susceptible to wear as this is the contact between the train and the track. CHAPTER 3, paragraph 2.18 discusses the wear characteristics of rail in detail.

On well-maintained rail and a well maintained track the natural wear of the rail due to wheel on rail is minimal. As a matter of fact, the maintenance of the rail, profiling through grinding or milling, is responsible for more artificial wear of the rail than the rail/wheel interaction. See CHAPTER 16 on rail profiling, especially paragraph 2 on the "magic wear rate".

Rail profiling is however one of the most important maintenance activities to avoid head checks and cracks developing due to rolling contact fatigue, hunting of the wheel across the rail due to poor rail profile, wheel flange contact in curves etc.

The rails in turnouts are especially prone to dynamic contact between rail and wheel and wear of the turnout components.

Fastenings and rail pads also wear over time and therefore has a finite life.

4. CONTAMINATION

The ballast bed can also wear due to stones grinding against each other causing them to turn to fine dust and or smaller pieces outside of the good ballast stone spectrum. The fine material is contamination of the ballast bed and has various other causes such as washed in during rains, wind-blown in etc. CHAPTER 3, paragraph 5 discusses ballast in detail.

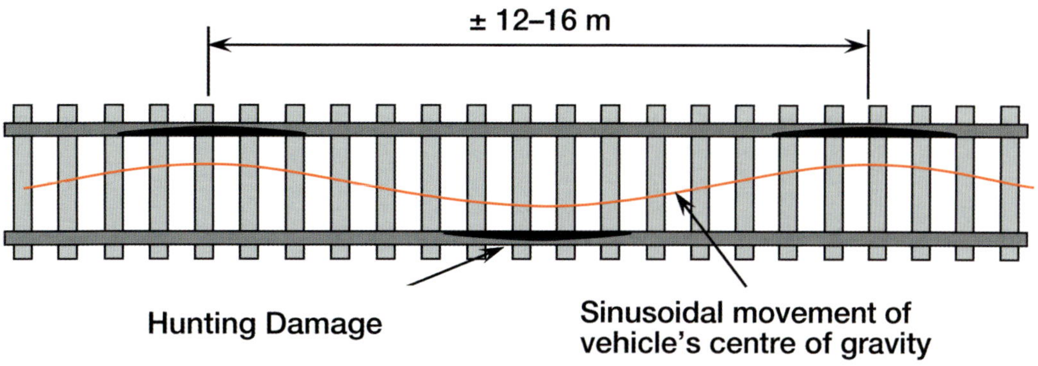

Figure 195: Hunting Following a Sinusoidal Damage Pattern

Chapter 6
Maintenance Strategy

1. INTRODUCTION

An efficient rail infrastructure is a prerequisite for any country's economic growth, development and prosperity to improve the general standard of living of all its people and to participate successfully in the global economy. An efficient rail infrastructure has undeniably been key to the success of most, if not all, affluent industrialised countries. An efficient railway system will stimulate the development of the national economy by transporting goods and passengers at minimum cost. Lower cost structures facilitate domestic competition and make exports more competitive on international markets.

However, for the railway system to be called "efficient" and to be able to transport freight and passengers at cost effective rates, the railway infrastructure must be reliable, available, maintainable, affordable and safe (RAMAS). This can only be achieved if an effective track maintenance strategy is followed which considers the entire track infrastructure, track life and entire lifecycle cost of the infrastructure. This chapter looks at maintenance as a process for which the strategy is the start of all.

2. PROCESS APPROACH TO TRACK MAINTENANCE

Track maintenance follows a process as illustrated by Figure 196. There are inputs into the process which is processed to provide a product which is a well maintained track. The process starts with a maintenance strategy.

2.1. Inputs into the Process

The inputs into the system as per Figure 196 fall mostly outside of this manual as a management function not directly involved with the day to day maintenance of the infrastructure though some aspects are addressed in various places in this manual especially the selection criteria for mechanised track machinery.

- **The maintenance strategy** has the objective to slow down the track deterioration (CHAPTER 5) by implementing various initiatives such high initial quality of construction and maintenance effort, setting intervention levels, etc. all which will reduce the track infrastructure lifecycle cost.

Input	Process	Output
Maintenance Strategy Finance Track Condition Info Maintenance Machinery Labor	Maintenance Planning Maintenance Practice	Reliable Available Maintainable Affordable Safe

Figure 196: Track Maintenance as a Process

- **Finance** (paragraph 6) must be available to maintain the track to the condition decided in the strategy, to procure the necessary machinery or contracts to carry out the maintenance work and to fulfil the budget requirements for employing qualified track staff.
- **Track condition information** to scientifically carry out maintenance must be established and the measured data analysed to produce defect reports and management reports. All possible parameters of the infrastructure must be measured. The measuring of the track infrastructure can only be done effectively with mechanised methods including the use of:
  - a track measuring vehicle (CHAPTER 9) for geometry, rail profile and overhead traction equipment measurements;
  - ultrasonic measuring vehicle and equipment (CHAPTER 10 paragraph 2) for detecting and measuring the size of internal rail defects;
  - eddy current measurements (CHAPTER 10 paragraph 3) for detecting and measuring the size of rail surface defects; and
  - various other mobile or hand-held measuring equipment to determine track condition wear or contamination, such as ground penetrating radar for the latter.
- **Mechanised track construction, maintenance and renewal machinery** of the required type, production, mix of machines, features etc. is essential to ensure that the track will be maintained to the required standard. This input into the process approach is the main subject of this manual and the types and selection criteria of machines will be discussed in all the following chapters.
- **Any company's** most valuable asset is its labour force (paragraph 8). Being short staffed or staff lacking in engineering and track maintenance qualifications, experience and knowledge can be one of the most costly mistakes a railway can make.

2.2. The Process

This section involves the actual planning and execution of the maintenance work.

- **Maintenance planning and scheduling** (paragraph 9) is an essential part of the process. It is also very scientific for which various methods, models and software packages are available. It is probably safe to say that most maintenance programs that fail would be due to poor planning. Despite its importance, this manual does not address maintenance planning due to its complexity and that a mere chapter on planning cannot do it justice.
- **Maintenance practice** involves all the procedures, processes and standards that a railway follows in day to day maintenance. There are many best practice guidelines and every railway should have its own maintenance manual and a set of standards that address every aspect of maintenance. In CHAPTER 25 there is a list of some of the most pertinent European Standards; there is also the AAR Manual of Standards and Recommended Practices, the American Railway Engineering and Maintenance-of-Way Association (AREMA), the International Heavy Haul Association Best Practice Guidelines, etc.

2.3. Output

RAMAS (reliability, availability, maintainability, affordability and safety) are the goals that every railway administration is seeking with regards to the track infrastructure as this would have a major influence on the ability of the railway to offer cost effective transport rates while still being profitable (private railways) or stay within government grants.

Better known is the acronym RAMS which does not include "affordability". RAMS is a European standard (EN 50126) which provides a process for the implementation of a system for the management of RAMS.

This manual does not have the same objectives or follow any of the principles of EN 51026-1:1999, however, the following definitions for the acronym were taken from the EN-standard:

- **Reliability:** The probability that an item can perform a required function under given conditions for a given time interval.
- **Availability:** The ability of a product to be in a state to perform a required function under given conditions at a given instant of time or over a given time interval assuming that the required external resources are provided.
- **Maintainability:** The probability that a given active maintenance action, for an item under given conditions of use can be carried out within a stated time interval when the maintenance is performed under stated conditions and using stated procedures and resources.
- **Affordability:** A measure of cost relative to the amount that the purchaser is able to spend.
  This definition is not part of RAMS EN Standard.
- **Safety:** Freedom from unacceptable risk of harm.

3. MAINTENANCE OBJECTIVE

The objective of maintenance is to lower the track geometry deterioration rate to preserve the service life of the track. Figure 197 schematically shows a hypothetical track deterioration curve and the influence of maintenance on track life.

The track is initially constructed using construction methods and materials that will place the track at a track condition "a" from where it will deteriorate in condition over time and traffic along the curve "ac" due to an increase in track roughness resulting in an increase in dynamic loads and accelerated rate of deterioration.

The curve "ac" represents deterioration of the track with no maintenance intervention. If the deterioration is left unattended, the functionality of the track will reduce until it can no longer be used safely.

However, for the safe passage of traffic at a realistic speed, the track cannot be allowed to deteriorate beyond a determined threshold for minimum allowable track condition. This threshold can be based on a number of measurements; it could be a number of defects per kilometre, a track quality index, etc.

To extend the life of the track, track deterioration must be slowed down through planned maintenance interventions (as depicted by the line "bd") which are executed on time. The track condition is therefore improved again to the point "d". However, the original as built functional condition cannot be regained by typical maintenance input due to the wear of the track components.

After maintenance input the track will continue to deteriorate along curve "de". Once again, as the deterioration approaches the threshold for minimum allowable track condition, maintenance input is required "ef" and, as before, the condition as achieved after the previous maintenance input cannot be regained.

This process continues with each maintenance input achieving a lower track condition than before and the interval between maintenance inputs reducing exponentially (refer to the distances between "be", "eg" and "gi"). This produces a new deterioration curve, curve "ak", which is much longer than the curve "ac" without maintenance input. The life of the track has therefore been extended through incremental maintenance.

This process will continue until the period between required maintenance inputs becomes uneconomically short. Complete track renewal e.g. formation rehabilitation, replacement of sleepers, rails, fastenings, ballast or any combination of these components will then be required during which the original as built condition can be reinstated (line "ij"). The whole process will repeat itself.

The longer the curve "ak", the longer the life of the track and the lower the track lifecycle cost will be resulting in lower freight costs and greater competitiveness on world markets for exports from the country in question.

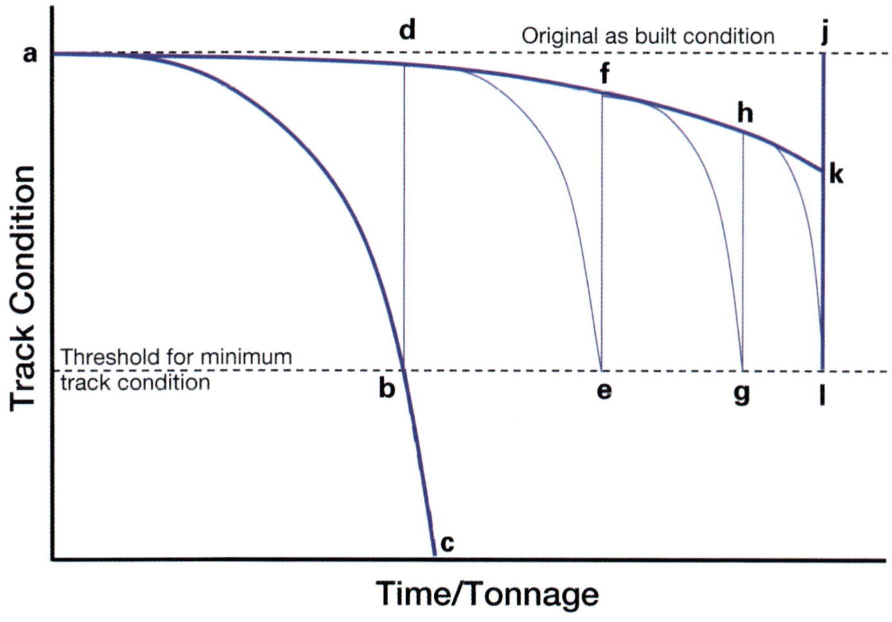

Figure 197 : Hypothetical Track Deterioration Curve

4. CONDITION PARAMETERS OF A TRACK MAINTENANCE STRATEGY

4.1. Initial Quality Related to Construction

Decisions made during the planning, design and construction phases of the track have far reaching consequences for expenditure later in the life of the track structure. This point is illustrated in Figure 198. If a lower initial quality of the track was selected with the use of lower standard track components or construction methods to save money in construction costs (the drop from "a1" to "a2"), the track will deteriorate quicker as illustrated by the curve "a2b2". The maintenance cycles will be shorter and track renewal will therefore also be required quicker (the difference between "j1" and "j2"), resulting in a reduced track life, higher freight costs and lower competitiveness. Therefore, the higher the initial quality the longer the track life will be and the lower the lifecycle costs.

4.2. Initial Quality Related To Maintenance

The principle of initial quality can also be extended to maintenance. Refer to Figure 199. All being equal, if the same initial quality of construction was used at point "a" and the deterioration is the same as before (deterioration curve "ab") but the maintenance methods employed does not bring the track back to the maximum condition possible (the difference between points "c1" and "c2"), the track will deteriorate much quicker causing premature track renewal (the difference between "j1" and "j2"), a shorter track life and a more expense lifecycle cost.

The selection of maintenance methods and maintenance machinery must therefore also receive careful consideration; for example, the use of some manual maintenance methods (labour intensive methods) cannot achieve the same level of precision or quality as mechanised methods and will therefore not be able

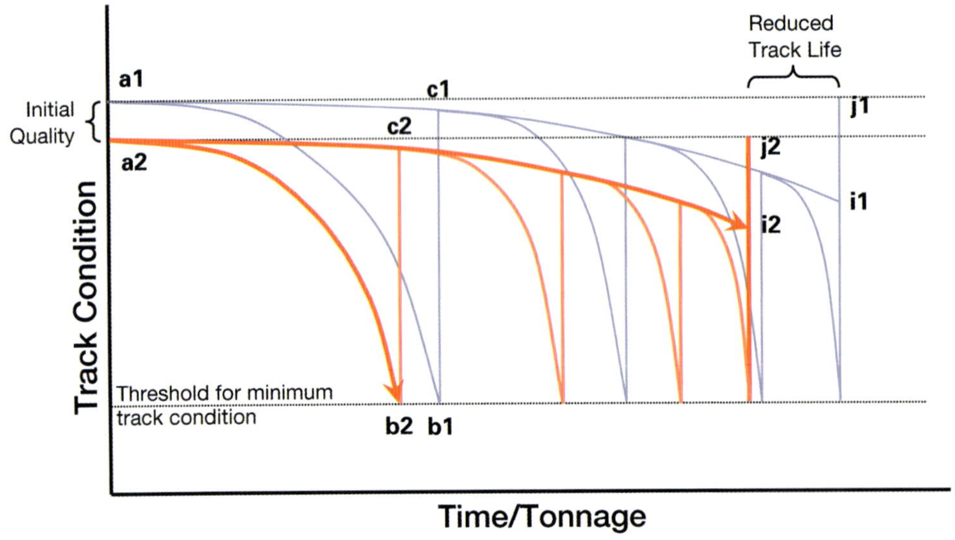

Figure 198: The Effect of Initial Quality in Construction on the Deterioration Curve

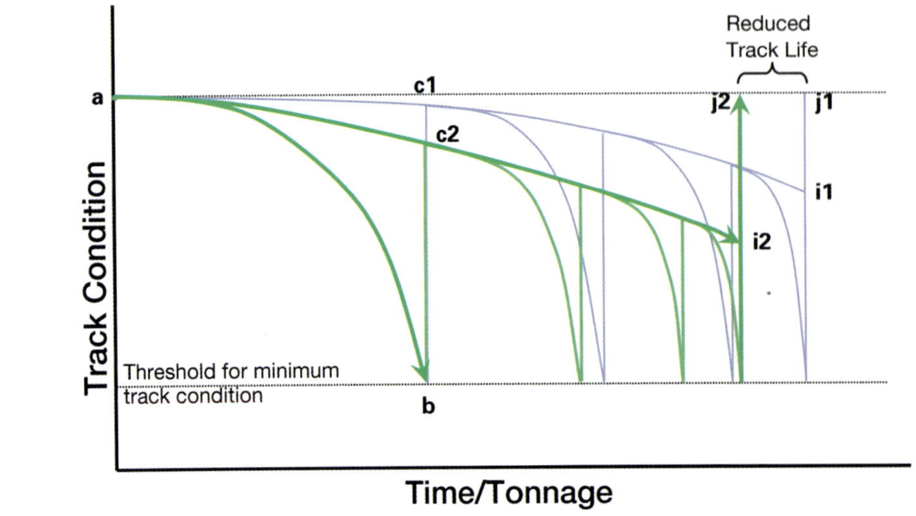

Figure 199: The Effect of Initial Quality in Maintenance on the Deterioration Curve

to bring the track condition back to the point "c1" and the same applies to mechanised machinery as they are not all equal in terms of processes used and quality of work.

4.3. Threshold for Minimum Allowable Track Condition

The maintenance strategy should also address the value at which the threshold for minimum allowable track condition is set as this will have a major influence on the track's lifecycle cost.

A paper by Prof Peter Veit "Outsourcing in Track Maintenance" provides some of the vital technical and economical correlations of the track based on a project adopted by the Austrian Railways since 1996. The paper and project concentrated on the optimisation of track maintenance and also addressed the issue of threshold.

If the threshold for maintenance intervention is set at a low track condition value as indicated by the line "threshold 1" on Figure 200, the effect of maintenance intervention at such a low value will be hardly noticeable at first but the track's service life will be reduced by the knock-on effect caused by the poor track condition. For example, if the track roughness index [track quality index (TQI) as measured and calculated by a track recording vehicle] is set too low before tamping takes place, the roughness will cause higher dynamic loading of the track which will cause crushing of the ballast, which in turn will result in fines in the ballast bed retaining moisture, etc.; eventually resulting in damage to the track material reducing the life of the track exponentially. Refer to CHAPTER 9 paragraph 5.2 for a description of how the TQI can be used as a threshold for maintenance intervention.

Prof Veit's research supported and proved this hypothesis. His evaluation showed relatively high initial improvements in quality values directly after maintenance, however, this level was not sustainable and the deterioration accelerated over time. The quality level achieved after maintenance was lower than when measures had been implemented sooner (such as at threshold 2). Intensive maintenance efforts at a later stage to extend the service life proved uneconomical.

It is however not only the level at which the threshold is set that is of significant economic importance. Research has also shown that the threshold should not be set at a fixed value but should rather be linked to the age of the track.

The general tendency by maintenance managers is to maintain a constant threshold for maintenance intervention throughout the life of the track (see curve "b" on Figure 201). With a constant threshold, maintenance intervention will always take place when the track condition has reached a predetermined level, irrespective of the age of the track.

However, in practice it is more likely to find an increasing threshold (curve "c" in Figure 201) which means that while the track is new, the available maintenance budget is concentrated on other sections where the

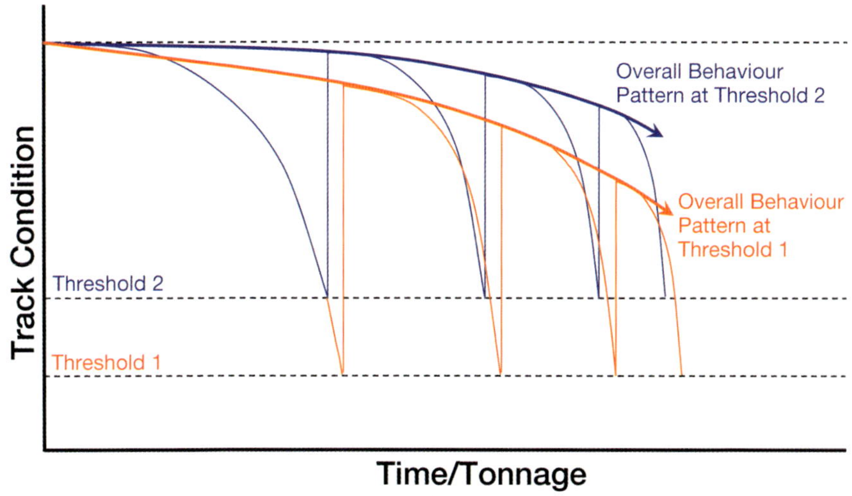

Adapted from Prof. Veit – "Outsourcing in Track Maintenance", OVG conference, Sep 2004

Figure 200 : Summary of the Effect of Maintenance Due to Threshold

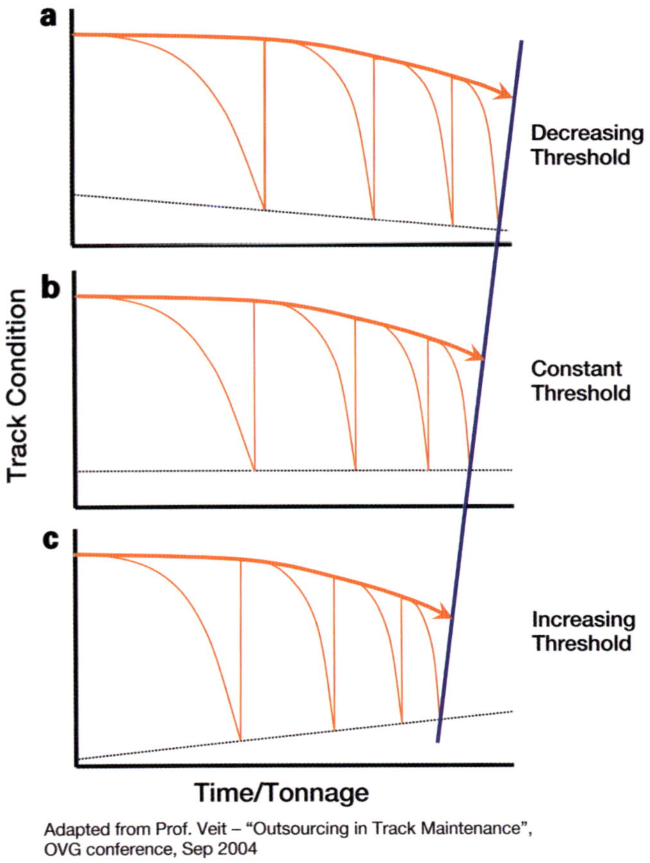

Figure 201: The Effect of the Various Thresholds on Lifecycle

track condition is poor. Initially the effect thereof cannot be seen. As the track ages, the threshold has to increase due to the increasing wear of all the track components, to maintain a reasonable availability and track life. However, the knock-on effect referred to earlier has already taken its toll and will negatively impact on the life of the track.

The Austrian project has however proved that the most economical approach to track maintenance is when resources are invested in a high initial quality level with corresponding maintenance. In other words, a long service life and lowest lifecycle cost can be expected when the threshold for maintenance intervention is at its highest while the track is still new. The threshold is then gradually reduced towards the end of the track life (curve "a").

5. MAINTENANCE TACTICS

The maintenance tactics that will be employed during the life of the track must form part of the maintenance strategy as this will influence the financial investment as well as the organisational structure required and will have an immense effect on the lifecycle cost and life expectancy of the track.

Figure 202 provides a diagrammatical layout of typical maintenance tactics. The two main tactics are either "planned" or "unplanned" maintenance. The danger of unplanned maintenance is that it grows and can easily consume all available resources over time. The railway can fall into a situation where all resources and finances are employed to address unplanned emergency work due to the unpredictability of the track which allows the track structure to deteriorate very fast due to a lack of preventive maintenance.

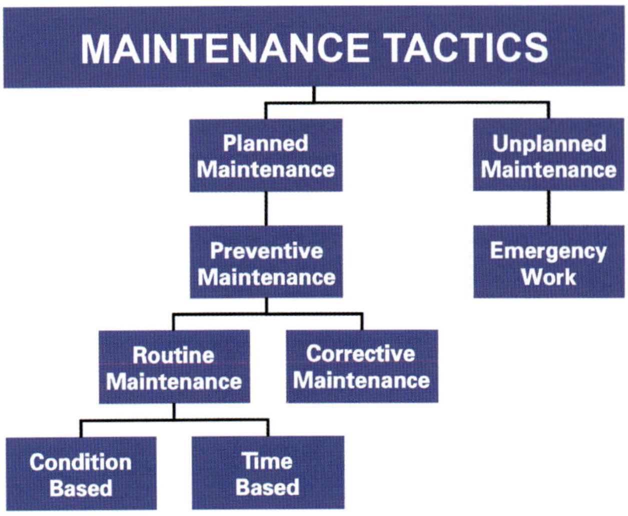

Figure 202: Maintenance Tactics Diagram

Ideally the majority of resources employed should be towards planned preventive maintenance activities as this is the only tactic that will extend the life of the track and will keep unplanned maintenance to a minimum, making the track more reliable, available and maintainable.

Preventive maintenance has two elements, routine maintenance and corrective maintenance. Corrective maintenance refers to work that is required to prevent failure after a defect has been identified and the date and time of the execution of such work can be planned i.e. it is not an emergency yet.

Routine maintenance is carried out to maintain the track before the threshold for minimum track condition referred to earlier has been reached which will prevent defects from occurring.

Time based routine maintenance is based on the elapsed time since the last maintenance action took place. Time based maintenance is usually carried out on off-track components such as drains, greasing etc. and is often seasonal.

Condition based routine maintenance will use the data from a track recording vehicle, trolley inspections, footplate inspections etc. to determine where the use of scarce resources can be optimised. Under most circumstances condition based maintenance will contribute towards the lowest possible lifecycle cost.

6. FINANCE

The level and consistency of financial input into track maintenance will have a major influence on the track life. Track maintenance comprises a large percentage of any railway's operational expenditure. When financial difficulty is experienced, the maintenance budget will usually be reduced first. The graph in Figure 197 was based on the hypothesis that the necessary maintenance input in financial terms has been allowed for. The question arises what the case will be if the maintenance input is inadequate i.e. both in terms of maintenance intervention at a point below the maintenance threshold (too late) and/or insufficient input (too little) to achieve the highest possible condition due to the unavailability of finance to carry out the work. Figure 203 illustrates how the deterioration curve will drop down much sharper than if timeous and sufficient maintenance was carried out.

The life expectancy of the track has been drastically reduced. In addition, the input required to renew the track will greatly exceed the input that would have been required if timeous and sufficient maintenance were carried out as depicted by the length of line "ab" as opposed to the line "cd".

The effect of such an underinvestment in maintenance was highlighted during the Regional Track Condition Assessment Project (RT-CAP) in the Southern African Development Community (SADC) which was carried out by Plasser South Africa and sponsored by Austria during the 1990's. [SADC consists of 14 member countries being Angola, Botswana, Democratic Republic of Congo, Lesotho, Malawi, Mauritius,

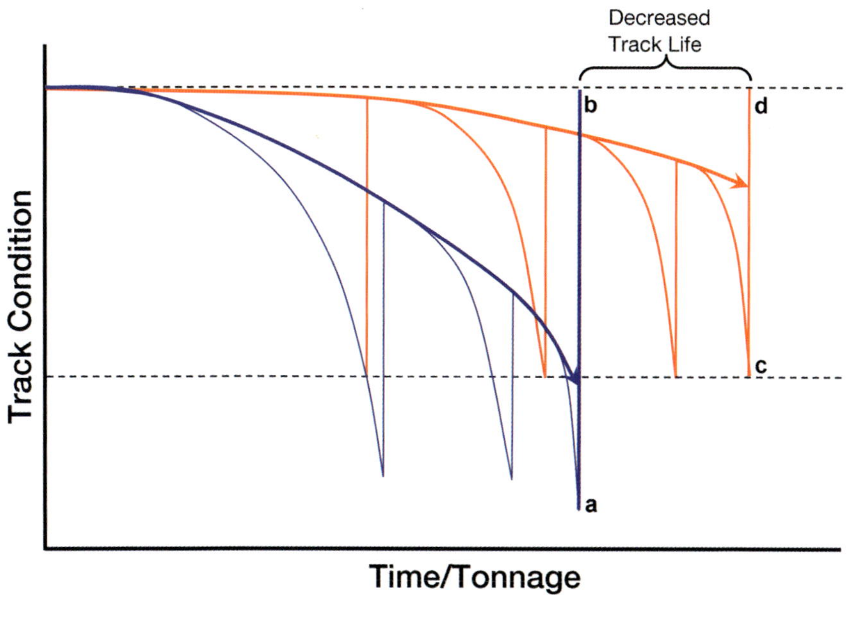

Figure 203: Track Deterioration Curve with Inadequate Financial Investment

Mozambique, Namibia, Seychelles, South Africa, Swaziland, Tanzania, Zambia and Zimbabwe though the assessment was not carried out on all of these countries.]

The maintenance cost curve of Figure 204 is based on the theoretical and practical relationship of under-investment in maintenance of the assets and the resultant intervention cost. The derailments and speed orders (speed restrictions) are the quality indicators which follow as a result of the above mentioned relationship.

A well planned and executed track maintenance strategy will aim to sustain the track condition at a predetermined equilibrium level. This equilibrium level is illustrated by the black horizontal line and can be measured in various different ways such as a standard deviation (measuring vehicle data) or number of workplaces per kilometre. It implies that the financial input is adequate to sustain the track condition at the predetermined equilibrium level.

An inadequate financial input is depicted by the blue maintenance cost curve "MC" which drops to below the equilibrium level, i.e. the financial input is inadequate to maintain the predetermined track condition. The deteriorating track condition will force speed orders (speed restrictions) to be imposed at an increasing number. The number of derailments will also increase.

The loss of customers due to late deliveries as a result of the speed restrictions as well as the cost of derailments will force the railway administration to reverse the negative trend to ensure the continued functioning of the track. This point of intervention is illustrated by "MC2". The shaded area between the equilibrium line and the maintenance cost curve depicts the accumulated backlog costs.

To bring the track condition back to the equilibrium level, a huge intervention cost is required over a short period. This becomes a capital expenditure, as opposed to operational expenditure, for which there is very seldom money available. The value of the intervention costs greatly exceeds the backlog value ("saved" amount) due to the knock-on damage to the infrastructure by the deteriorating track. If the indirect costs of derailments, reduced traffic due to speed restrictions and loss of business are added to the intervention costs, the increase in the lifecycle cost of the track is astronomical. This could have been prevented by small incremental maintenance expenditure.

Re-investment in maintenance will be depicted by the line "MC2" to "MC3" which indicates an acceleration in expenditure and renewal cost. After the peak maintenance cost expenditure level ("MC3") has been reached, maintenance expenditure can be allowed to slow down until it levels off at the equilibrium level, at which point the level of derailments and speed orders will also follow to reach an acceptable level.

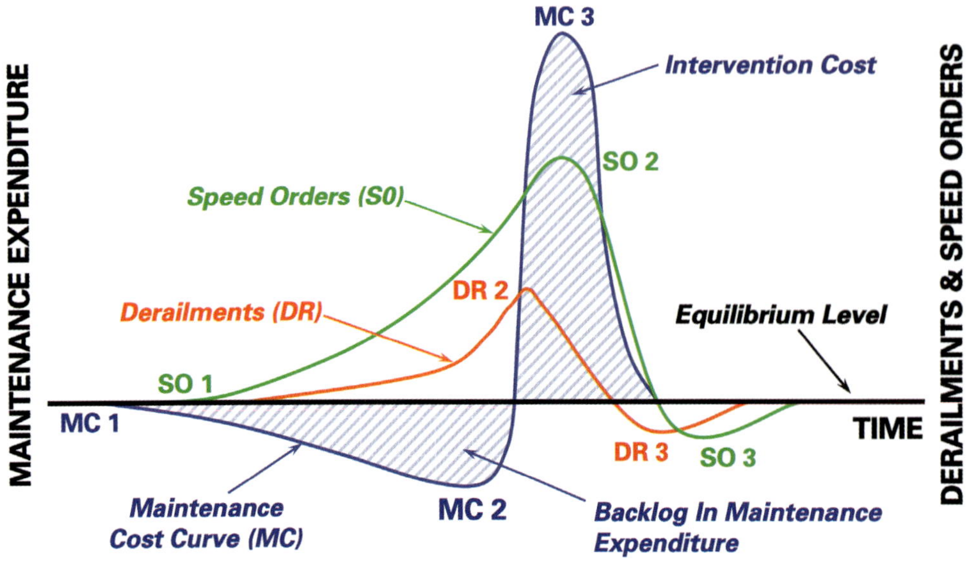

Figure 204: Maintenance Cost Curve (From RT-CAP Project, Courtesy Plasser South Africa)

7. TRACK INFORMATION

The fact that maintenance intervention is required before the track deteriorates to a point below a minimum threshold (quality standard) and for the track to be maintained to an equilibrium level implies that the track condition must be determined on a regular basis.

There are various methods for gathering track information and a combination of all of these methods will contribute towards adequate track information for decision making. These include footplate inspections and measurements that should take place at regular intervals, maintenance managers that do trolley inspections to gain first-hand experience of the track condition and even observations by train drivers can be valuable. All the observations must be recorded and the data must be analysed to produce track information.

Various parameters of measure can then be used to decide where the threshold for minimum track condition lies, such as number of workplaces and/or speed orders per kilometre. For scientific management the data obtained from a track recording vehicle can be manipulated to produce graphs that will provide very accurate information regarding the required level of financial investment to maintain the equilibrium.

Infrastructure measuring vehicles measure and record, among other, the track geometry, overhead contact wire geometry, rail profile and ballast profile and produces various different reports which ranges from lists of track geometry defects by type and location to reports intended for scientific financial management.

For detailed information about infrastructure measuring and recording cars and the parameters they measure as well as the reports generated, refer to CHAPTER 9.

Figure 205: Plasser & Theurer Offers a Vast Range of Infrastructure Measuring Vehicles

8. LABOUR FORCE

Even when maintenance contractors are extensively used, the railways will always need a well-qualified and experienced labour force.

The fit between the needs, skills, experience and values of the employees and the objectives, needs, values, norms and procedures of the railway company (Figure 206) will have a deciding effect on the efficiency and economic viability for the railways.

A bad fit due to low salaries, irregular payment of salaries, lack of training, inadequate equipment etc. will lead to a negative perception among the employees. This will lead to mistrust in the ability of management, low commitment and low work output. Maintenance will eventually suffer to the detriment and demise of the track infrastructure. This is very common in some developing countries.

A concerted effort should be made by any railway to ensure that a good fit between the employee and the company is achieved which will lead to commitment and higher productivity, the result of which will eventually be seen in the track.

9. MAINTENANCE PLANNING AND SCHEDULING

Effective maintenance planning will also affect the track life and lifecycle costs. Planning refers to more than just deciding when and where to work. Planning also includes addressing root causes of problems as opposed to addressing the symptoms, understanding trends and being able to recognise built-in defects such as rail joints (where defects are detected by a recording vehicle). There are just too many variables for any human to manage and cannot be efficient or effective without the use of a proper maintenance management system.

Various systems are available, ranging from highly sophisticated computer systems to a simple line layout diagram. The maintenance planning system should show where what type of maintenance needs to be done and where it was done, what the interval of the maintenance is, where recurring defects which require more investment are, etc. Such a system will allow the maintenance manager to manage his resources effectively.

Planning takes place at various levels in the organisation. Over and above managerial planning in terms of financial investment and the maintenance planning as described above, the day to day logistical planning around the maintenance site is also important for the efficiency and cost effectiveness of the maintenance input.

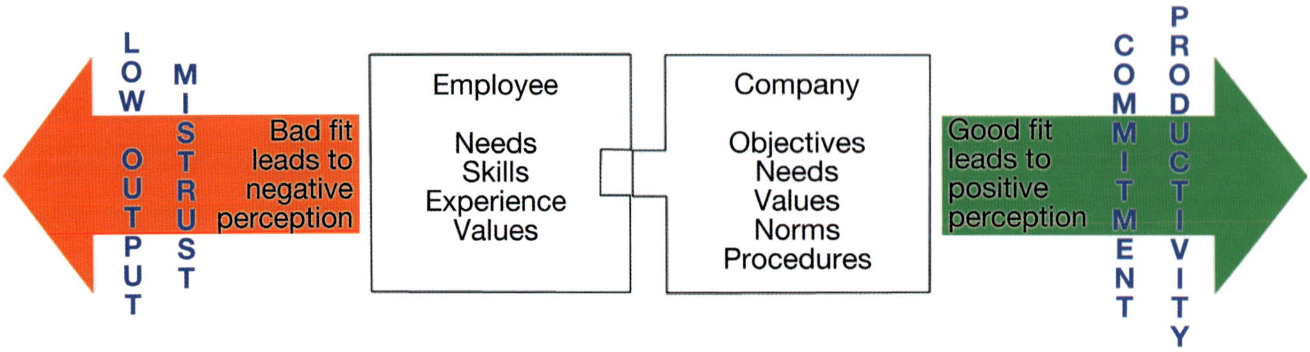

Figure 206: Employee and Company Match

10. HOLISTIC APPROACH TO TRACK MAINTENANCE

The maintenance strategy should look at the entire infrastructure when planning either preventive or corrective maintenance activities. The reason would be that a defect in one component could be seen as a defect in another. For example, an out of stagger measurement on the overhead equipment (OHE) contact wire could actually be because of cant deficiency in a curve, or a severely side worn rail could give the impression of wide gauge.

It is therefore possible for track recording vehicles of high quality to relate all measurements to the same GPS coordinates and planning software to plot all defects including rail wear, OHE, ultrasonic measurements, eddy current measurements, ground penetrating radar measurements, etc. on one report to relate one defect to another to establish a cause and a plan of action.

All track components need to me maintained to the same standard. If grinding for example takes place but the ballast is fouled, much of the benefit of grinding will be lost very quickly since corrugations are likely to return quickly. Similarly, if no or inadequate grinding takes place the dynamic loading will cause pulverisation of the ballast drastically shortening the ballast cleaning benefit.

All components are interrelated since a defect in one will cause a defect in another with overall track condition deteriorating exponentially. The same applies to rolling stock maintenance, especially wheels, due to the effect that the wheels have on the track and vice versa.

11. CONCLUSION

A railway line is an expensive and valuable asset especially in the developing world where it provides vital infrastructure for sustainable social development. The railway is a very reliable and economic means of transporting goods and passengers. Wisely planned and implemented maintenance at regular intervals will extend the life of the railway to remain economically viable for many years to the benefit of economic growth of the country involved and will ensure a track infrastructure that is reliable, available, maintainable, affordable and safe. The old adage "a stitch in time saves nine" rings very true in track maintenance.

Chapter 7
Deciding between mechanised and labour-intensive Track Maintenance

1. INTRODUCTION

The debate around mechanisation versus labour is as old as mechanisation itself and dates back to approximately 1780's. Though mechanisation propelled the average income of ordinary people to unprecedented sustained growth, it is often blamed for job losses and unemployment.

In the railway track maintenance industry, mechanisation really only started gaining momentum in the 1950's but grew very fast and is in full swing today with virtually every aspect of track maintenance being mechanised.

In first world countries the question of mechanised versus labour-intensive track maintenance is no longer debated due to the lack of precision and quality that labour methods is capable of, the tight geometric precision that is required from high speed and high axle load lines and the physical hard work associated with track work which very few people are willing to do today. In addition, the labour cost and the opportunity cost of traffic on lines in first world countries excludes labour-intensive methods completely. It is however not as straight forward in developing countries since labour is much cheaper and traffic is far less to the point where there might be little opportunity costs at all. There are also the issues of unemployment, social responsibility of state owned enterprises and the unavailability of mechanised maintenance machinery.

However, African countries as an example, are rich in valuable minerals and tourist potential in terms of its shear natural wonders. It is also very populated with millions of people requiring affordable and reliable transport. The business is certainly there and can be illustrated by South Africa's two heavy haul lines that carry almost 150 million net tons of freight per annum and international mining groups that are spending billions of Dollars to upgrade and build new sections of railway line in Africa.

Despite this, Figure 208 shows a container terminal in Africa where hundreds of containers are stacked for 20 to 30 days before they can be transported, almost exclusively by road using totally un-roadworthy trucks (Figure 209) on over-congested roads. They pay heavy toll fees and are involved in many accidents, often fatal; all because the maintenance on the railway infrastructure that is still there has been neglected to the point where it is no longer safe to use.

Figure 207: Nearly Every Aspect of Track Maintenance can be Mechanised

Figure 208: Many Containers Stuck in a Harbour Waiting to be Transported to its Final Destination

Figure 209: Un-Roadworthy Trucks Used to Transport Containers Over Long Distances

There certainly is a need to increase rail traffic volumes with heavier axle loads and higher speeds. There is also the issue of interoperability between African countries, especially South Africa where strict minimum standards for track condition are maintained.

All of the above influences the debate around mechanised and labour-intensive track maintenance methods. This chapter investigates if labour-intensive track maintenance methods are still an option today.

2. THE TRACK MUST BE MAINTAINABLE TO MAKE A DECISION

The track must be in a condition whereby it can be maintained otherwise the argument of mechanised versus labour-intensive methods would be futile. This may sound like an obvious comment but many lines in many countries are in a very poor state and are no longer maintainable.

Track maintenance has many aspects but maintaining the track geometry can be singled out due to the destructive effect poor geometry has on the rest of the track structure, train energy consumption and the danger of derailments. The main track geometry parameters are vertical alignment also known as top or profile (Figure 210), horizontal alignment (Figure 211), cant (Figure 212) and gauge (Figure 215). See also CHAPTER 8.

Figure 210: Vertical Alignment Defect

Figure 211: Horizontal Alignment Defect

Cant is the difference between the vertical height of the left hand and right hand rail crowns (Figure 213) at a coincident point to assist steering around curves (Figure 212). In the absence of cant, especially around very short radius curves and the lack of transition curves, poor steering will result in side wear on the high leg rail (Figure 214). See CHAPTER 4 paragraph 2 on the principles of steering. Once side wear has reached this advanced stage, the track is no longer maintainable and too dangerous for traffic. Rail replacement must first be carried out.

Ballast tamping is the process used to correct geometry defects in vertical alignment, horizontal alignment and cant and thereby placing the track skeleton in a new and correct geometric position. This requires sufficient ballast to be packed underneath the sleeper. If the track has insufficient ballast, it is not maintainable.

Figure 212: Cant Visible of Track Around Curve

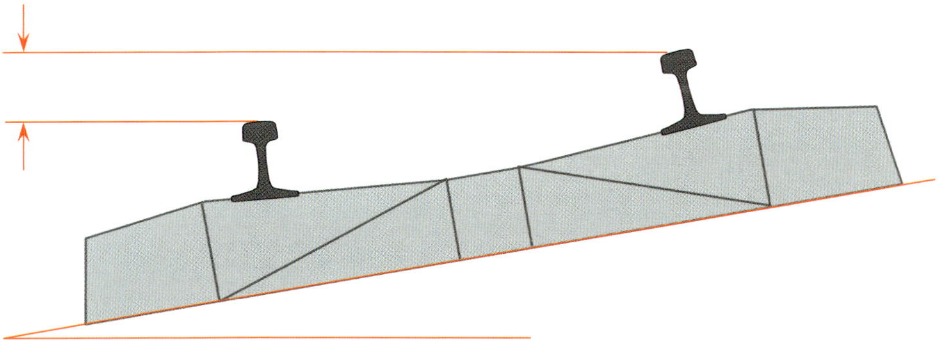

Figure 213: Cant

Figure 214: Severe Side Wear on a Curve

Gauge is also a geometric parameter which is the distance between the insides of the rail crown (Figure 215). Broken sleepers and missing or broken rail fastenings (Figure 216) will not be able to hold the gauge and may result in a derailment. The practice of fitting other sleepers in-between broken sleepers to hold gauge is not acceptable. Broken fastenings will also prevent the track geometry to be adjusted since when the track is lifted, the sleepers will stay behind preventing ballast from being packed underneath the sleeper. Broken sleepers and fastenings must be replaced before the track will be considered maintainable.

All the other track components are equally important though. If, for example, the rail surface is not smooth and profiled, the dynamic loads from passing trains will damage the entire structure as can be clearly seen in Figure 217 where a rail surface defect (locomotive wheel burn mark) caused the sleeper underneath to break. The formation must also be stable since this is the platform on which the track is built. Once the formation has failed, the track will not be maintainable as it would not be able to hold geometry.

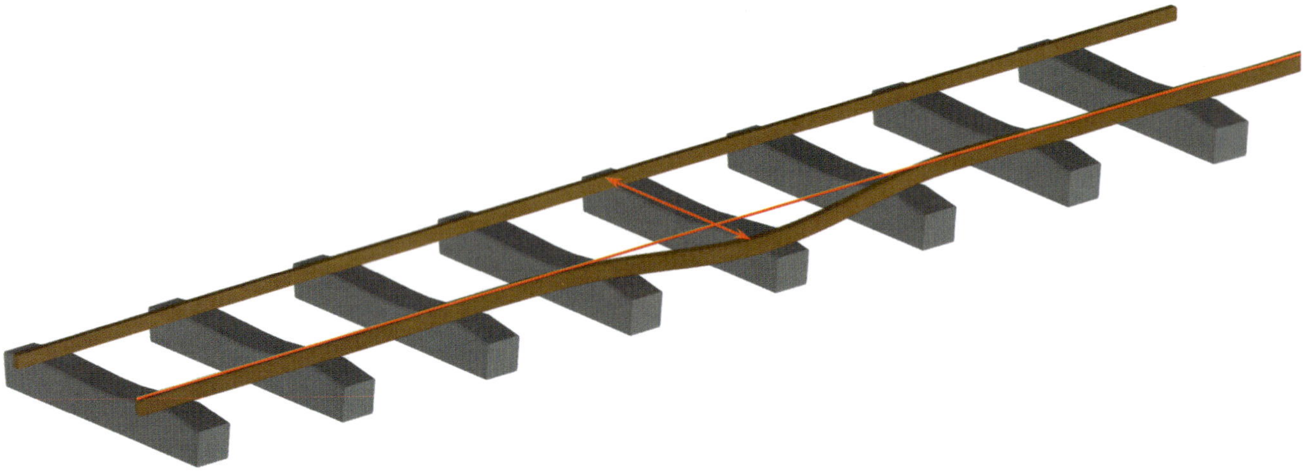

Figure 215: Gauge Defect

Figure 216: Broken Sleepers and Fastenings

Figure 217: Rail Surface Defects

3. MECHANISED VERSUS LABOUR-INTENSIVE METHODS FOR VARIOUS MAINTENANCE ACTIVITIES

3.1. Tamping

Tamping is the process whereby the geometry of the track is corrected. During tamping (Figure 218) the track is lifted, levelled, aligned and ballast squeezed into the void underneath the sleeper to correct the vertical alignment (Figure 210), horizontal alignment (Figure 211) and cant (Figure 212) of the track. Tamping is arguably the most important and most frequent of all track maintenance activities due to the dynamic impact loads caused by vertical and horizontal alignment defects. See also CHAPTER 11 for a more detailed description of tamping.

Many different tamping machine models are available with different specialised features. These features include specialised tamping units for tamping turnouts, continuous action tamping machines for high production tamping, small tamping machines for cost effective spot tamping, etc. When these machines are used for the function they have been designed for, they will produce the lowest possible unit cost of tamping.

The fastest tamping machine in the world is the Plasser & Theurer 09-4X continuous action tamping machine with a production rate exceeding 70 sleepers per minute. This machine is designed for high traffic density lines where available maintenance windows and the tamping cycles are very short. The application possibilities of this machine are probably very limited but the advanced technology it employs, the continu-

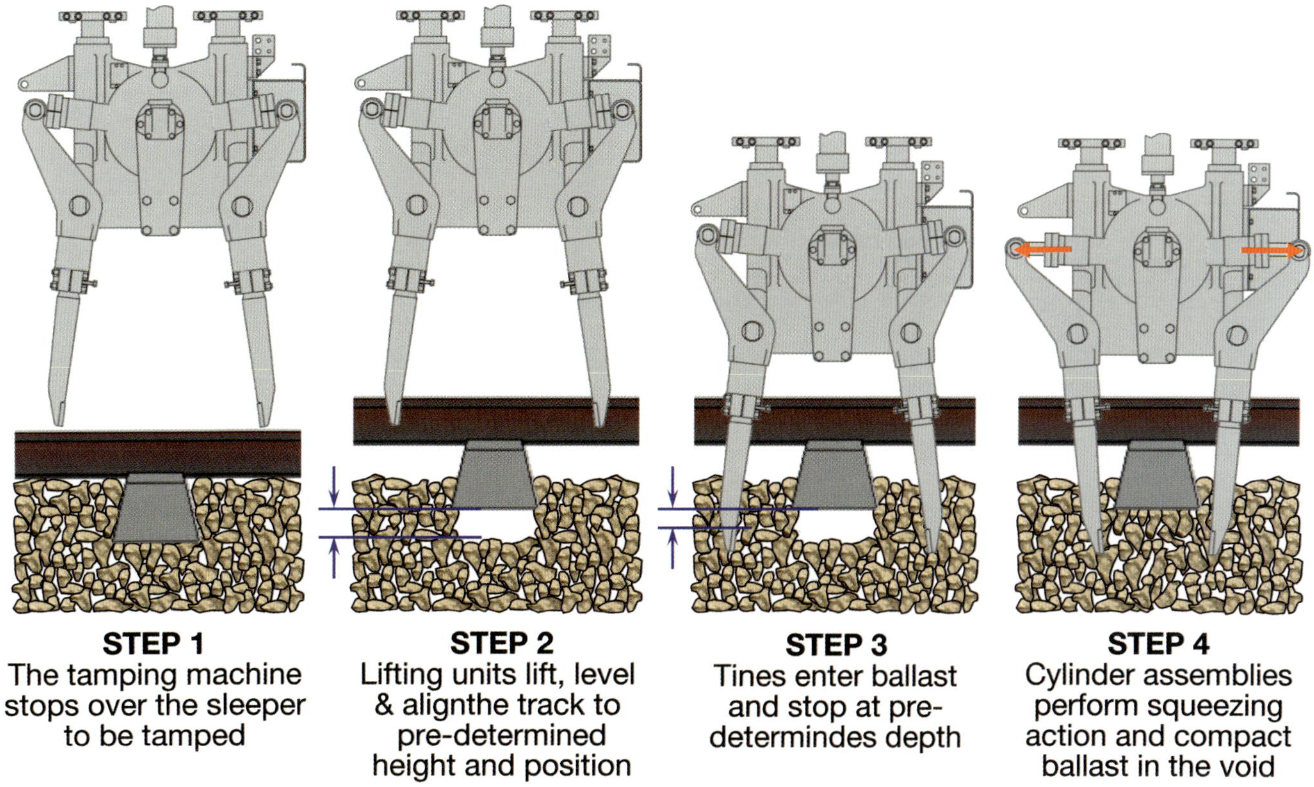

Figure 218: Tamping Process

ous action principle, is not. Continuous action tamping machines separate the cyclic (also called indexing) movement of the tamping unit/s from the machine's mainframe. The mainframe where the operator is sitting therefore moves at a constant speed whereas the cyclic braking and acceleration of the tamping process is made possible by mounting the tamping unit/s to a much lighter satellite frame, thereby producing the maximum possible production. Continuous action tamping machines can tamp from one (see Figure 219) to four sleepers per cycle, depending on the machine model.

Smaller tamping machines are also available which can produce similar quality but at a much lower tamping rate. Examples of such machines are the 08–16 Split Head tamping machine (Figure 220) or the 08–16 cyclic acting single sleeper open line tamping machine; the latter producing a nominal tamping rate of approximately 18 sleepers per minute. These machines are designed for lower density main lines or branch lines which are relatively short in length and/or where the tamping cycles and the maintenance windows are longer. They are also ideal behind lower production ballast cleaning machines.

If a high production tamping machine is utilised on a branch line or siding, it will complete the tamping cycle very quickly resulting in idle time for the machine and increasing the unit cost of tamping substantially, unless the machine can be redeployed. Branch lines also have less ballast and are more likely to have uneven sleeper spacing, all of which will reduce the production ability of the high production tamping machines and increasing the unit costs of tamping.

On the other hand, if the 08–16 cyclic acting single sleeper open line tamping machine is used on very busy lines where high production is necessary, it will be unable to maintain the required tamping cycle. Four 08-16 tamping machines with their operators and support teams will therefore be required to achieve the same results as one 09-4X for example. Calculations have shown that the unit costs of production of four slower machines will be more than double than that of one high production machine and this does not even include the cost of administrating four track possessions in four different maintenance windows which may not even be possible.

Figure 221 is an example of a very small self-propelled tamping machine which is designed for tamping over very short distances with a high level of mobility due it being able to fit on small trucks. This machine cannot lift and align the track and tamps at a maximum rate of approximately 2 to 4 sleepers per minute. Lifting and

Figure 219: 09–16 Single sleeper Continuous Action Tamping Machine

Figure 220: 08-16 Split Head Tamping Machine

aligning is carried out ahead of the machine using track jacks and a level. In comparison with the 08-16, nine of these small tamping machines with nine operators and 90 labourers to mechanically lift the track ahead of the machines should also be able to tamp 18 sleepers in one minute but tamping at a continuous 18 sleepers per minute will not be attainable or practical.

This machine will only be cost effective when one or two of them are used for spot tamping to correct short distance track defects or short sidings. Similarly, if a high production machine is used under the same conditions, the unit cost of tamping will be very high due to the establishment costs of the machine and the machine not being able to work at its optimum design capacity.

Figure 221: Tamping Machine without Lifting and Aligning Unit

Vibratory hand tamping units such as those shown in Figure 222 can be considered a semi-mechanised tamping method since relatively good results can be achieved with these machines under specialised circumstances. Take note that the units shown in Figure 222 do not use the percussion or impact principle which some handheld units employ which is damaging to the ballast and should be avoided. However, continuous tamping with handheld vibratory tampers should not be attempted due to questioning quality that can be achieved over longer distances and low production that will be achieved.

It will require a team of 360 tamper operators and 180 labourers to lift the track to also achieve 18 sleepers in one minute. But, such a large group of people will be in one another's way and attaining a continuous 18 sleepers per minute will not only be impossible but unpractical too. Such a team will cost at least 8 times more than one 18 sleeper per minute tamping machine. And this does not even consider the huge difference in actual production per day, quality and precision between a machine and hand tampers. These hand tampers are designed to support turnout tamping machines in areas where the tamping machine cannot access the turnout. To a lesser extent it can also be used to correct short distance defects until the root cause of the defect can be addressed.

Electrical Horizontal Tampers using a Team of Four People Simultaneously

Vertical Tamper Available in 2-stroke, 4-Stroke, Electrical or Hydraulic

Figure 222: Vibratory Hand Tamping Machines

Figure 223: Manual Tamping Using Pick Axes

At the lowest end of the scale a very large team in excess of 1,000 labourers with pick axes to shovel pack the track will be required to achieve 18 sleepers in one minute but such a large labour team will be totally unpractical if one considers their logistical, transport, accommodation and supervision requirements. However under the right circumstances to fix small isolated faults on the line the unit costs of a small team will be more affordable than large mechanised machines.

As can be seen from the above, there is no definitive point where mechanisation ends and labour intensity starts; it's a matter of "horses for courses" and affordability and practicality of the specific circumstances. The unit costs, production requirements and minimum precision and quality required will dictate the most appropriate method to be used.

To summarise the comparison between mechanised and labour-intensive tamping: For fixing short distance geometry defects, especially lines in a poor but maintainable condition, labour-intensive methods could well be more effective and affordable than machines. But as higher density traffic and type of traffic demands a better track condition, labour-intensive methods becomes increasingly ineffective and unpractical as more and more labourers are required to maintain the line. Furthermore, hand tamping methods using percussion or impact vibratory tampers (not to be confused with vertical or horizontal vibratory tampers in Figure 222) or pick axes to tamp the track are likely to cause damage to concrete sleepers (Figure 224) and the unevenness in the longitudinal direction due to inaccuracy and inconsistency produced by manual tamping will have a negative effect on train energy requirements and the dynamic loading on the track structure will cause accelerated wear.

As the required track condition increases, small mechanised tamping machines will become more affordable and as the track condition enters the sphere of high speed and heavy axle loads, higher production tamping machines will reduce the unit costs of tamping while labour-intensive methods become impossible.

Figure 224: Damaged Sleepers due to Repetitive Hand Tamping

3.2. Ballast Regulating

For the ballast bed to perform its very important functions, it has to comply with a specific ballast profile (Figure 226). However, over time the ballast bed will deviate from this profile due to vibrations from traffic, people and animals walking across the track etc., spreading the ballast over a wide area, exposing the sleeper shoulders and leaving little ballast in the cribs (Figure 225). This will require the ballast bed to be re-profiled to reinstate the longitudinal and lateral stability of the track due to traffic induced forces (Figure 227).

Mechanised regulating using a ballast regulating machine (Figure 228) cannot be used if there is no ballast to regulate but from the moment the minimum volume of ballast is available, the production and precision of mechanised ballast regulating cannot be equalled. A ballast regulating machine can bring about savings in ballast cost by transferring ballast from an area with surplus ballast to an area with insufficient ballast (model depending) which would not be possible with labour-intensive methods. Ballast regulating machines can also transfer ballast across the track to ensure an even ballast profile, optimising ballast usage. See also CHAPTER 13 for a more detailed description.

On low traffic density lines and lines that are in a poor but maintainable condition, labour-intensive ballast regulating can be done affordably and practically, especially if local communities along the line are used for this purpose. Even on lines with slightly higher traffic volumes, good quality of work can still be achieved with labour-intensive methods but proper supervision would be advisable. On high traffic volume lines careful consideration of unit costs would be required when deciding between mechanised and labour-intensive ballast regulating methods.

Figure 225: Poor Ballast Profile

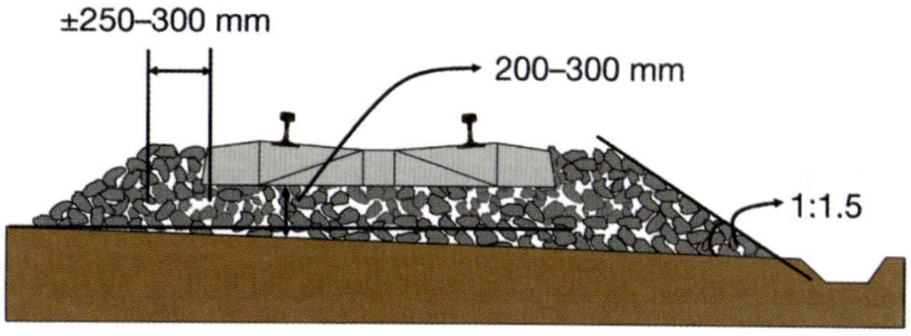

Figure 226: Design Ballast Profile

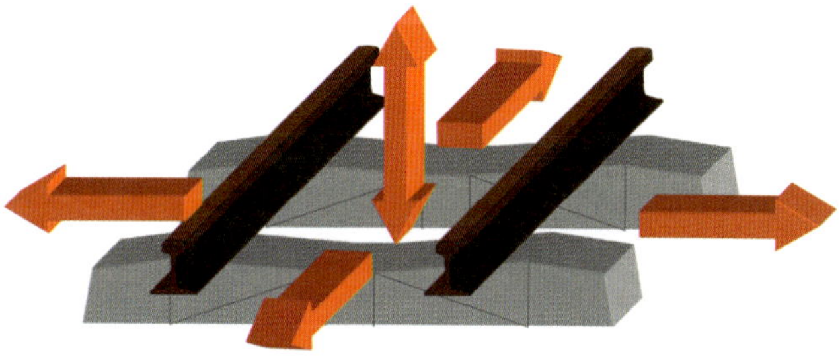

Figure 227: Ballast Resists Forces

Figure 228: Ballast Regulating Machine

3.3. Ballast Cleaning

Ballast cleaning refers to the process of removing the fine material from the ballast bed. The fine material fills the void space between the ballast stones, making the ballast bed inelastic and the fines also retain moisture, leading to rapid deterioration of the entire track structure (Figure 229). In the past this used to be carried out with ballast forks, but this process is extreme hard work and the most important area of the ballast bed to be cleaned, the sleeper bearing area (the area directly beneath the sleeper which is hard to reach), was therefore often omitted making the whole process a pointless exercise. A bigger problem however is that manual screening of the ballast does not re-instate the formation cross fall (Figure 230) and it creates unevenness in the formation which are areas where water can collect leading to accelerated deterioration of the formation (Figure 231).

A ballast cleaning machine (Figure 232) on the other hand uses a continuous chain which removes the entire ballast bed and at the same time re-instates the cross fall of the formation (Figure 233). Any unevenness in the formation is also removed automatically to ensure a smooth formation free from any collecting areas for water (Figure 234 and Figure 235). See also CHAPTER 14.

Depending on the machine type and track gauge, a ballast cleaning machine can screen between approximately 200 and 500 metres per hour which will never be possible with a labour team and ballast forks. Only ballast stones that conform to specification will be returned to the track producing high quality, long ballast cleaning intervals and savings in ballast.

Figure 229: Excessively Fouled Ballast Bed

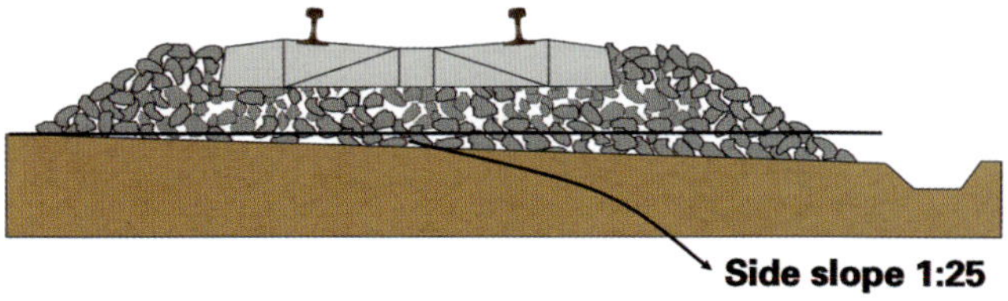

Figure 230: Required Side Slope (Cross Fall)

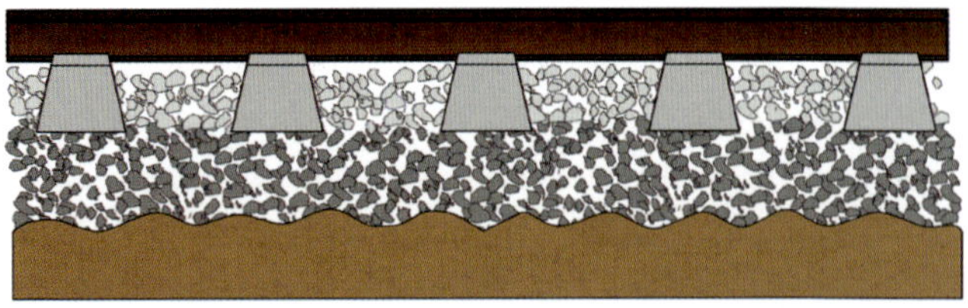

Figure 231: Damage Caused by Hand Ballast Cleaning Creating Areas for Water to Collect

Figure 232: RM74 Ballast Cleaning Machine

Manual screening using ballast forks is almost never affordable no matter what the track condition is and the damage caused by manual screening makes this method totally undesirable where higher track conditions are required.

Figure 236 graphically shows a cost comparison for 1 km of ballast cleaning between fully mechanised, partially mechanised and manual (labour-intensive) ballast cleaning methods. Separate calculations for industrialised countries with a high wage level as well as for low wage level countries are also illustrated for comparison.

Even with low wage costs, mechanised ballast cleaning is the most cost-efficient form of track maintenance compared to partial mechanisation or complete manual ballast cleaning. By altering the strategy the total costs for a track network can be lowered considerably.

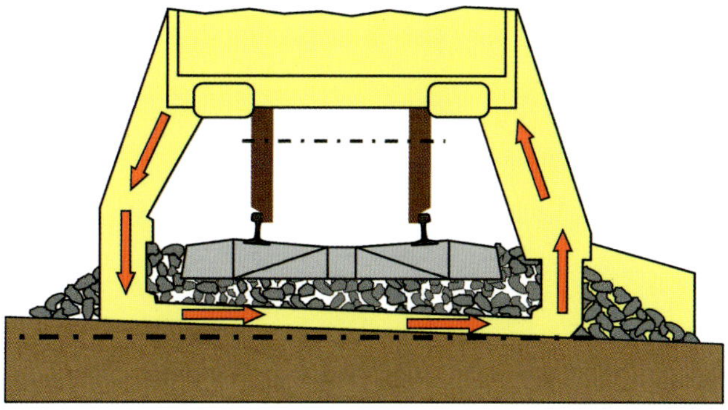

Figure 233: Reinstating Cross Fall

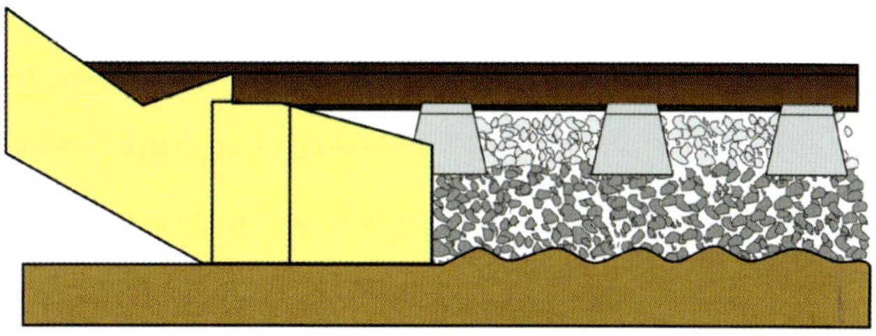

Figure 234: Removing Unevenness on The Formation

Figure 236: Comparison of the Costs for 1 Km of Ballast Cleaning

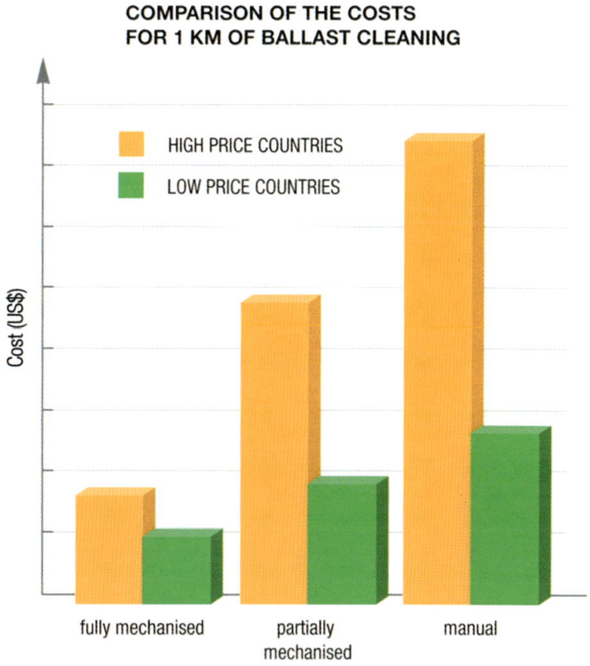

Figure 235: Ballast Cleaning Machine Removes Entire Ballast Bed, Screens out Fine Material and Returns Ballast that fits in Ballast Envelope Specification

Figure 237: Flat Rail Profile

Figure 238: Head Checks and Flaking

Figure 239: Damaged Switch Blade due to Lack of Grinding

Figure 240: Small Switch Grinding Machine

3.4. Rail Grinding

Over time and traffic the rail crown profile becomes flat (Figure 237) which reduces the rail and wheel interaction negatively affecting steering which further accelerates the wear of the rails. Rolling contact fatigue also hardens the crown which has the tendency to develop head checks, cracks, flaking (Figure 238) and can eventually lead to rail breaks and derailments.

Rail grinding cannot be done by manual labour methods and is a very important track maintenance activity if any reasonable rail life wants to be achieved. Mechanised railbound grinding machines have to be used. Refer also to CHAPTER 16. However, when it comes to the grinding of switchblades and crossings which are an equally important activity to prolong the life of the very expensive turnouts, mechanised methods are no longer a prerequisite and hand grinding using smaller machines (Figure 240) can achieve equally good results which would be more affordable.

3.5. Wayside Maintenance

Drain cleaning is a maintenance activity that is neglected by most railways with severe consequences to the track infrastructure, especially the life expectancy of the formation. This is one activity that can be very effectively and affordably done by labour-intensive methods (Figure 241). Mechanised methods would however be more efficient and durable. See more on drain cleaning methods in CHAPTER 3, paragraph 9.

Weed control (Figure 242) is another maintenance activity which is neglected by many railways, especially those in developing countries. Semi-mechanised methods using road/rail vehicles with spraying equipment or even labour intensive methods can be cost effectively alternatives.

3.6. Track Construction and Renewal

Track construction and track renewal can be done very effectively with a combination of semi mechanised and labour-intensive methods (Figure 243 and Figure 244). In this regard semi-mechanisation refers to the use of smaller machinery such as gantry cranes, rail threaders, etc. which still require a large labour force.

However, to ensure a railway line of good quality and long life expectancy, a ballast tamping machine and regulator must be used to final the line and a flash butt welding machine preferably used to weld the rail sections into continuously welded rail.

See also CHAPTER 20 for a more detailed description of track construction and renewal machines and processes.

Figure 241: Drain Cleaning using Labour

Figure 242: Poor Weed Control

Figure 243: Labour-Intensive Track Construction

Figure 244: Semi Mechanised Track Renewal using Gantry Cranes and other Machinery (Visible in the Background

4. DECISION MAKING CRITERIA

The previous paragraphs provided an indication of which maintenance activities can be carried out cost effectively and practically using labour-intensive methods. It is however clear that it is only cost effective and practical under certain circumstances. The same applies however to mechanised maintenance methods as well. Two of the main circumstances that influence the cost effectiveness and practicality of a chosen maintenance method are the traffic density on the line and the design specification of the line. These circumstances are variables and can therefore best be illustrated with a graph. The decision to use mechanised or labour-intensive maintenance methods will however also have a major influence on the expected track life and lifecycle cost.

4.1. Traffic Density

At the one end of the scale there are low density lines and branch lines with perhaps one train per day or even per week which are sometimes only kept open for strategic reasons. The required track speeds would be low, the permissible tolerance of track defects relatively high, maintenance cycles quite long and availability of the line for maintenance will also be high.

High traffic density main lines, high speed passenger lines and heavy haul lines have several trains per day such as the 160 km/h commuter Gautrain and the Sishen to Saldanha heavy haul iron ore line in South Africa; the latter with 30 ton axle loads and 342 wagons per train. The required track speed will be at least 80 km/h, the track defects will have very small permissible tolerances for safety and productivity reasons, the maintenance cycles will be short and the available windows for maintenance will also be short and few.

The effect of traffic density on the cost effectiveness and practicality of mechanised or labour-intensive methods is illustrated in Figure 245. Note that the curves are illustrative and not based on actual figures.

At very low traffic volumes, the cost effectiveness of mechanised methods will be low initially but will grow quicker than the growth in traffic due to very small and low production machines being available which are very cost effective.

As the traffic increases, mechanised methods will become exponentially more cost effective as higher production machines are able to work at their maximum production therefore reducing unit costs.

Labour-intensive methods on the other hand will be much more cost effective at low traffic volumes by using small labour teams with vibratory tamping units or hand tools for shovel packing to address only the worst defects. As the traffic volume grows, the cost effectiveness of labour-intensive methods will reduce quicker than the growth in traffic.

When the traffic reaches a certain level, labour-intensive methods become completely unpractical and unaffordable due to the size of labour teams that will be required, the lack of quality and precision they can produce, damage they can cause to the track structure and due to their low production.

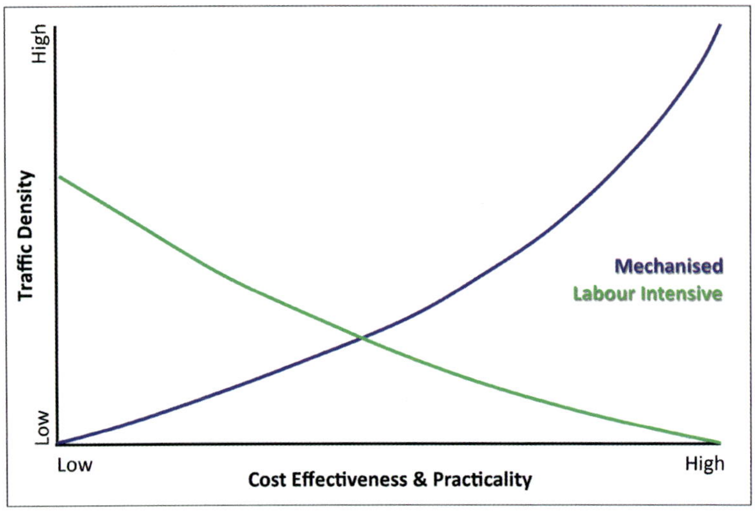

Figure 245: Effect of Traffic Density on the Cost Effectiveness and Practicality

4.2. Design Specification

Low traffic density branch lines are often constructed of a light structure which may have short and light rails of 12 metre lengths and profiles of 30 kg/metre; with fishplate joints; steel or wood sleepers and a shallow ballast bed. These lines are often in a poor condition as well which adds other criteria to this argument of cost effectiveness and practicality.

However, to benefit from the economy of scale that higher axle loads of at least 18 to 20 tons and longer and faster trains offer, a heavier track structure of at minimum 48 kg/metre continuously welded rail, heavy concrete sleepers and a ballast thickness of at least 200 to 300 mm under the sleeper now becomes necessary for thc safety and stability of the track against increased traffic induced forces. For effective passenger traffic, section speeds of at least 80 km/h would be required.

The effect that the track design specification has on the cost effectiveness and practicality of mechanised or labour-intensive methods is illustrated in Figure 248.

Track condition depending, even a light track structure is very practical and cost effective to maintain from a mechanisation point of view since smaller and lighter machines can be used. However, as the structure becomes heavier, mechanised maintenance machines will become increasingly cost effective due to their ability to effortlessly deal with heavier track structures.

A light structure is also practical and cost effective to maintain with labour-intensive methods especially if one considers the mobility of a labour team and the high repetitive maintenance that fish-plated joints require. However, at some point the heavy structure becomes increasingly unpractical for labour teams to handle.

Figure 246: Light Track Structure

Figure 247: Heavy Track Structure

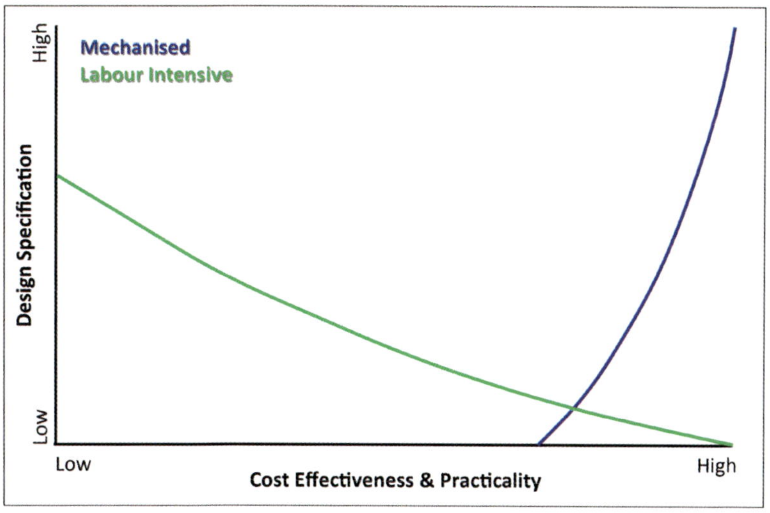

Figure 248: Effect of Design Specification on the Cost Effectiveness and Practicality

4.3. Track Life and Lifecycle Cost

The objective of track maintenance is to attain the lowest possible lifecycle cost by extending the service life of the track. It is therefore clear that the maintenance strategy and effectiveness of the maintenance will influence the service life and ultimately the lifecycle cost of the track.

To relate this back to the debate between mechanised and labour-intensive methods:

The precision and quality of a completely mechanised construction process will produce the highest possible initial quality track for a given quality material and construction standards therefore producing the longest track life and lowest lifecycle cost with regular maintenance.

A combination of mechanised and labour-intensive methods for construction can produce similar quality, as long as the earthworks is carried out using at least earthmoving machines to produce the best possible platform for the track structure and the line is finalised using ballast regulating and tamping machines to produce the best possible uniformity in the line in terms of horizontal and vertical alignment for a high initial quality.

If however, track construction is carried out using only or predominantly labour-intensive methods, the lack of precision and quality will produce a line with a lower initial quality, reduced track life and higher lifecycle costs.

In terms of maintenance, fully mechanised maintenance methods will produce the precision and quality required to always bring the track condition back to the highest possible level considering the wear in the track components.

Fully labour-intensive maintenance will achieve quite the opposite. Depending on the type of maintenance activity carried out, the lack of quality and precision will produce a curve similar to line “bc2” and even lower (CHAPTER 6, Figure 199), depending on the type and traffic density of the line. However, if the decision between mechanised or labour-intensive maintenance is chosen carefully, the point “c2” can be brought very close to “c1” despite some activities being labour-intensive.

5. CONCLUSION

There certainly is a place for both mechanised and labour-intensive track maintenance methods and in many cases they can be complementary and not contradictory as long as they are directed towards the conditions where the most economic benefit will be achieved from both.

However, as a line moves into the sphere of fast (greater than 120 km/h) and heavier axle load traffic (greater than 22 ton per axle) on heavy concrete sleepers, continuous welded rails and sufficient ballast, labour-intensive methods becomes increasingly uneconomical due to their low production and damage caused to the track structure and soon reaches a point where labour-intensive work is just not possible to produce the quality, precision and consistency required. Labour-intensive methods is then limited to only some wayside maintenance.

Chapter 8
Track Alignment and Geometry

1. INTRODUCTION

The alignment of a railway consists of straight sections (tangent track) connected by curves in the horizontal plane (curved track) to move around obstacles and vertical curves (gradients) to follow the topography.

Geometry is defined as the science of the properties and relationships of sizes and distances in a given space or the relative arrangement of objects or parts of an object in that given space (Concise Oxford Dictionary). If you apply this definition to railway maintenance, geometry means the relative relationships between the two rails or two points on one rail. The same also applies to the overhead electrical contact wire geometry.

Track geometry and alignment are interrelated, for example, the radius of the curve influences the geometry parameter superelevation (cant); for track geometry condition management the track is divided into sections, i.e. related to alignment in this context; etc.

The following paragraphs provide the theory behind track alignment and geometry. The term "track geometry" is used very loose in this context as this chapter also explains geometry of the rail and overhead traction equipment.

For additional reading refer to European Standards EN 13803, Track Alignment Design Parameters and EN 13848-5, Track Geometry Quality.

Note: Standards mentioned in here are mostly those that have been adopted by the South African railways in the past. Where mentioned in this chapter it acts mostly as a guide and as a reminder that standards need to be considered and established for the railway in question.

2. TRACK ALIGNMENT

2.1. General Terms and Definitions

- **The path of a railway line –** The course the railway line follows across terrain which is determined by:
  - Economic requirements – e.g. shortest possible journey time between 2 points.
  - Topography – e.g. mountains, valleys, built-up areas (fixed structures).
  - Operational requirements – e.g. traffic type (freight, heavy haul, commuter, high speed).
- **Railway axis (centreline of the layout) –** The centre line between the track axes (track centrelines) of double-track lines (Figure 249).

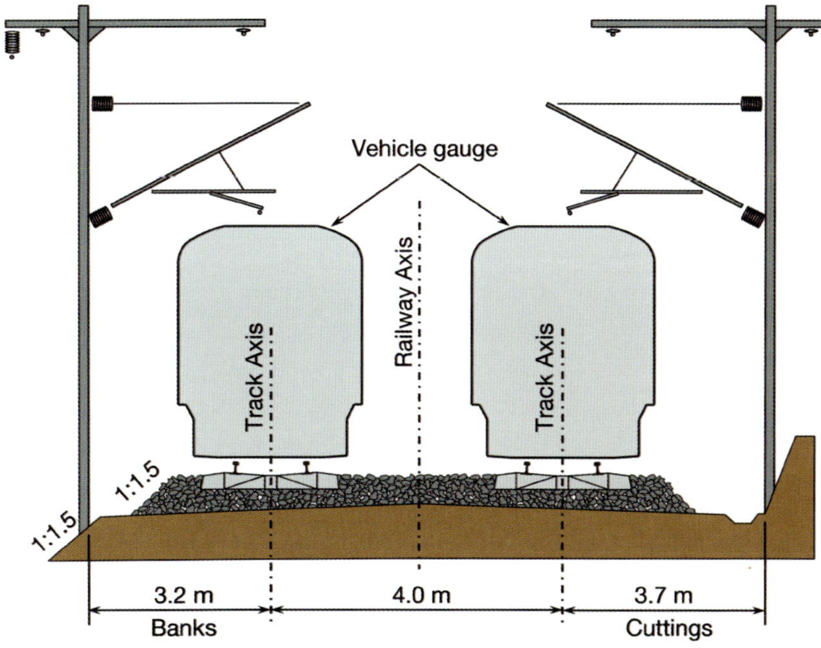

Figure 249: Double Line Cross Section (Minimum Dimensions)

- **Track axis (centreline of the track) –** The track axis or centreline of the track is that line in longitudinal direction of the track which has the same distance to the two running edges of the rails in crosswise direction regardless of any gauge widening (Figure 249).
- **Track plane –** The track plane is the plane formed from the track axis (centreline of the track) and the crosswise direction of the track in the observed point of the track axis. It lies level with the top edges of the rails.
- **Line layout –** The geometrical course of the track axis or railway axis either measured in the longitudinal direction or crosswise to the track.
- **The course of the track axis or railway axis in the longitudinal direction –** This is referred to as a gradient, vertical curve or slope.
- **The course of the track axis or railway axis in the cross wise direction (horizontal to the track)** – This is referred to as alignment and consists of straight sections and curves.
- **Main Lines –** The principal running line or lines of a section or lines between stations, junctions, interloops or halts. It connects two destinations.
- **Secondary lines –** Connect two destinations on lines of lessor importance and/or lower traffic density. Also referred to as branch lines.
- **Passing or crossings loops –** These are installed on single lines to allow trains traveling in opposite directions to pass one another. These lines are classified as running lines.
- **Spurs and sidings –** These are constructed on the line to allow for maintenance machines to make way for traffic. These may also be used a staging of machines or rolling stock over night and where track material can be offloaded before transportation to the installation site.
- **Private siding –** A railway track owned by the local authority, a company or a private person and is connected to the track of a railway network owner.

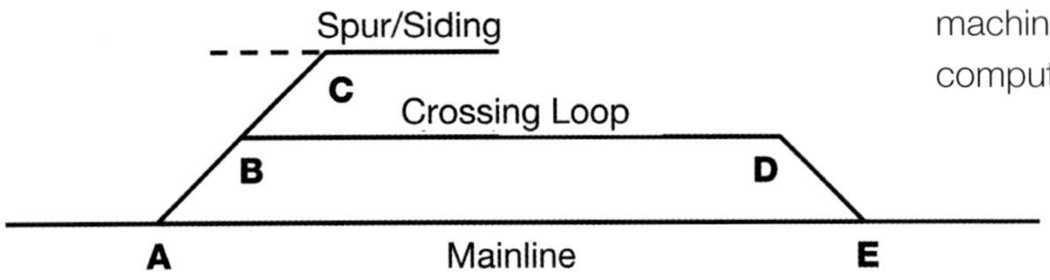

Figure 250: Mainline, Crossing Loops and Spurs/Sidings

2.2. Curved Track (Horizontal Plane)

Curves are pretty complicated in their design but the objective here is not to dwell into curve design but rather to understand curve geometry since that is what is important from a mechanised maintenance point of view. In the simplest terms curves connect two sections of tangent (straight) track to go around obstacles or change direction.

2.2.1. Curve Parameters

A train travelling at a speed "v" entering a curve with a radius R will experience a centrifugal lateral acceleration of a=v2/R. This lateral acceleration may have the effect to displace wagon loads, risk the overturning or derailment of wagons and have high lateral forces on the track, which increases wear of the rails and wheel flanges, lateral displacement of the whole track, etc.

It is therefore desirable to limit these effects without negatively impacting on the track speed. The lateral acceleration can be partially compensated for through the use of superelevation (cant) in the circular curve (refer to paragraph 3.1.3 of this chapter), the use of a maximum possible curve radius and the use of transition curves. Curves are therefore generally divided into three sections, namely, a transition curve at the beginning and end of the curve and a circular curve between the two transition curves. See Figure 251.

2.2.1.1. Curve Radius

A curve is identified by the radius measured in metres (see Figure 251) of the circular section of the curve. The bigger the radius the "flatter" the curve is and the lower the centrifugal forces will be going through the curve. This will allow higher speeds through the curve. The radius is however linked to the line layout determined by economic and operational factors and topography.

Very small radii curves must be avoided for their limiting speed restrictions, high rail wear and maintenance requirements, however, a radius exceeding 1,600 metres should also be avoided due to the additional maintenance requirements such as pre-measuring and calculating offsets during tamping, especially with machines that are not equipped with automatic guiding computers.

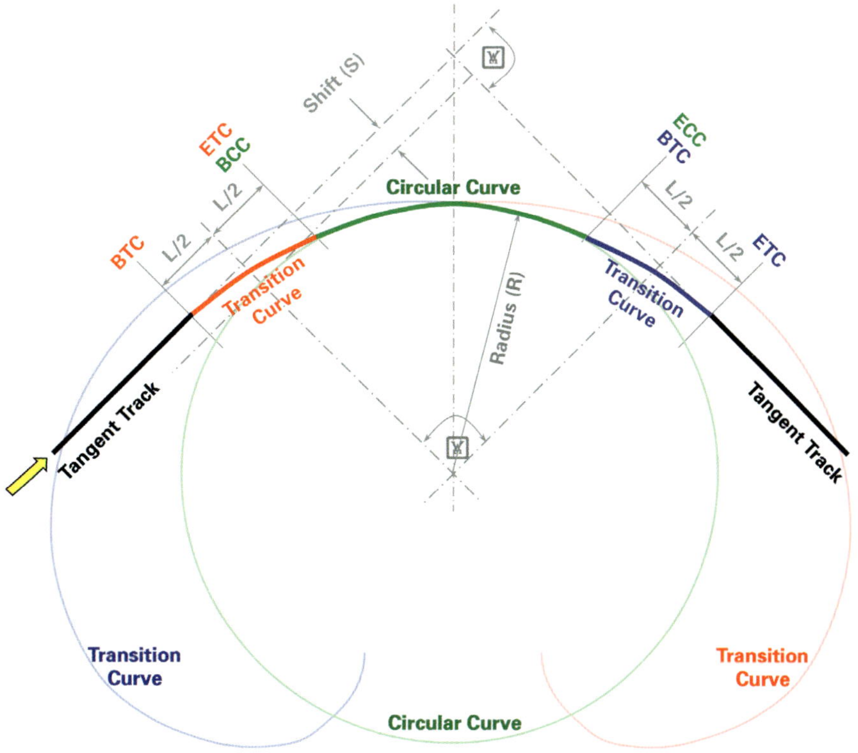

Figure 251: Curve Design

2.2.1.2. Transition Curve

The transition curve allows the centrifugal force experienced by a train travelling at a constant speed to increase uniformly from the end of the tangent track to the maximum force in the curve. This is achieved by gradually decreasing the curve radius from the tangent track to the maximum radius of the curve in the form of a spiral.

Transition curves are also referred to as a spiral curve because that section of the curve has an ever changing radius and offset in the form of a spiral. Another transition curve follows after the circular curve, which gradually increases the curve radius again to the tangent track.

2.2.1.3. Circular Curve

The circular curve follows on the transition curve and has a constant radius and constant offset.

The transition between these sections is denoted by the following fixed points depending on the side from which the curve is entered (see Figure 251):

- the beginning of the transition curve (BTC),
- the beginning of the circular curve (BCC),
- the end of the circular curve (ECC), and
- the end of the transition curve (ETC).

The maximum cant of the circular curve (paragraph 3.1.3) must also be ramped up from zero on the tangent to the maximum of the circular curve generally in equal steps over the length of the transition curve.

2.2.2. Curve Layouts

2.2.2.1. Compound Curve

Compound curves are formed by two single circular curves of different radii that have a common tangent with the end of one coinciding with the beginning of the other.

Both these curves are on the same side of a common tangent. Compound curves should be avoided as far as possible.

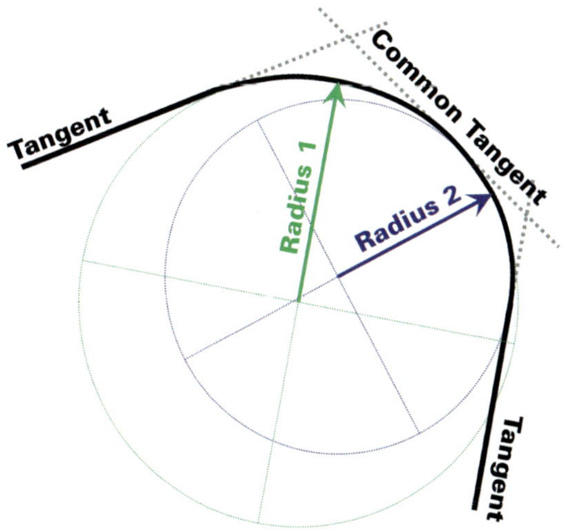

Figure 252: Compound Curve

2.2.2.2. Reverse Curve

Reverse curves are defined as two curves on opposite sides of a common tangent and where with a common point of tangency are undesirable in track layouts. In a reverse curve the end of the one transition curve coincides with the beginning of the transition of another curve.

In reverse curves, where the transition curves meet without intermediate straight sections, the transition of the cant from the one curve to the next curve must be carried out gradually. The connection of such superelevated reverse curves is called „track scissor" or „crossed superelevation ramps".

2.2.3. Curve Direction

The direction of a curve is identified as to the "left" or to the "right" respectively looking in the direction of increasing kilometres. See also paragraph 4 on sign conventions.

2.2.4. Measuring of Curves to Establish Curve Information

When a curve is "measured", it is actually the offset from the middle of a 10 metre cord to the rail crown that is measured at 5 metre intervals. This is the same measurement as horizontal alignment. The measuring position is called a station. The measuring instrument (a versine) is a nylon string of 10 metres which is used in conjunction with a special ruler in the middle of the string (see Figure 255) to establish the distance between the string (the cord) and the rail.

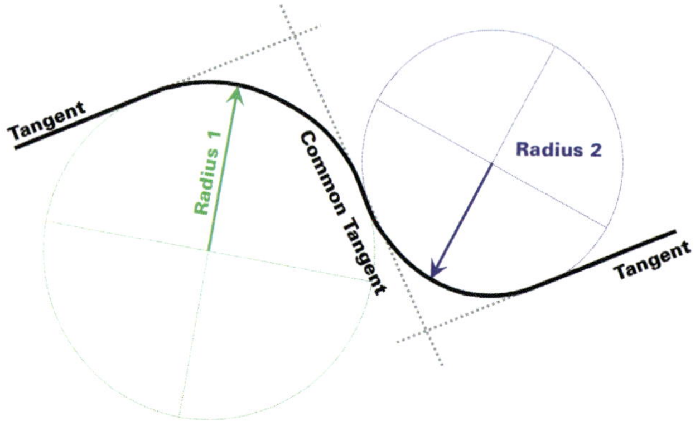

Figure 253: Reverse Curve

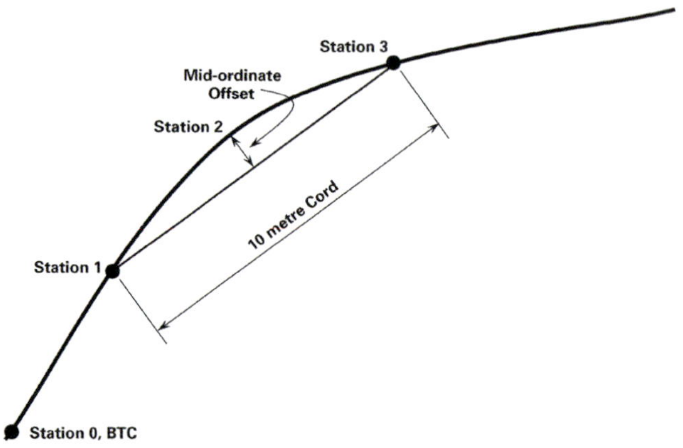

Figure 254: Measuring the Offset Distance Between a 10 Metre Cord and the Rail Crown

Figure 255: Measuring the Curve with the Manual Method and a 10 m String

2.2.4.1. The Hand Measuring Process

- First the beginning of the transition curve (BTC) must be established. This point should be sign posted without which the beginning of the curve is, at best, an estimate.
- A tape measure is used to measure the position of every station from tangent to tangent at 5 metre intervals. Three stations X, Y and Z are also measured before the beginning of the transition curve. A chalk mark is made on the crown (or on the foot of the rail) to indicate the position of the station. The station number is written on the foot of the rail with white chalk (see Figure 256).
- The mid-ordinate offset can now be measured using the string held by two people at the chalk marks and a third person to measure the offset in the middle of the 10 metre string at each station using a special ruler (see Figure 255). Readings are taken 14 mm below the crown.
- The measured offset is written on the foot of the rail in white chalk next to the station number.
- A form similar to the example in Figure 257 can be used to write the offset under the column "MO" (mid-ordinate) against the corresponding station number. A dot is made on the graph paper at the offset value.
- The string is moved on by 5 metres and the next station is measured, the value is written on the foot of the rail and recorded on the form.
- This process continues until the whole curve has been measured and recorded on the form.

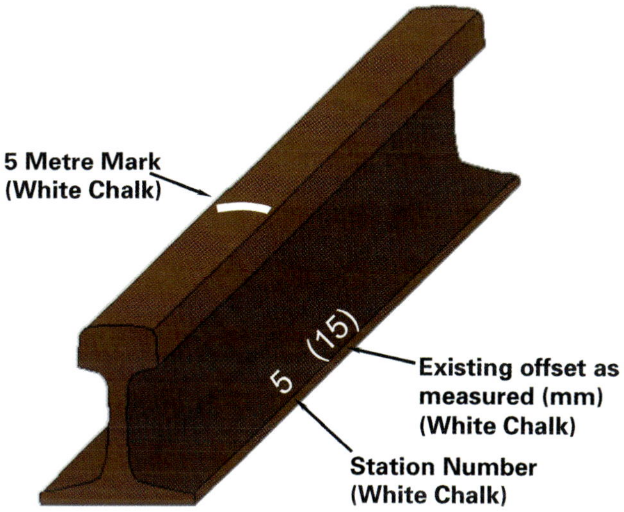

Figure 256: Marking Made on the Rail

The data is then used to calculate the transition lengths, the curve radius, the curve offset and the required cant. Accurate information is required to ensure that, when the curve is tamped, the tamping machine places the curve in the desired position since errors in measuring may result in curves being pulled completely out of geometry, with disastrous consequences.

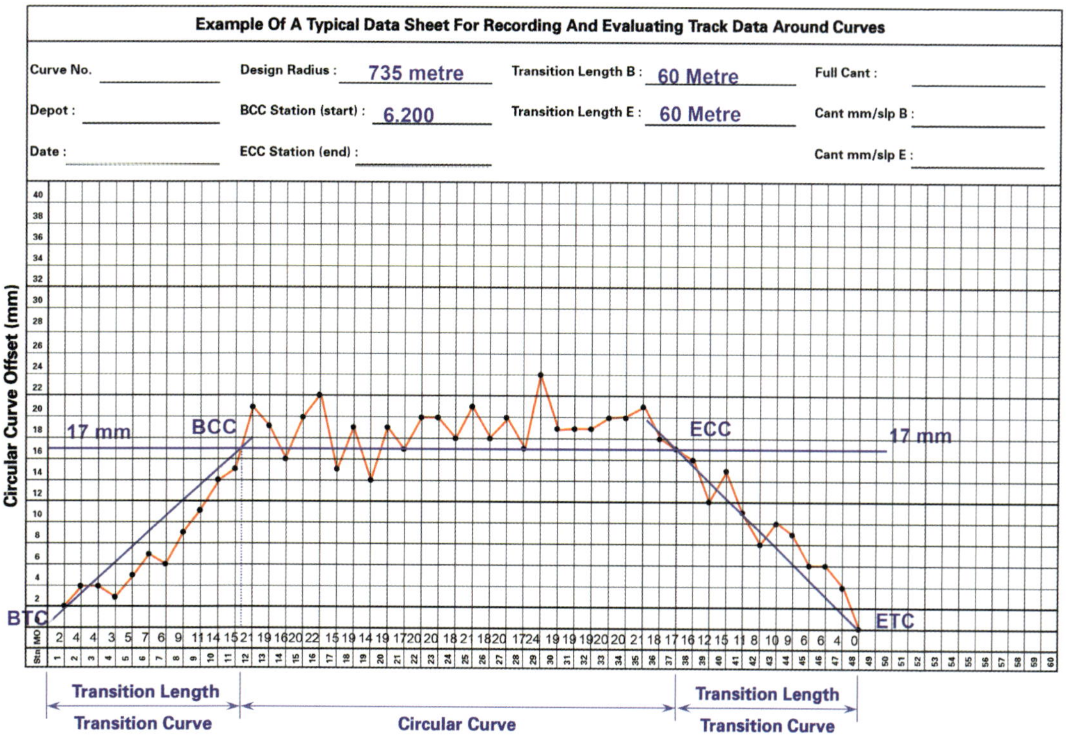

Figure 257: Example of a Form to Dot down the Measured Offsets

2.2.4.2. Establishing the Curve Data from the Measured Information

The form in Figure 257 can now be used to plot the curve and to establish the position and values of the curve data. The following curve data is calculated:

- **Offset of the Circular Curve**
  First the average offset for the circular curve must be established. This is done by calculating the average of the offsets that are clearly in the circular curve by adding all the mid-ordinate offsets and dividing it by the number of offsets used. A line is drawn at that offset which is at 17 mm in the above example.

- **Offsets and length of Transition**
  A line is drawn from station 0 (at the BTC) to where the line crosses the circular curve offset line (at 17 mm in the example). This line represents the transition curve. The same is done for the transition curve at the end of the curve.

  The transition lengths can be calculated by adding the number of stations between the beginning of the transition curve and the end of the transition curve and multiplying this number by 5 metres (the distance between stations).

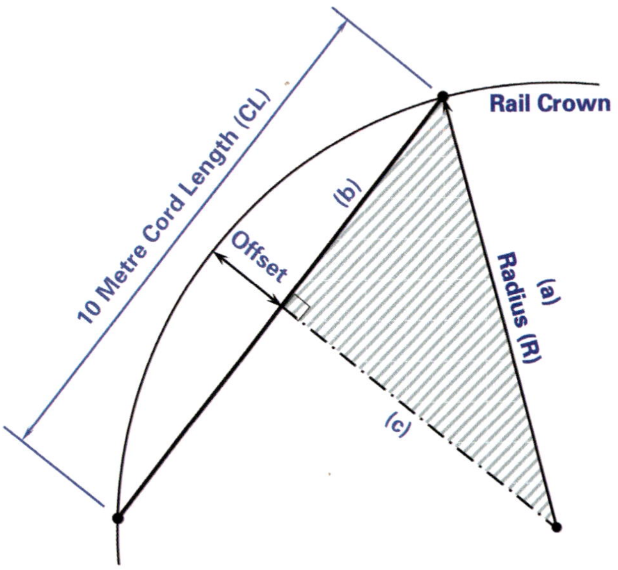

Figure 258: Calculating the Curve Radius

$$a^2 = b^2 + c^2$$

$$\textbf{Curve Radius (R)} = \sqrt{\left(\frac{CL}{2}\right)^2 + \left(R - \text{Offset}\right)^2}$$

$$= \frac{CL \times CL \times 125}{\text{Offset}}$$

Where : CL = Cord length in metres (10 metres)
R = Radius of the circular curve

On main lines the transition lengths should normally be:

– 60 metres long for circular curves of 300 metres or sharper radius
– 80 metres long for circular curves flatter than 300 metres radius

- **Curve Radius**
  The radius of the curve can now be calculated from the average offset using the Pythagorean formula and the right angled triangle shown in Figure 258:

- **Desired Superelevation (Cant)**
  Once the curve radius has been established, the radius and the section speed are used to establish the maximum superelevation for the circular curve from the track standards of the particular railway.

  The specified superelevation applies only to the circular curve and will remain constant through the circular curve.

  The rate of superelevation increase in millimetre per sleeper must be calculated over the transition curve. The superelevation starts at 0 mm at the beginning of the transition curve and will reach the required value at the point where the transition curve joins the circular curve (end of transition curve/beginning of circular curve).

2.2.5. *Compensation Methods for Curves*

Depending on the curve parameters, the following compensation may be made in curves for safety, comfort of passengers and wear of track and rolling stock material:

- Check rails are provided on the low leg of curves to prevent excessive wear of the high leg but they increase the curve resistance. See CHAPTER 3 paragraph 2.16.
- Gauge widening.
- The moving vehicle structure gauge must also be increased to allow for superelevation.
- Permanent speed restrictions may be imposed which should be sign posted before the curve.

2.2.6. Maintaining Curves

Over time trains may move curves to a new position where it generally becomes stable. This position may have different curve parameters to the original design.

When a curve is tamped, the curve parameters such as the radius and the length of the transition curves are used to establish the offset and superelevation for the tamping machine.

If the original design data for the curve is used (where available), the curve will be returned to its original design position. This is not always possible or desirable and the curve is likely to move back to its stable position again. It is therefore necessary to measure and establish the new curve data before tamping commences.

Measuring curves by hand using a string (the principle is explained above) is not accurate enough for main lines, heavy haul lines and high speed lines.

It is therefore recommended that the parameters of all the curves on a line are established using a track measuring vehicle (CHAPTER 9) and by applying a best fit for the curves using software available from the track recording vehicle. The fixed points of the curves must then be marked on the track and a curve list kept with all the curve parameters to be used by the tamping machine for all the curves on the line. This will produce a smooth geometry for curves.

Refer to CHAPTER 3, paragraph 2 and, in particular, paragraphs 2.14, 2.18.2 and 2.20.3 for more maintenance requirements of the rail on curves.

2.3. Gradient, Vertical Curve Or Slope

- The longitudinal gradient at any random point is the angle of gradient Θ (Figure 259 and Figure 260) of the track axis to the horizontal plane (inclination from the horizontal). Longitudinal gradient is illustrated in the vertical plan and is therefore called a vertical curve.

A vertical curve should be in the form of parabola (mirror-symmetrical curve) to ensure an even rate of change of grade.

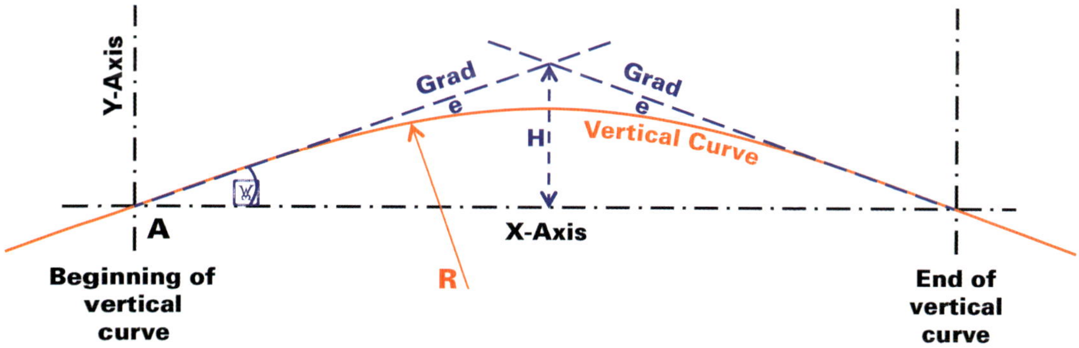

Figure 259: Parabolic Shape of a Vertival Curve

Where:

x = Distance along the x-axis in 20 metre intervals

y = Distance along the y-axis in metres

G = Grade in metres per 20 metre of the grade line passing through “A”
(G is positive on an up grade and negative on a down grade)

H = Elevation of point “A” in metres

R = Radius (R is negative on a summit and positive on a sag)

The equation of the vertical curve is:

$Y = ½Rx2 + Gx + H$

The grade at any point “x” is:

$\frac{dy}{dx} = Rx + G$

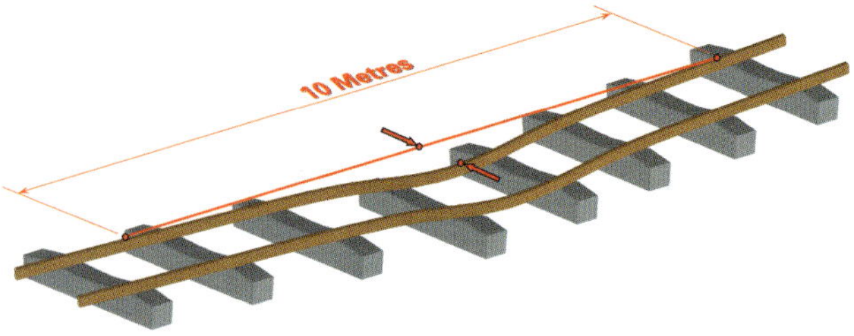

Figure 262: Symmetrical Horizontal Alignment Measurement

Figure 263: Symmetrical Measurement in the Middle of a 10 Metre String

Figure 264: Horizontal Alignment Defect

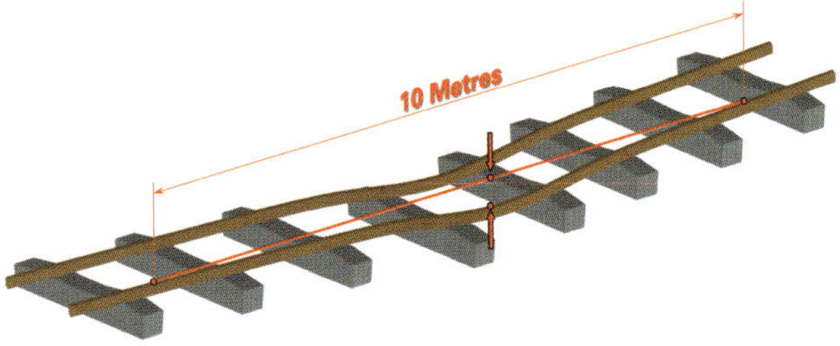

Figure 265: Symmetrical Vertical Alignment Measurement

3.1. Track Geometry

3.1.1. Horizontal Alignment

Horizontal alignment (also known as versine) is measured in the middle of a 10 metre cord (Figure 262 and Figure 263). This measurement shows a deviation from an offset measurement in the horizontal plane (Figure 264).

Horizontal alignment defects are often caused by a lack of ballast under and/or at the sleeper shoulder. The track therefore has little resistance against lateral displacement due to expansion on continuously welded rail (kick-outs) or due to traffic-induced forces. Horizontal alignment defects show both rails having a deviation from a straight line, otherwise it would be defined as a gauge defect.

Deviations in the horizontal plane are particularly dangerous for train derailments and are very uncomfortable to passengers due to the lateral acceleration experienced. It may also cause freight to shift or break loose from their positions.

3.1.2. Vertical Alignment

Vertical alignment (also known as top, profile or longitudinal profile) is similar to horizontal alignment in the way it is measured; i.e. in the middle of a 10 metre cord (Figure 265). This measurement shows the vertical relationship between three points on one rail.

Vertical alignment problems are encountered where the structure under the rails cannot carry the load distributed from the rails and sleepers because of formation failure or because of high dynamic loads caused by rail joints and other rail surface defects (Figure 266).

Vertical alignment defects have a similar danger to derailments, passenger discomfort and freight shifting as horizontal alignment.

Figure 266: Severe Vertical Alignment Defects

3.1.3. Superelevation (Cant or Cross Level)

Cant is a measurement showing the difference between the vertical height of the left-hand and right-hand rail crowns at a coincident point (Figure 267). Calculating the required cant for the section speed, type of traffic and gauge is not within the realms of this manual.

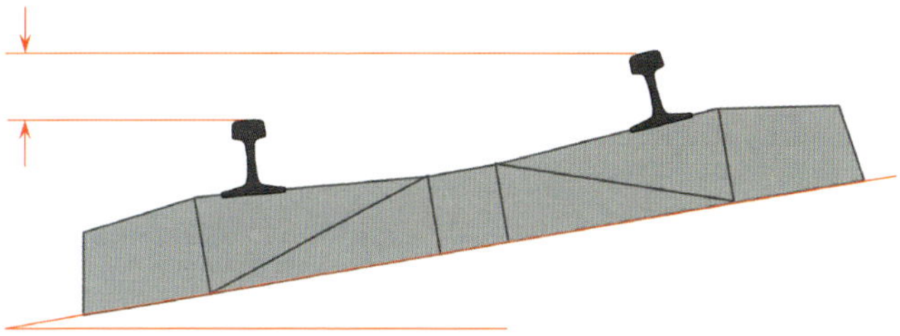

Figure 267: Cant Measurement

Suffice to assume that every railway will have a set of standards that provide the required cant for a particular radius of a circular curve and section speed. This information is required to be available either with the tamping machines in the form of a curve list or should be provided on a sign board at the curve.

The principles of cant can be explained as follows: When a vehicle moves in a curve, centrifugal acceleration (lateral acceleration) will occur that acts in the opposite direction of the curve direction (Figure 268 and Figure 269). Cant is required to:

- aid the steering of trains around curves;
- compensate for the lateral acceleration of trains; and
- limit a change in wheel loads between the inner and outer rails.

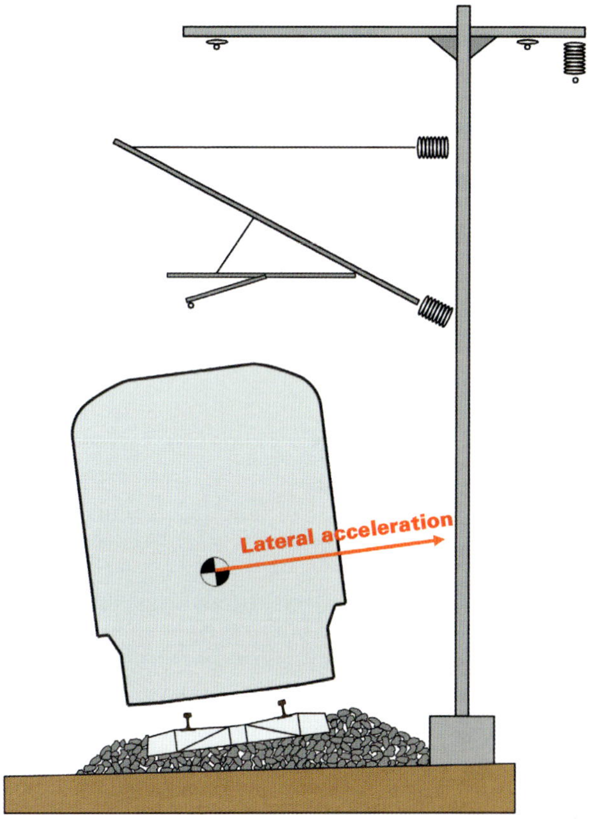

Figure 268: Lateral Acceleration

The maximum lateral acceleration will be in the circular curve (see section 2.2.1.3 above). When a train enters a curve without a transition curve, the lateral acceleration will change from zero to a maximum in an instant which brings in another force, the roll angle acceleration in extreme cases. Therefore the cant is gradually introduced from zero to the maximum in the circular curve over the length of the transition curve.

In essence, cant balances the centrifugal forces associated with the speed of the train with gravitational forces (Figure 268). This is referred to as curving at the balance speed. At the balance speed the wheel loads are equal on the high and low rails. This is however not an exact figure due to, for example, potential different speeds of trains due to curves in grades (going up or down the grade) and gravitational load due to empty or loaded trains.

Figure 269: Direction of Centrifugal Force on a Left-Hand Curve

Unbalanced conditions are associated with either cant deficiency or excess cant which results in a transfer of the vertical load between the high and the low rail and increased lateral loads on both rails. On a deficient superelevation the train moves laterally toward the high rail transferring vertical load from the low rail to the high rail and vice versa.

In general the aim would be for a balance speed for the greatest load as this will cause the greatest rail damage. It is also suggested that curving with superelevation deficiency is probably a better option than that of curving with excess superelevation (International Heavy Haul Association's "Guidelines to Best Practice for Heavy Haul Railway Operations – Management of the Wheel and Rail Interface").

3.1.4. Gauge

Gauge is the distance between the insides of the rail crowns and is measured in millimeters at a point 14 mm below crown level.

There are several different gauges used all over the world including:

- Broad gauge
  - 1,676 mm – Portugal, Spain
  - 1,600 mm – Ireland, Brazil
  - 1,524 mm – Former Soviet Union, Finland
- Standard gauge which is the most common gauge around the world
  - 1,435 mm – All central European countries, USA, China, etc.
- Cape gauge
  - 1,067 mm – South Africa, Australia and secondary lines in several other countries
- Narrow gauge
  - 1,000 mm – Countries in Africa and various secondary lines in Europe

Gauge problems may be encountered where there is excessive rail side wear, deformation (metal flow) of the head, where an incorrect clip combination is used, the sleepers are damaged or the fastening system is broken or loose. Figure 216 was a good example of gauge problems due to a lack of track maintenance.

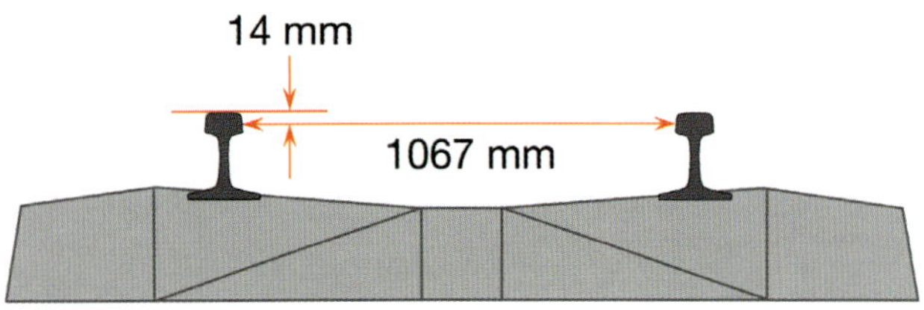

Figure 270: Gauge Measurement

Gauge widening can be artificially built into curves of small radii to assist with rolling stock steering to limit wear of wheels and rail alike. Gauge widening will normally start at radii less than 200 metres starting at approximately 10 mm widening up to as much as 20 mm.

Gauge widening on curves with transition curves will be increased gradually with the offsets.

Gauge tightening may also occur where there is metal flow on the rail head.

3.1.5. Twist

The twist parameter is determined by calculating the algebraic difference between two superelevation measurements taken 2.75 metres apart. Twist is encountered where the vertical alignment in both rails is such that one wheel of a vehicle with a rigid frame will lose contact with the running surface of the rail with the obvious danger of a derailment. Figure 266 above would result in a twist defect.

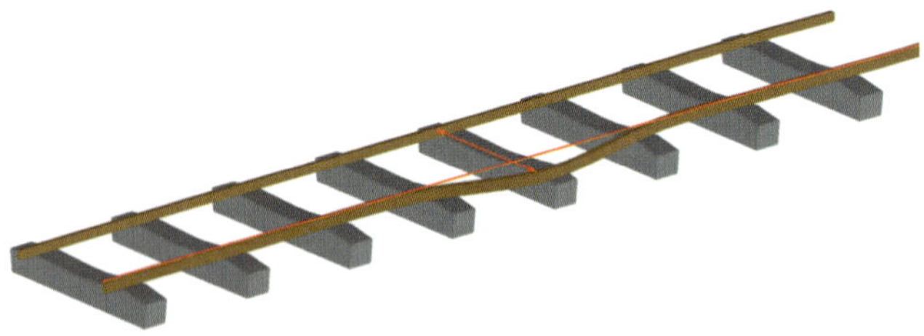

Figure 271: Gauge Defect

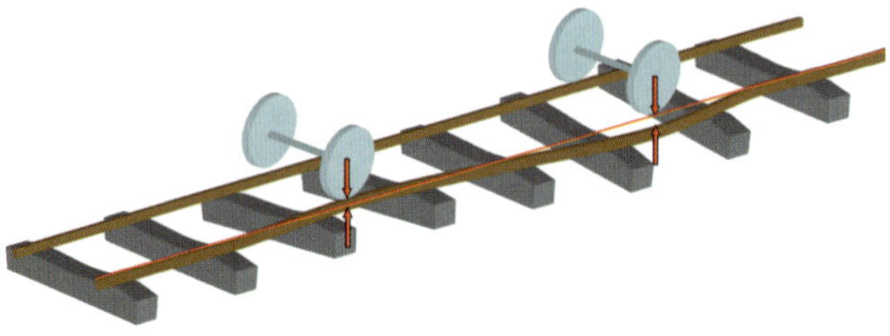

Figure 272: Twist Defect

3.1.6. Maintaining Track Geometry

The first step in maintaining track geometry is to determine the geometry condition using a proven track measuring and recording car (CHAPTER 9). Where the geometry exceeds a set threshold, maintenance is required. Refer again to CHAPTER 6 paragraphs 3 and 4.3.

Vertical alignment, horizontal alignment, superelevation and twist defects are all rectified using a tamping machine. However, it is important to use a tamping machine which can lift, level and align the track within narrow tolerances. Refer to CHAPTER 11. Often the defect may be related to other problems such as fouled ballast or worn track components which would require other maintenance interventions such as ballast cleaning or material replacement. Gauge is an example of a defect which can generally only be rectified with material replacements, such as rails with side wear and broken fastenings or sleepers.

3.2. Rail Profile Parameters

Measuring rail profile parameters will be explained at the hand of the method used by Plasser & Theurer infrastructure measuring and recording vehicles. Detail regarding the system used is explained in CHAPTER 9, paragraph 4.2.

3.2.1. Rail Height and Type

In order to measure the rail height, the measuring system determines the radius at the base/web area of the rail (Figure 273). A circle is drawn according to this radius at the fillet area and the centre points of the circles are connected. The length of this line is unique for each rail section and can therefore be used to determine the rail type. A line drawn perpendicular to the midpoint of the first line to the top of the rail crown represents the rail height.

3.2.2. Rail Width

The rail width is determined by drawing two lines perpendicular to the rail center line at 14 mm below the crown (Figure 274). One line extends to the gauge side of the rail crown and the other to the field side. These two lines provide gauge and field side wear, as well as the total crown width.

3.2.3. Rail Inclination

Rails are inclined at an angle of 1:20 to the vertical on sleepers in order to provide the 1:20 contact surface with the running surface of the wheel (Figure 275). This is a very useful parameter to determine broken sleepers, worn or loose fastenings and pads and inadequate fastening strength.

3.2.4. Crown Wear

Crown wear is the vertical difference in height between a new rail profile and a worn profile measured in the centre of the rail crown. The crown wear is also expressed as an overall head loss percentage of the total crown area (Figure 276).

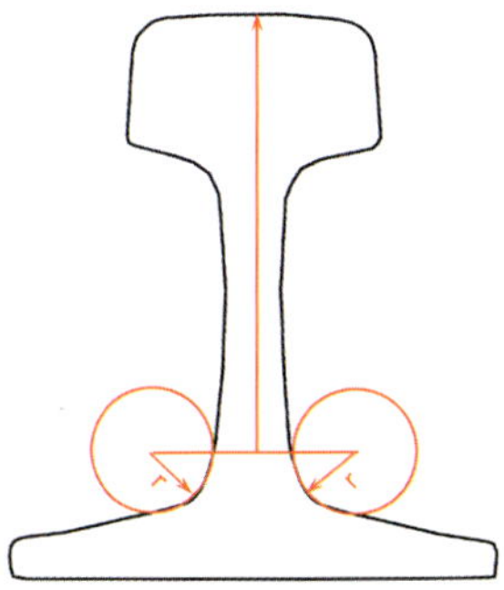

Figure 273: Rail Height and Type

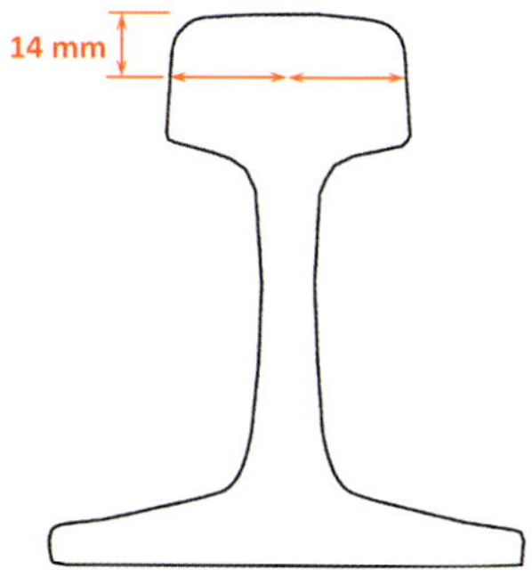

Figure 274: Rail Width

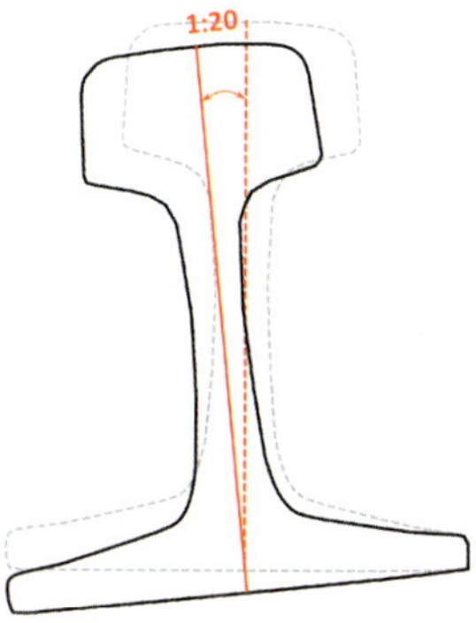

Figure 275: Rail Inclination

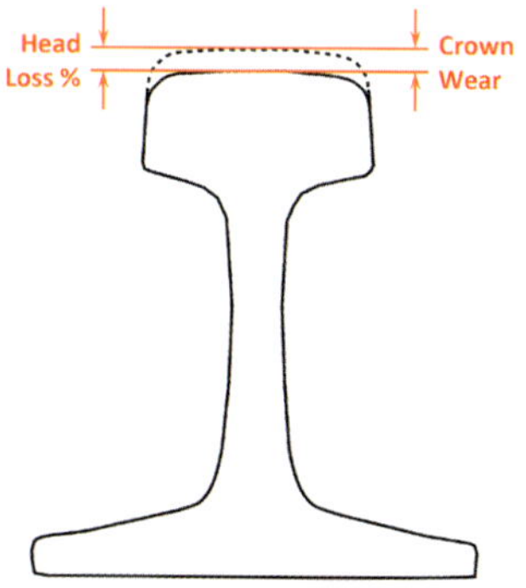

Figure 276: Crown Wear

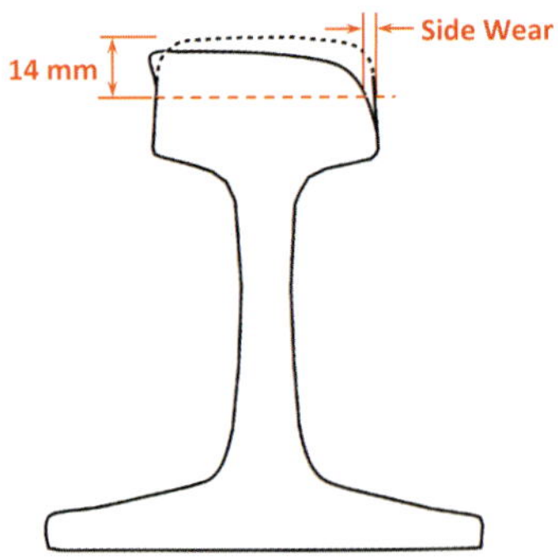

Figure 277: Side Wear

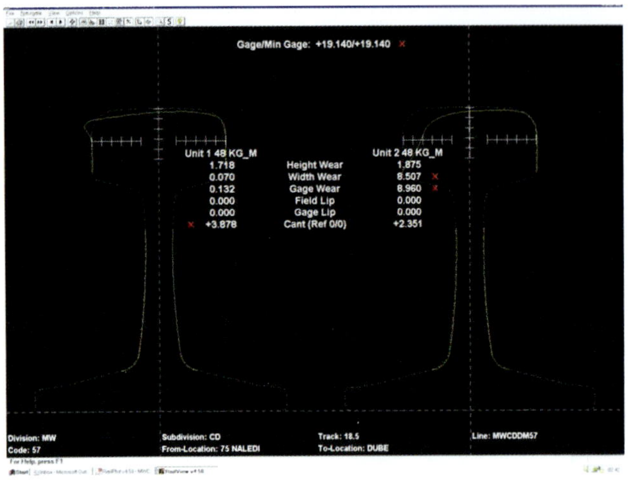

Figure 278: Measuring of Rail Profiles

3.2.5. Side Wear

Side wear is the horizontal difference in width between a new rail profile and a worn profile measured 14 mm below the rail crown level on the gauge side or inside of the rail (Figure 277).

3.2.6. Maintaining Rail Profile

The most important aspect of rail profile management is determining the existing rail profile through measuring and recording. Infrastructure measuring and recording vehicles can be equipped with rail profile measuring equipment (Figure 278). Refer to CHAPTER 9. Refer also to CHAPTER 3, paragraph 2.2 for more detail on rail profiles and management.

3.3. Overhead Electrification Contact Wire Geometry

Similar to the track, the overhead electrification components must also comply with certain geometric requirements to ensure their reliability and to avoid the danger of hook-ups and other damage to the electrification system and locomotives alike.

Figure 279: Overhead Electrical Structure on the Sishen to Saldanha Iron Ore Line, RSA

3.3.1. Vertical Contact Wire Force

Vertical contact wire force is a measurement of the deviation of the actual perpendicular vertical force from 80N between the measuring pantograph and the contact wire (Figure 280).

High vertical contact forces may indicate abnormalities on the line, like kinks, low-tension problems, excessive down slope in the direction of travel or the locality of incorrectly set equipment (section insulators, steady arms or other suspension points) which might cause damage to pantographs.

Low vertical contact forces may indicate loss of contact and might cause arcing. Repeated arcing on the same point might cause the contact wire to burn off. Contact loss also causes poor current supply to locomotives. This might cause surging in the system and damage to the locomotive motors.

3.3.2. Longitudinal Contact Wire Force

This is the frictional force experienced by the pantograph in the direction of travel (Figure 280) between the pantograph measuring head and the contact wire. High relative longitudinal forces are caused by lack of lubrication (where applicable), quick changes in slope and incorrectly set equipment.

3.3.3. Contact Wire Height and Slope

Contact wire height is a measurement with reference to the rail level and is approximately 5 metres above rail (in South Africa). See Figure 280.

Contact wire slope is calculated as the change in contact wire height over a distance of 50 metres. The contact wire generally slopes towards locations where the contact wire height might be limited, such as bridges or tunnels.

3.3.4. Contact Wire Stagger

Contact wire stagger is the horizontal "zig-zag" of the contact wire between suspension points (mast poles) and is measured in millimetres relative to the track centreline (Figure 281). The contact wire is staggered on purpose in order to cause a more even distribution of wear on the pantograph current collecting pan. In areas where there are no stagger this would generally be the cause of a broken suspension point at the mast pole.

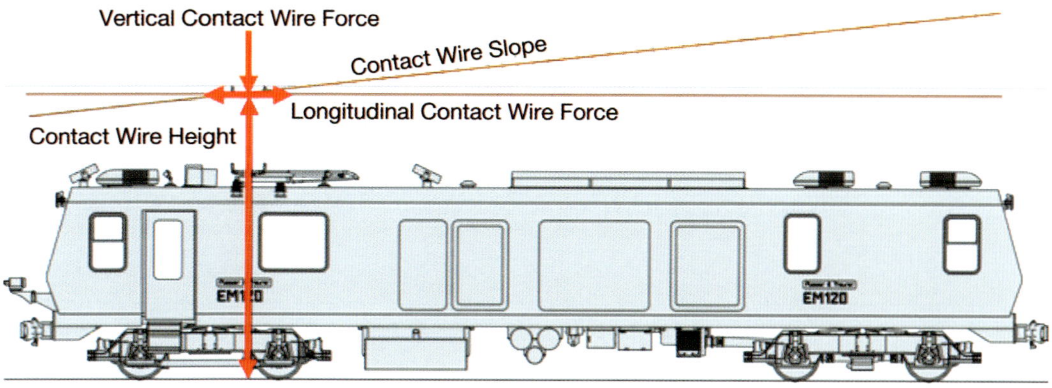

Figure 280: Measured Contact Wire Geometry

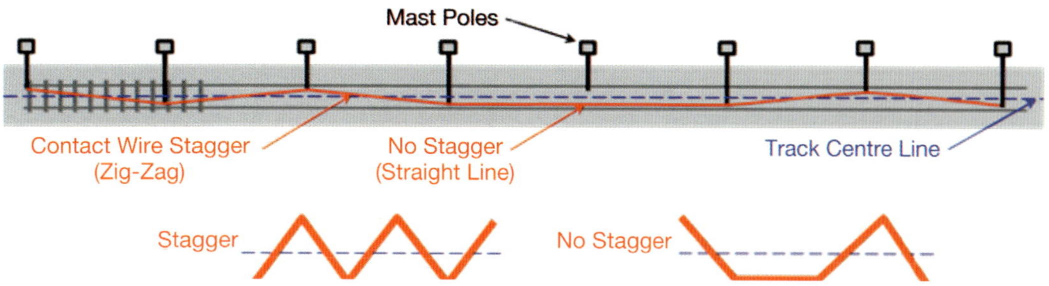

Figure 281: Contact Wire Stagger/No Stagger

Figure 282: Contact Wire Illustrated Sliding Down Pantograph Horn

Figure 283: Notice Contact Wire Moving Out from the Left at a Turnout as well as a Tensioning Device

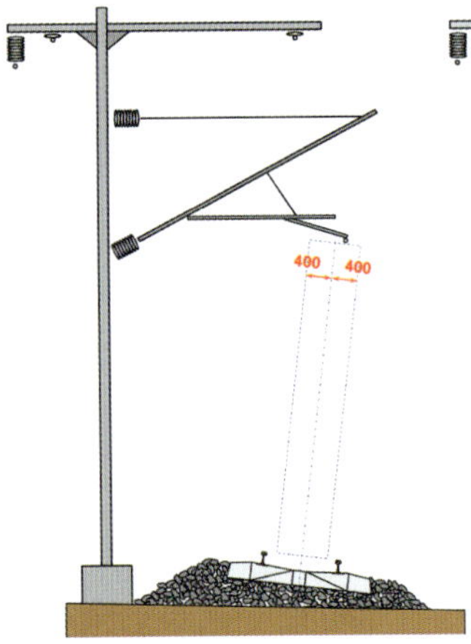

Figure 284: Stagger Set for Track in Curve

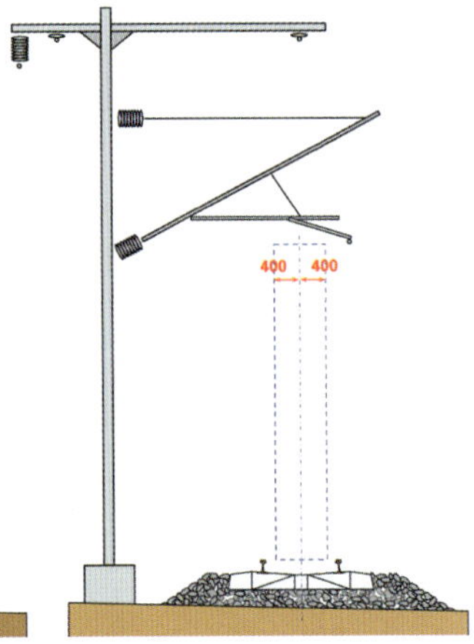

Figure 285: Stagger Exceed Limit if no Cant

3.3.5. Horn Contact

Horn contact is detected at a point where the contact wire makes contact with the pantograph ends or horns and is found at locations where out of specification stagger is encountered or at points where another contact wire slides in too low next to the running wire, such as at overlaps, striking points and at turnouts, for example (Figure 283).

3.3.6. Maintaining Overhead Electrification Systems

Similar to the track, the overhead electrification system geometry must also be measured to determine where maintenance work is required. However, often a recorded overhead contact wire geometry defect is actually a track geometry defect. For example, the track on curves must have a given superelevation (cant) and the contact wire will be positioned according to that cant (Figure 284 showing stagger halfway within maximum limits, exaggerated for illustration purposes). If the cant is however lost (Figure 285), the contact wire may not be positioned within specification and, when measured, could be recorded as a defect. However, if the cant is reinstated, no contact wire geometry defect will be recorded.

It is therefore beneficial from a maintenance efficiency point of view to measure the track and overhead electrification system geometry at the same time at a coincidental point and in so doing able to determine where the actual defect lies by analysing both reports at the same time.

Maintenance to the overhead electrification system consists of replacing insulators and other equipment, the cleaning of insulators, greasing and protecting of components etc. Maintenance to the overhead electrification system can generally not be mechanised but access to the overhead electrification structure can be with the use of overhead electrification maintenance machines (Figure 286). These machines may be equipped with specialised equipment such as catenary support masts, earthing and measuring pantographs, electrical or pneumatic hand tools, cranes, workshop space, etc.

One of the largest maintenance activities of an electrical maintenance department is replacing the contact wire when the maximum wear has been reached. This is an extensive and time-consuming process during which no traffic can run. Mechanised contact wire replacement machines (Figure 287) rolls up the old contact wire (model depending) and installs the new contact wire at the final tension and stagger, saving valuable time and potential injury. Refer to CHAPTER 22.

3.4. Clearances

A moving vehicle structure gauge (Figure 505, South African cape gauge) is generally set by railways to ensure that minimum clearances are maintained for trains to safely pass one another on double lines and in loops, safely cross bridges and tunnels and fit within the electrical overhead structure, where applicable. Branches encroaching into the vehicle structure gauge must also be monitored.

Infrastructure measuring vehicles can be equipped with instrumentation to establish fouling of the structure gauge by branches or even fixed structures such as tunnels due to the track moving.

The structure gauge must also be considered for the purchase of rolling stock and must therefore be included in the specification.

Figure 286: Overhead Electrification Maintenance Machines

Figure 288: Moving Vehicle Structure Gauge for 1067 mm Gauge in RSA

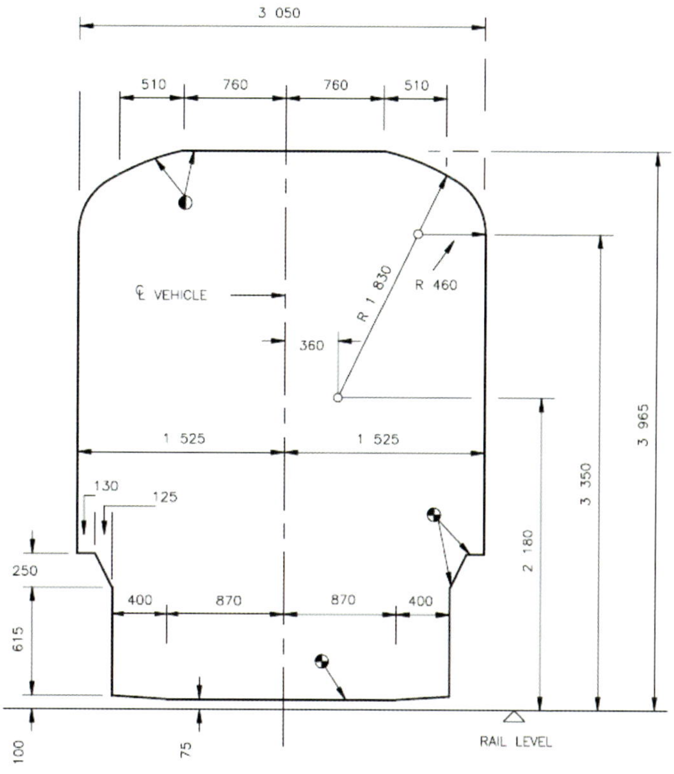

Figure 287: Contact Wire Installation Machine

4. +/– SIGN CONVENTIONS

4.1. Conventions "True to Coordinates" for Track Geometry

When the parameters are defined true to coordinates, there should be no need for further explanation such as left-hand curve or right-hand curve, trough or peak, etc.

- Angles, curvature and radii are plus (+) in a right-hand curve and in a trough and minus (–) in a left-hand curve and on a peak in the direction of increasing km position.
- Longitudinal gradient is plus (+) when the line runs uphill in the direction of increasing km position and minus (–) when it runs downhill.
- Superelevation is plus (+) when the left-hand rail is higher than the right-hand rail in the direction of increasing km position and minus (–) when the right-hand rail is higher than the left-hand rail.
- Lateral acceleration to the left in the direction of increasing km position is plus (+) and minus (–) to the right. The true-to-coordinate vertical acceleration is plus (+) in a trough and minus (–) on a peak.
- Twist is plus (+) when the superelevation increases with the km position and minus (–) when the superelevation decreases–
- All jolts are plus (+) when the acceleration from which they are derived is increasing in the direction of rising km position and minus (–) when it is decreasing.

4.2. Absolute Values

When existing values have to be compared with permissible values, this can be done only with absolute values, i.e. the minus signs must be removed. With conventionally defined static and dynamic values, therefore, an immediate comparison can be made with the permissible values.

Chapter 9
Infrastructure Condition Measuring and Recording

1. INTRODUCTION

A strategy of condition based preventative maintenance will ensure the longest life from the track at the lowest possible lifecycle cost. This requires maintenance intervention before the track deteriorates to a point below a minimum condition threshold (refer again to Figure 197) which implies that the condition of the track must be determined on a regular basis through inspection, measuring and recording and that threshold levels are set.

The European Standard EN 13848 can be consulted for more information on track recording vehicles.

2. TRACK INSPECTIONS

Though this chapter focusses on mechanised track measuring and recording methods, it is important to mention here that regular, scheduled trolley, footplate and patrolled inspections of the track must not be neglected since measuring vehicles do not replace these very important visual inspections. Measuring vehicles only replace manual measuring and surveying of the track geometric condition.

3. INFRASTRUCTURE MEASURING VEHICLES

Measuring the infrastructure manually by surveying gangs is time-consuming, cost-intensive and not accurate enough for any main line today. Infrastructure measuring vehicles can be equipped with a combination of various available measuring systems to establish the condition of the whole infrastructure, depending on the requirements of the railway. Plasser & Theurer as the largest manufacturer of infrastructure measuring vehicles offer a wide range of machine layouts to suit any requirements.

These measuring vehicles can either be self-propelled or a suitable passenger rail coach can be equipped with all the instrumentation and form part of a train consist. However, it should be expected that predominantly self-propelled vehicles will be used (Figure 289) in most countries. Self-propelled vehicles can measure at a speed of up to 160 km/h depending on the machine type in use.

Figure 289: Plasser & Theurer's Range of Infrastructure Measuring Vehicles

Unlike manually-measuring methods, measuring vehicles can measure the track geometry under load as the measuring equipment is positioned on bogies under load. The advantage of measuring under load is that so-called blind vertical defects will become visible under load. As the recording vehicle travels over the track, the measuring system on the recording vehicle calculates the track geometry parameters, displays them in graphical form for the operator, provides information on exceptions and stores the data for filing purposes, later processing or more detailed analysis.

4. GEOMETRY PARAMETERS MEASURED AND RECORDED

The following measuring systems can be installed on infrastructure measuring vehicles:

- Contactless inertial navigational track geometry measuring system
- Rail profile measuring system
- Software package for calculating the equivalent conicity (possible only in combination with the rail profile measuring system)
- Axle box acceleration measuring system
- Corrugation measuring system
- Tunnel clearance, ballast profile, track centre distance measuring system
- Contact wire geometry measuring system with mast pole detection system
- Pantograph for dynamic catenary wire geometry measurements
- Catenary parameter measuring system (voltage, pantograph pressure, etc.)
- Monitoring system for contact wire wear
- Traction plate position measuring system
- Ultrasonic rail flaw detection system

The above measuring systems can be supplemented by the following video systems for a more accurate evaluation of the measurements:

- Driver's view video system (in direction of travel)
- Video system to monitor contact wire position
- Rail component video monitoring system (rail surface, rail fasteners, rail fishplates, sleeper condition)
- Third rail video system
- Thermal imaging video system
- Head check monitoring system (cracks at the gauge side of the rail head)

In this chapter only the most frequently used systems will be discussed.

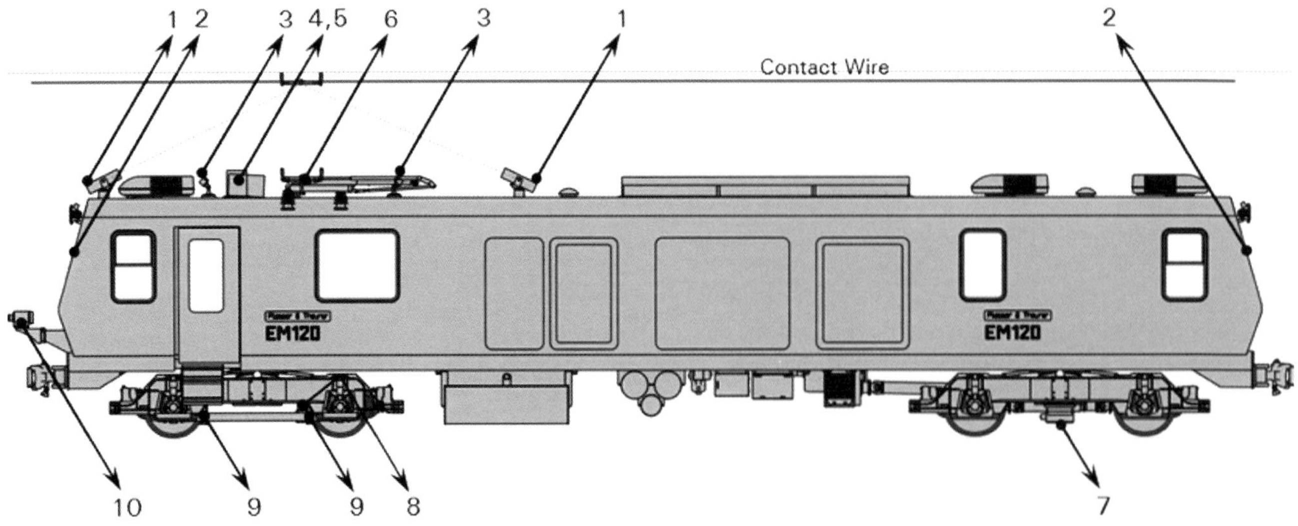

1 Contact Wire Video Recording System
2 Driver's View Video Recording System
3 Overhead Video Lighting System
4 Contact Wire Position Measuring System
5 Mast Pole Detection System
6 Pantograph (Dynamic Measuring)
7 Rail Profile and Wear Measuring System
8 Inertial Measuring Unit (IMU)
9 Optical Gauge Measuring System (OGMS)
10 Laser Mirror System

Figure 290: Location of Measuring Systems on Track Recording Vehicles (EM120)

4.1. Track Geometry

The Plasser & Theurer contactless inertial navigational track geometry measuring system is standard on its measuring vehicles and measures the following geometric parameters at 250 mm sampling intervals (refer again to CHAPTER 8, paragraph 3):

- Horizontal alignment of the left and right rail (also known as alignment or versine) with simultaneous display of three space curves and three reference lengths (chord lengths), selectable by the user
- Vertical alignment of the left and right rail (also known as top or profile) with simultaneous display of three space curves and three reference lengths (chord lengths), selectable by the user
- Cant (also known as superelevation)
- Gauge
- Twist
- Curve radius
- Gradient
- Position using differential GPS
- Speed
- Distance
- Events (e.g. tunnels, bridges, etc.)
- Ambient temperature

Two separate systems are used for these measurements, namely, the Position and Orientation System for Track Geometry (POS/TG) and the Optical Gauge Measuring System (OGMS).

- **The Position and Orientation System for Track Geometry (POS/TG)**
  The POS/TG consists of a rack-mounted navigation computer system which receives data from the Inertial Measurement Unit (IMU), Differential Global Positioning System (DGPS) and the Distance Measuring Indicator (DMI) to process accurate measurements. The POS/TG provides track geometry measurements in millimetres as well as standard deviation calculations and a full set of navigation outputs, consisting of position, speed, altitude, angular rate and bogie acceleration.

  The IMU is mounted to the measuring frame on the bogie [Figure 290 (8) and Figure 291 (iii)] and consists of silicon linear accelerometers and fibre optic gyros that sense three dimensional acceleration and rotation of the IMU. By picking up the changes in angle and movement of the IMU, the IMU supplies earth coordinates of a space curve (blue lines in Figure 291) as they deviate from the track design line (green lines). The red horizontal arrows represent the alignment defects and the orange vertical arrows represent top or profile defects calculated from the space curve. The IMU space curve represents the exact position of the track in space.

i) Measuring frame with fixed vertical distance to the rail surfaces
ii) The Optical Gauge Measuring System (OGMS)
iii) Inertial Measurement Unit (IMU)

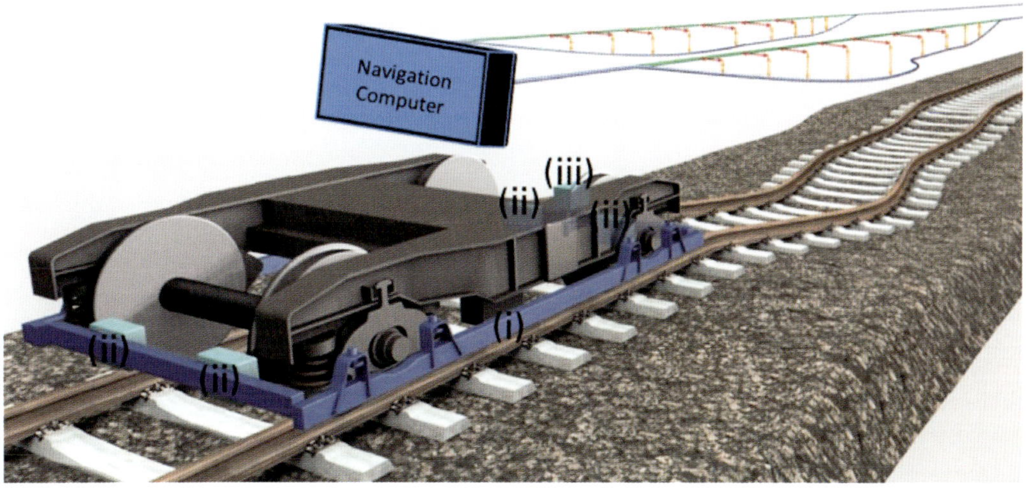

Figure 291: The Measuring Bogie with Instrumentation

The DGPS provides an accurate localisation in terms of latitude and longitude at each of the recorded measurements which is absolutely essential for useful track recordings. It automatically adds the exact location data acquired to the measured data. Normal GPS broadcasts from satellites are not accurate enough; the raw GPS data is therefore corrected using the Differential Global Positioning System which provides a maximum accuracy deviation of 400 mm. The DGPS coordinates come from the GPS receiver or, in areas with no reception, such as tunnels, from the IMU until DGPS reception is resumed.

The DMI is an external sensor that is attached to the vehicle wheel which converts fractional wheel revolutions to increments of distance.

- **The Optical Gauge Measuring System (OGMS)**
  The OGMS sensors (four of them) are also mounted to the measuring frame [Figure 290 (9) and Figure 291 (ii)] and consist of a set of lasers, mirrors and cameras and measure the horizontal displacement from the centre of the bogie to the left and right rails (Figure 292). These units tell the IMU where it is in relation to the rail through triangulation.

4.2. Rail Wear

Rail wear is also measured with an optical-measuring system mounted on the infrastructure measuring car bogie [Figure 290 (7)]. The system measures the full cross-sectional rail profile and compares the information with that of a new rail profile in order to produce a rail wear exception report in real time.

The rail measuring equipment contains lasers and cameras which create a plane of light that surrounds almost the full rail profile. See Figure 293. Using a high resolution camera, the image of a full rail profile is acquired at 250 mm sampling intervals. The computer converts the picture to digital format with 700 xy co-ordinates per rail (Figure 294). This image is then superimposed over a new rail profile, digitally analysed, and the rail wear calculated at speeds up to 300 km/h and accuracy of ±200 µm.

The following rail profile parameters are measured (refer again to CHAPTER 8, paragraph 3.2):

- Rail height
- Rail weight (rail profile)
- Rail head width
- Rail inclination

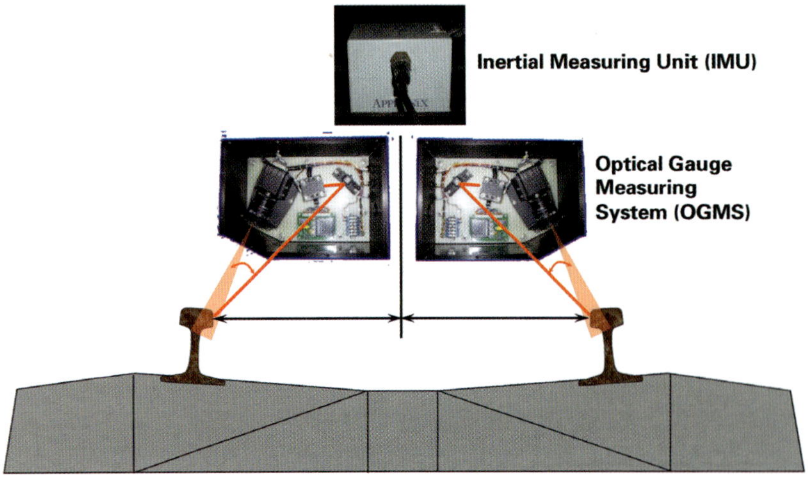

Figure 292:
IMU and OGMS

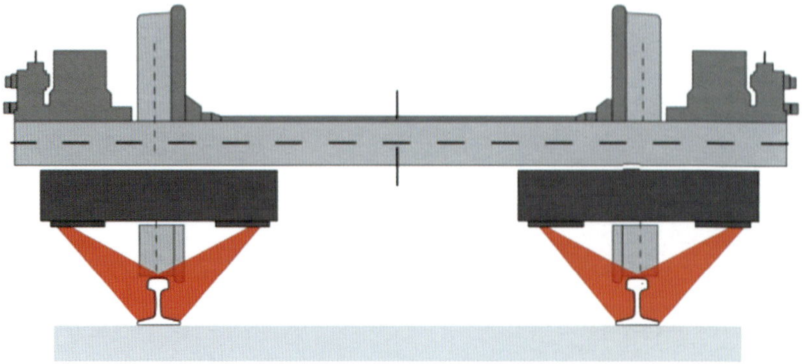

Figure 293:
The Rail Profile Measurement Sensor Heads

By comparing these measurements in real-time with a defined target profile the following parameters are calculated for both rails:

- Rail height and width wear for both rails
- Wear of the rail crown area, given in square millimetres or percentage
- Wear of the rail for up to nine user-defined positions across the rail head such as side wear
- Gauge and field side over burr for both rails

The recordings of those points on the rails outside the pre-selected tolerance are marked by the system and listed in a rail profile exception report. The measured parameters are transmitted to the on-board computer where they are stored together with the measured track data.

4.3. Head Checks, Small Fissures and Cracks

External rail defects can propagate into the rail causing a rail break under ideal conditions with potential disastrous consequences. Refer to CHAPTER 3 paragraph 2.19.1 for more detail on external rail defects.

Head checks, small fissures and cracks on the rail gauge side are detected using monochromatic line scan cameras mounted to the bogie frame of the track recording vehicle monitoring both rails.

Each line scan camera is linked to an individual image acquisition and processing computer. The line scan cameras sample the track longitudinally with a resolution of 0.5 mm along the rails and 1.0 mm across the rail edge.

The high resolution video images are stored on the hard drives of the image acquisition computers. Each image is tagged with differential GPS and track location information. The images are stored and synchronised with all the other measured track geometry, rail data, driver's view video and track component video system to allow evaluation of rail surface defects with other geometry or rail defects.

Rail surface defects can also be detected using eddy current technology. Refer to CHAPTER 10 paragraph 3.

4.4. Rail Corrugations

Corrugations are wave marks on the rail crown which are the source of noise and increased stress on the track components due to dynamic loading. Refer to CHAPTER 3 paragraph 2.18.3 which discusses corrugations as an external rail defect.

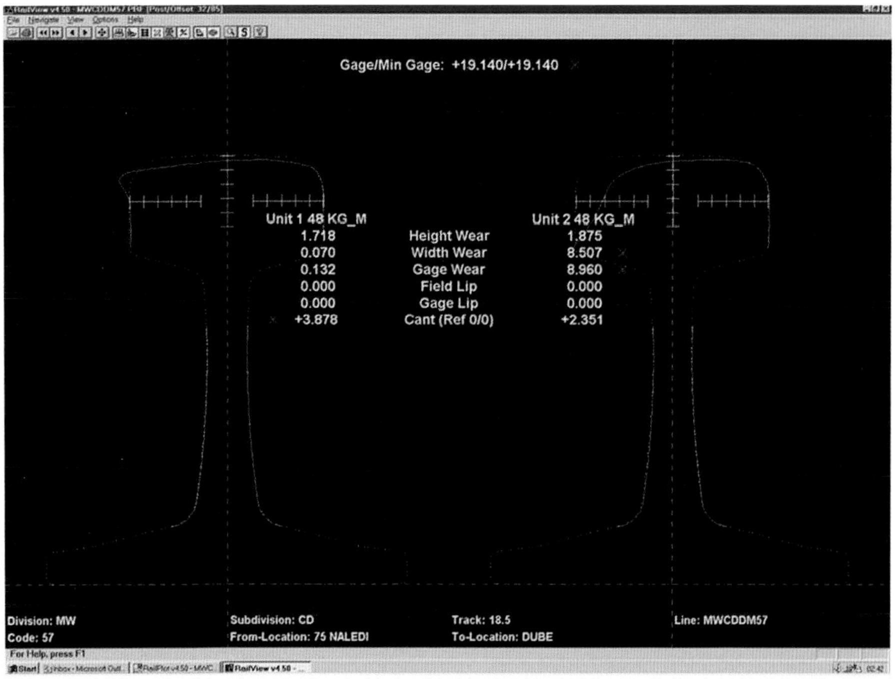

Figure 294: Display of Rail Image

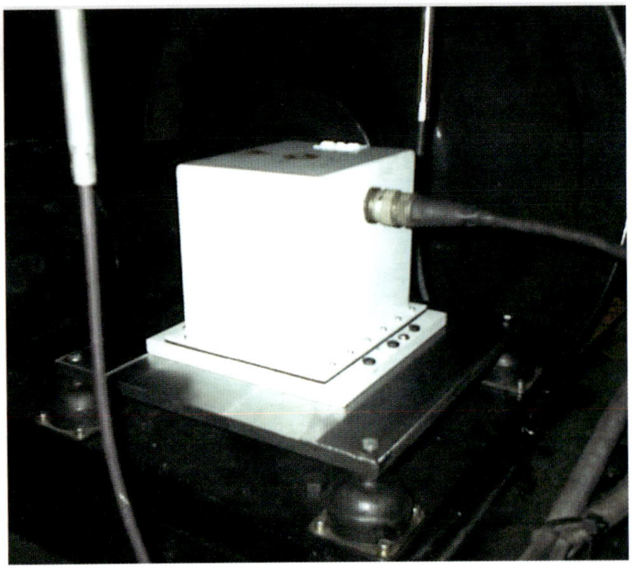

Figure 295: Corrugation Measuring System

Infrastructure measuring and recording vehicles can be equipped with a corrugation measuring system (Figure 295) mounted to the measuring frame guided vertically parallel to the rail surface to measure and record the existence, severity and location of corrugations.

The corrugation measuring system is designed to measure and analyse corrugated rails in different wave length ranges. The corrugation system makes use of a short, three-point asymmetrical chord. The measured chord offset is a representation for the condition of the rail surface. With an optimum chord ratio and a sampling distance of 5 mm, an excellent transfer function for the measured rail surface is accomplished.

The corrugation measuring system is designed to measure at speeds up to 300 km/h, maintaining a sampling rate of one sample per 5 mm of travelled distance. No minimum measuring speed is required. The detectable corrugation wave lengths are between 10 mm and 1,000 mm.

The following parameters are measured:

- Corrugation amplitudes for both rails, either as RMS value or PEAK value for up to four freely definable wave length ranges
- Corrugation wave lengths for both rails for up to four freely definable wave length ranges
- Raw data for chord measurements for both rails
- From the chord measurements a corrugation space curve is calculated and displayed as a PEAK to PEAK value within a user defined band width (corrugation amplitude).

Figure 296: Plasser & Theurer Ultrasonic Rail Flaw Detection Vehicle

4.5. Internal Rail Flaws

The ultrasonic rail flaw detection system is used to detect internal rail flaws such as horizontal, vertical and transverse fissures (cracks) in the crown, foot or web, wheel spin fractures, bolthole cracks, defective welds, rail surface defects, etc. The system is mounted on a telescopic axle and is suitable for speeds of up to 60 km/h. The fourteen transducers that measure one rail are located in two wheel probes (per rail) which are centred above and running on the crown of the rail (Figure 297). A fine mist of water is used to create a contact fluid between the probe wheels and the rail for ultrasonic sound transmission.

Figure 297: Ultrasonic Rail Flaw Detection Wheels

Refer to CHAPTER 3 paragraph 2.19.2 for more detail on internal rail flaws and CHAPTER 10 paragraph 2 for more detail on ultrasonic rail flaw detection.

A separate workplace inside the ultrasonic measuring vehicle is used to analyse the ultrasonic signals. The ultrasonic data is used to generate a side profile of the actual rail.

Pattern recognition software is used to identify and classify the rail flaws. The measuring system supplies location and track information to the ultrasonic rail flaw detection system. Off-line multiple run comparison histograms can be created and analysed.

4.6. Overhead Contact Wire Geometry

A contactless, laser-based, optical-measuring system is utilised to measure the horizontal and vertical positions of the contact wire [Figure 290 (4 & 5) and Figure 298 (ii)]. The system is designed to measure and process the height and stagger of up to four overhead wires simultaneously in real-time at track section speeds.

The measuring sensor is mounted on top of the roof of the infrastructure measuring vehicle. The measured data is transmitted to the processing unit inside the car via an oscillating mirror. A high frequency modulated, temperature-stabilised laser beam scans a range of ±35° above the track recording vehicle lateral to the travelling direction at 100 times per second (Figure 299). When the laser beam hits a wire, the light is reflected and the receiving unit assesses the reflection. Analysing the phase shift between outgoing and reflected light determines the distance from the mirror to the catenary wire. The measured polar co-ordinates are converted into x-y co-ordinates by taking the angle of the mirror into account.

The measured stagger and height values are referenced to the track plane (a plane through the top of both rails) by compensating for vehicle body movements and bogie rotation. Utilizing optical gauge measurement data, the stagger values can also be referenced to the track centre line.

The conditioned measured data is transmitted via Ethernet to the main server of the measuring vehicle and stored together with the other measured data. During the measuring run, the catenary parameters can be shown, together with other parameters such as superelevation or curvature. The main advantage of this coincidental measurement is to separate seemingly overhead contact wire defects which are actually caused by track defects. Refer again to Figure 284 and Figure 285.

The laser beam is eye-safe for distances greater than 2.5 m. As an additional laser safety measure, the scanner cover is closed automatically at speeds below 3 km/h. The laser scanner is fitted with continuous, automatic calibration at intervals of 20 milliseconds. The system is not influenced by ambient light.

Figure 298: Overhead Contact Wire position and Mast Pole Detection System

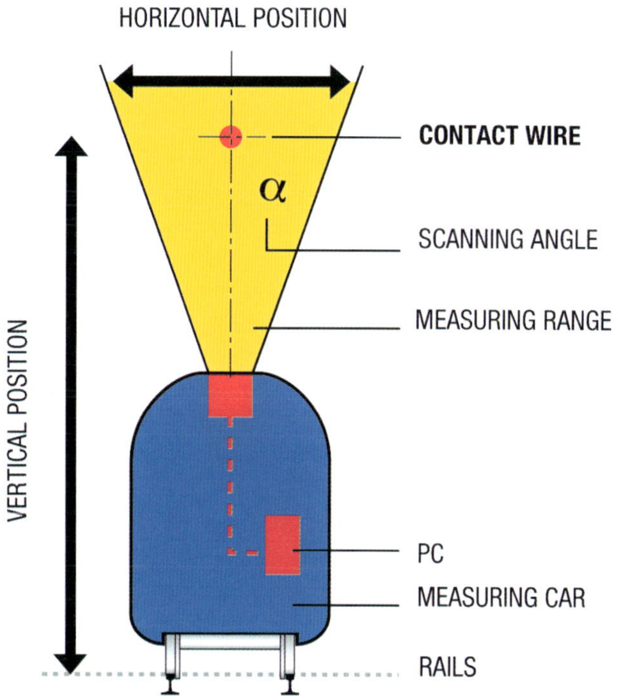

Figure 299: Working Principle (need high resolution images from Roman)

The following contact wire parameters are measured (refer again to CHAPTER 8, paragraph 3.3):

- Vertical contact wire force
- Longitudinal force
- Contact wire height and slope
- Horn contact
- Contact wire stagger

The measuring system measures the position of the contact wire at rest. With the pantograph [Figure 290 (6)] the contact wire position can also be measured under dynamic conditions. The pantograph simulates the same dynamic loading normally applied to the contact wire by a locomotive travelling along the track. The actual pressure applied by the pantograph to the contact wire and the accelerations are measured. Additionally the contact wire voltage is measured. The loss of voltage also indicates when the dynamic forces will cause the pantograph to loose contact with the catenary wire. All measured data are transmitted to the central server, synchronised with the distance measurement and stored.

The pantograph is instrumented with an acceleration measuring system consisting of a two axis fibre-optic sensor to measure the vertical and lateral pantograph accelerations. Accelerations are measured by interpreting laser light intensity fluctuations by comparing the intensity of the laser light emitted by the fibre into the sensor with the light returned by the second fibre from the sensor.

A remote unit, which is emitting and receiving the laser light, is installed inside the measuring car. The fibre-optic is routed through an isolator in order to separate the electrical potential of the pantograph from the car inside. The remote unit outputs two analogue voltages, one representing the vertical, the other the lateral accelerations.

The analogue output signals are connected to an application specific computer plug-in board for sampling and data processing. The board samples the signals with a frequency of up to 25 kHz. After performing digital low pass filtering on each input channel, the board analyses the recorded accelerations on a distance based interval (user defined, generally 250 mm) by calculating the acceleration RMS and extracting minimum and maximum PEAK values per channel.

The RMS and PEAK values are then stored lined-up with the other measuring data for displaying, printing and further analysis.

4.7. Mast Pole Detection System

The Mast Pole Detection System automatically locates the position of the overhead electrification mast poles along the track as an input into the OHE geometry analysis. The mast pole position is also used for exact synchronisation of track locations. The system performs data acquisition and processing in real-time at track section speeds.

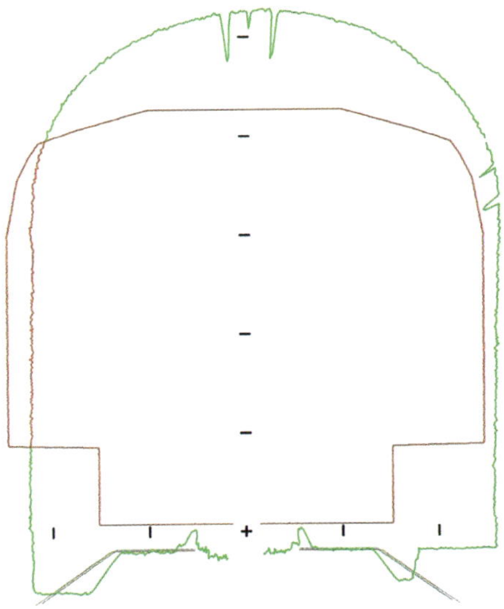

Figure 300: Illustration of Clearance and Ballast Volume Measurement

Figure 301: Working Principle Get high res picture from Roman

4.8. Structure Clearances and Ballast Profile

Infrastructure measuring cars can be equipped with a clearance and ballast profile measuring system to establish clearances to tunnels, platforms, adjacent lines and other wayside structures to the vehicle structure gauge (Figure 300). Any object that fouls the vehicle structure gauge is recorded as an exception at a specific track kilometre point. Refer again to CHAPTER 8, paragraph 3.4.

The system also measures the calculated excess or deficit of ballast for the left and right, inner and outer shoulder. The measurement is in square millimetres but is converted to cubic metre per 200 metre break length for reporting.

This system utilises a laser mirror scanner (Figure 302) which is a high speed profile-measuring device using an electro-optical range detection method and a beam scanning mechanism. The laser mirror scanner delivers profile data at almost 360° per scan, at up to 100 revolutions per second and performs a maximum of 300,000 distance measuring points per second. The system has a measuring range of 1.3 to 30 metres.

The system uses the "time of flight" principle. An electrical pulse generator periodically drives a semiconductor laser diode to send out infrared light pulses. The echo signal reflected by the target hits a photodiode on the Laser Mirror Scanner, which generates an electrical signal. The time intervals between transmitted and received pulses are counted, sent to the microcomputer and transformed into the measured distance between the laser system and the target.

4.9. Video Recording Systems

Two high resolution video recording systems are available; a contact wire and a driver's view video recording system. The camera can record up to 15 images per second. At a speed of 100 km/h this is equivalent to a frequency of approximately one image every 2 m.

The digital image data stream is transformed, compressed online, and stored on the hard disk of the video recording computer. The recorded video images can be viewed on the infrastructure measuring vehicle and, with the off-board analysis tool, on any PC work station, after installation of the software. Images can be retrieved quickly due to linked track and location information with every image. The off-board analysis tool also allows the recorded JPG images to be converted into AVI video files.

The video recording systems can be supplemented by powerful lighting systems.

- The contact wire video system is directed at the pantograph and takes snap shots of the overhead structure every 2 metres. A defect detected by the contact wire measuring system can then be viewed on a photo from the video system to establish or eliminate causes of the defect.
- The driver's view video system (Figure 303) consists of two cameras, one for each direction of travel. The camera used for recording is automatically selected depending upon the direction of travel of the vehicle.

Figure 302: Laser Mirror Scanner Fitted to the Machine

Figure 303 : Driver's View Video Image

5. ANALYSIS OF MEASURED RESULTS

The objectives of measuring the track geometry are to provide track information to:

- locate and rectify emergency defects in order of priority; and
- to assess the condition of the infrastructure to:
  - prepare the preventive maintenance plan, and
  - to establish the effectiveness and adequacy of financial expenditure.

5.1. Prioritising Defects

The amount of data produced by an infrastructure measuring vehicle is vast and consists of various tables of deviation measurements in millimetres. For example, the gauge measurement will not be given as 1,070 or 1,064 mm, for example, but rather as a +3 mm or –3 mm deviation from the required gauge.

In order to prioritise such deviations, railways and international standards organisations have applied thresholds to the measurements. These thresholds can typically be indicated as "within the required standard", a "deviation from the standard" and a "deviation constituting an urgent repair". The table in Figure 304 provides an example of typical threshold values for track geometry measurements as specified by Transnet in South Africa.

Exception reports will list all deviations from a required threshold and may be produced in separate reports, one for exceeding "deviations from a standard" and another for exceeding "deviations constituting an urgent repair". These reports can be converted to a spread sheet format and sorted from highest to lowest exceedence by parameter.

This information can then be used to rectify defects. However, this is spot maintenance which should have been avoided in the first place by a programme of preventative maintenance. If such a programme is, however, in place, spot maintenance would indicate an underlying problem which should be rectified to avoid the same defect from recurring and wasting valuable resources that could have been used for preventative maintenance.

5.2. Assessing the Condition of the Infrastructure

The Track Master of a specific track section needs the actual data of defects measured in millimetres at a specific location on the line for carrying out corrective spot maintenance activities in order of priority as described above.

Spot maintenance is, however, very costly and inefficient. To prolong the life of the track and in so doing achieve the lowest possible lifecycle cost, a programme of preventative maintenance must be followed as described in CHAPTER 6. This requires the track section to be maintained when a predetermined threshold for minimum allowable condition has been reached (paragraph 4.3 of CHAPTER 6) and involves the whole track section to be tamped, for example, which is probably several kilometres in length.

The Maintenance Manager therefore requires a summary, a single figure, of the condition of each track section under his authority to prioritise workload and scheduling of available resources. This single figure must be compared with a single figure threshold to decide which track sections must be maintained next.

Parameter	Required Standard (mm)	Deviation From Standard (mm)	Deviation Urgent Repair (mm)
Vertical Alignment	±3.5	±14.0	±19.0
Horizontal Alignment	±2.5	±10.0	±14.0
Cant	±3.0	±12.0	±16.9
Narrow gauge	-3.0	-7.0	-10.0
Wide gauge	+5.0	+12.0	+25.0
Twist	± 2.5	±6.9	±9.5
Curves	(0.05 x Mo) + 2.5	(0.2 x Mo) + 2.5	(0.3 x Mo) + 2.5

Figure 304: Example of Thresholds for Track Geometry (1067mm Gauge)

The measurements in millimeters per parameter at a specific location are, however, not of much use to establish general condition assessments of various track sections or to establish a threshold for minimum track condition. Using an average figure of a parameter's or combination of parameters' measurements per track section would be statistically inconclusive.

For this purpose a standard deviation (a statistical formula) is calculated of the measured values bundled in specific break lengths (generally 200 metres) per track section. The software on the Plasser & Theurer measuring car can calculate standard deviations for each of the following geometry parameters:

- Average vertical alignment (PRA)
- Average horizontal alignment (ALA)
- Twist (TWT)
- Superelevation or cross level (SUP)
- Track gauge (GAU)

The table in Figure 305 serves as one example of a comprehensive report used by South Africa for a short line from kilometre 105 to 106.4. Columns A to E provide the average standard deviations for each of the parameters per 200 metre break length. If the average of each of the parameters is calculated for the whole track section a single figure, which is statistically correct, will then quantify the track condition per parameter.

The standard deviations allow for trend analysis and the highlighting of problem parameters. So, for example, a standard deviation smaller than 2 is acceptable for each parameter whereas a standard deviation of 1 is considered to be very good.

To obtain a single figure to quantify the track condition for all parameters for this example, the averages of all the parameters can be added (Columns E+D+C+B+A=column X) to provide a Track Quality Index (TQI). The TQI of this track section is therefore 6.91 and can now be compared with other track sections to determine where maintenance priorities are and to determine deterioration rates. A TQI of less than 10 would be considered acceptable whereas a maximum TQI of 7.5 would be allowed for high speed passenger lines and heavy haul lines. The maximum values will be decided by the management of the particular line.

A further index, the Tamp Track Quality Index (TMPTQI, column Y) is also calculated to quantify tamping requirements and to evaluate tamping performance. The gauge parameter (column A) cannot be rectified with tamping and is therefore excluded from the TMPTQI. In the example this track section has a TMPTQI of 5.54. This section can now be compared with other sections and the tamping priorities determined by the section with the highest TMPTQI. There are, however, other considerations such, as sections with many turnouts or other built-in defects or underlying root causes of defects.

			X	A	B	C	D	E	Y
KM FROM	KM TO	BREAK LENGTH	TQI	GAGE	TWIST	SUPR	PRAVG	ALAVG	TAMP TQI
105.000	105.200	200	12.12	3.23	2.03	2.31	2.70	1.88	8.92
105.200	105.400	200	6.28	1.03	1.09	1.40	1.49	1.29	5.27
105.400	105.600	200	5.01	1.05	1.05	0.97	0.91	1.06	3.99
105.600	105.800	200	5.37	0.73	0.98	1.01	1.11	1.56	4.66
105.800	106.000	200	8.47	2.25	1.47	1.54	1.66	1.57	6.24
106.000	106.200	200	5.93	0.58	1.27	1.54	1.68	0.90	5.39
106.200	106.400	200	4.97	0.65	0.70	0.79	1.85	1.00	4.34
		Average	**6.91**	**1.36**	**1.23**	**1.37**	**1.63**	**1.32**	**5.54**

Figure 305: Example of Track Quality Index Report

The TQI, TMPTQI or even standard deviation values are therefore also ideal to set threshold values. If, for example, the TMPTQI for a track section reaches 7.0, tamping would be required (Figure 306). Instead of TMPTQI, some maintenance managers prefer to use the single parameter of top or vertical alignment.

Various reports can be prepared using the TQI values to compare track sections with one another to establish priorities in terms of the maintenance program and for general investment decision-making.

6. REPORTING OF MEASURED RESULTS

The measured data is made available in the form of on-board network viewing of the measurements in real time, printed real time reports and off-board post-processed reports.

6.1. On-Board Network Viewing of Measurements

The infrastructure measuring vehicle has an on-board area with LAN points where representatives of the railway can view the measured data on screens in real time.

6.2. On-Board Real Time Reports

The following two reports are printed in real time which the representatives of the railway take with them as they disembark the machine to immediately address emergency defects:

6.2.1. Strip Chart

The strip chart (Figure 308) graphically displays all the parameters measured, with lines indicating the "deviation constituting an urgent repair" threshold (paragraph 5.1). The scale of the graph is also provided, which allows maintenance staff to establish the size of the defect in millimetres. A separate strip chart is printed for each of the measuring systems installed.

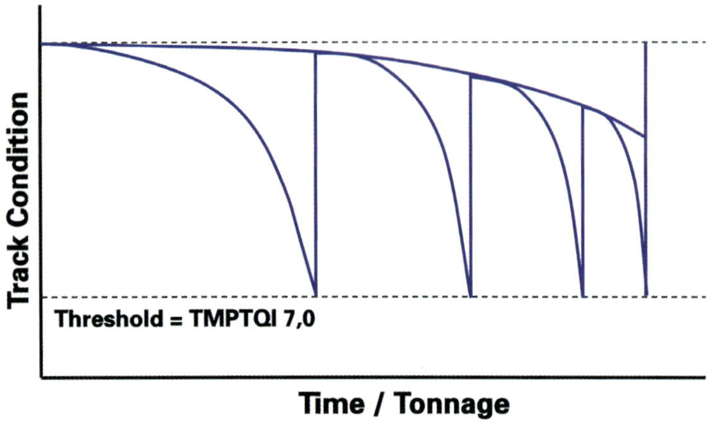

Figure 306: Using TMPTQI as a Threshold for Minimum Track Condition for Maintenance Intervention (Tamping)

Figure 307: On-Board Network Viewing Screens behind the Machine Operator and Pilot

6.2.2. Exceptions Report

Whenever a measured parameter exceeds the pre-set limits (normally the "deviation constituting an urgent repair" threshold) an exception will be listed in an exceptions report (Figure 309). The report reflects, among other, the exceeding parameter (i.e. alignment, twist etc.), the maximum deviation in millimetres and the location in a table format. A separate exceptions report is printed for each of the measuring systems installed.

Exception reports will generally not feed into a preventative maintenance programme but is rather a preventative corrective maintenance activity to avoid an infrastructure catastrophic failure.

6.3. Off-Board Post Processed Reports

All the measured data are also post-processed into various standard reports presented in table and graphic formats for each of the measuring systems installed. Many reports are customer-specific and expert advice and training would be required when a new machine is commissioned.

The reports can be converted to a spread sheet from which various manipulations of the information can be made and various graphic displays can be prepared to assist decision making.

The following are just an introduction into the type of reports that are available to get an understanding of the value of an infrastructure measuring vehicle for an efficient and cost-effective maintenance strategy.

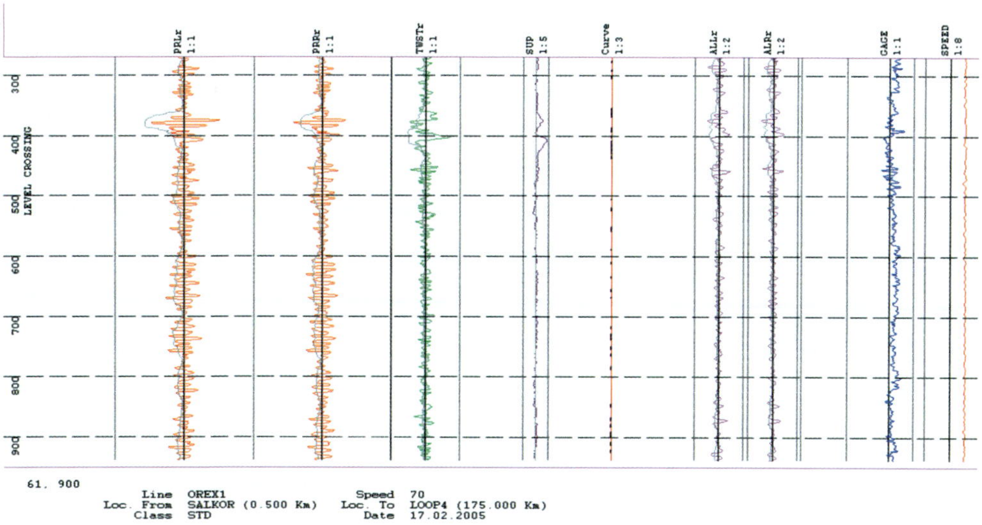

Figure 308: Example of a Strip Chart

EXCEPTION REPORT PAGE 1

2,500

Line C1XBAH Speed 80
Loc. From GMR OOS KAS Loc. To ELSBURG
Class STD Date 16.12.2009

TRK --FROM-- KM M	--TO-- KM M	LENGTH M	PARAMETER		MAXIMUM VALUE	EXCEPTION LOCATION	CLASS CL	SUB			
2,505	2,506	1	PROFILE LEFT	-21	2,505	STD	16.12.2009	[-26.21506412	28.18585054	1672.26]
2,515	2,517	2	TWIST 2.75m	-13	2,516	STD	16.12.2009	[-26.21513615	28.18591871	1672.09]
3,314	3,316	2	TWIST 2.75m	-15	3,315	STD	16.12.2009	[-26.22068329	28.19100720	1664.15]
3,197	3,513	316	SUPERELEVATION	21	3,494	STD	16.12.2009	[-26.22217006	28.19172224	1662.51]
3,537	3,546	9	SUPERELEVATION	14	3,539	STD	16.12.2009	[-26.22255532	28.19184919	1662.10]
3,560	3,566	6	SUPERELEVATION	4	3,561	STD	16.12.2009	[-26.22274348	28.19190137	1661.81]
4,391	4,393	3	SUPERELEVATION	25	4,392	STD	16.12.2009	[-26.23005621	28.19377433	1654.72]
4,417	4,420	3	TWIST 2.75m	-14	4,418	STD	16.12.2009	[-26.23028494	28.19380719	1654.68]
4,420	4,422	2	PROFILE LEFT	-26	4,421	STD	16.12.2009	[-26.23030651	28.19381193	1654.63]
4,419	4,422	3	ALIGNMENT RIGHT	22	4,421	STD	16.12.2009	[-26.23030651	28.19381193	1654.63]
4,420	4,422	2	ALIGNMENT LEFT	17	4,421	STD	16.12.2009	[-26.23030877	28.19381224	1654.63]
4,422	4,424	3	TWIST 2.75m	17	4,422	STD	16.12.2009	[-26.23032215	28.19381400	1654.63]
4,418	4,423	4	SUPERELEVATION	12	4,422	STD	16.12.2009	[-26.23032215	28.19381400	1654.63]
5,92	5,96	4	SUPERELEVATION	-21	5,95	STD	16.12.2009	[-26.23639178	28.19405112	1651.73]
5,288	5,289	1	PROFILE LEFT	-22	5,288	STD	16.12.2009	[-26.23813866	28.19410078	1650.87]
5,292	5,293	2	PROFILE LEFT	23	5,292	STD	16.12.2009	[-26.23817476	28.19410176	1650.87]
5,314	5,316	2	TWIST 2.75m	14	5,315	STD	16.12.2009	[-26.23838194	28.19410573	1651.32]
5,317	5,318	1	PROFILE LEFT	21	5,317	STD	16.12.2009	[-26.23839998	28.19410616	1651.34]
5,331	5,336	5	TWIST 2.75m	-14	5,335	STD	16.12.2009	[-26.23856227	28.19411003	1651.26]
5,377	5,379	3	TWIST 2.75m	-27	5,378	STD	16.12.2009	[-26.23894892	28.19411977	1650.73]
5,400	5,405	5	ALIGNMENT LEFT	-19	5,401	STD	16.12.2009	[-26.23916098	28.19412454	1650.43]
5,417	5,419	2	PROFILE LEFT	-25	5,418	STD	16.12.2009	[-26.23930894	28.19412487	1650.08]
5,409	5,420	11	ALIGNMENT LEFT	23	5,411	STD	16.12.2009	[-26.23924926	28.19412619	1650.16]
5,418	5,420	2	ALIGNMENT RIGHT	16	5,418	STD	16.12.2009	[-26.23931347	28.19412477	1650.07]
5,424	5,425	1	PROFILE LEFT	-23	5,425	STD	16.12.2009	[-26.23937015	28.19412285	1650.02]
5,379	5,685	306	SUPERELEVATION	26	5,525	STD	16.12.2009	[-26.24025695	28.19393708	1649.07]
5,704	5,709	5	ALIGNMENT LEFT	-21	5,707	STD	16.12.2009	[-26.24153423	28.19283151	1648.65]
6,604	6,607	3	TWIST 2.75m	-14	6,604	STD	16.12.2009	[-26.24685340	28.18605470	1640.94]
6,645	6,648	3	TWIST 2.75m	11	6,648	STD	16.12.2009	[-26.24711079	28.18572803	1640.30]

Figure 309: Example of an Exception Report

6.3.1. Track Geometry

6.3.1.1. Track Quality Index (TQI) Report

The TQI Report is provided in a table format (see Figure 305 for an example) and can be used to determine where maintenance should be done. Refer again to paragraph 5.2 for an explanation of the use of the Track Quality Index.

The data can be manipulated by sorting on different parameters to prioritize where the most critical areas exist. The maintenance program can be drawn up by using cut-off values (thresholds) for the different parameters.

6.3.1.2. ADAII Report

The ADAII calculation derives a travel comfort quality index for selected parameters for pre-selected distances. A 50 m break length is used as default. The maintenance quality indexes (ADAII indexes) of the report are based on the horizontal and vertical accelerations the un-sprung vehicle would experience when traveling with maximum allowed track speed over the measured track geometry.

6.3.1.3. Number of Exceedences per Threshold Report

This report contains the number of exceedences per thresholds per track section as described in paragraph 5.1, which provides a different analysis of the condition of the track section.

6.3.1.4. Curve Report

The curve report is a listing of the following geometry information:

- The start and end kilometer values of the curve transitions and circular curve
- The differential GPS coordinates of the curve fixed points. Refer again to CHAPTER 8, paragraph 2.2
- The lengths of the transition curves and circular curve
- The average mid-ordinate offset value over the circular curve
- The radius calculated from the average offset value
- The average cant on the circular curve
- The speed as input for the specific curve
- The average gauge on the circular curve

The measured parameter of "horizontal alignment" is further manipulated to produce curve data based on a 10 metre cord. The accuracy for a 10 metre mid-cord offset measurement is 1 mm RMS or better.

The computer system uses the measuring car's measurements (at 250 mm sampling intervals) of the previous 40 metres to calculate a running mean for the average horizontal alignment of the left and right rail. See Figure 310. The running mean is then used to establish the exact point at which the offset exceeds the running mean by a set figure. This is the beginning of the transition curve. The other fixed points are similarly located.

The curve report, together with ultrasonic measurements and historical rail break data, is used to plan the grinding program and the rail replacement program on curves.

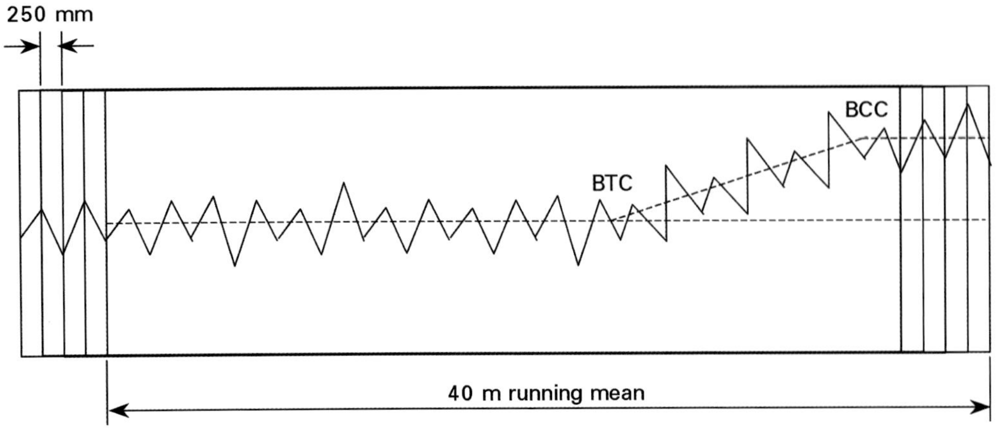

Figure 310: Using a 40 Metre Running Mean to Establish the Curve Fixed Points

6.3.1.5. Event Report

Turnouts, bridges, tunnels and wayside structures can be chosen as a "device" between any points and or for any distance which provides the following information:

- The line information similar to other examples above
- The chosen kilometer start and end points with the associated distance
- The TQI value and standard deviation values for all the parameters
- The number of exceedences of the thresholds
- The GPS co-ordinates of the device

6.3.2. Rail Profile Reports

The following electronic data is provided after post-processing for every section measured:

- Rail wear maximums and averages for 20 metre break lengths
- Rail wear maximums and averages for every straight and curve
- All recorded rail wear values
- Rail wear maximums and averages for the section as a whole

Software is available for post processed rail grinding analysis based on rail profile data collected.

The software uses user defined rail grinding templates for the analysis process. It calculates the profile deviations over the defined locations on and displays them on a strip chart. Profile deviation plots (Figure 311) and enlarged profile deviation plots (Figure 312) can then be printed.

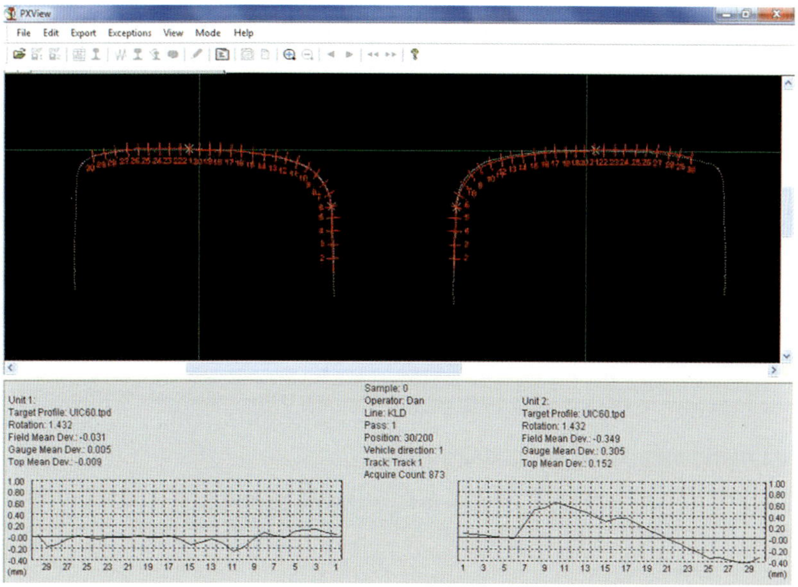

Figure 311: Profile Deviation Plot

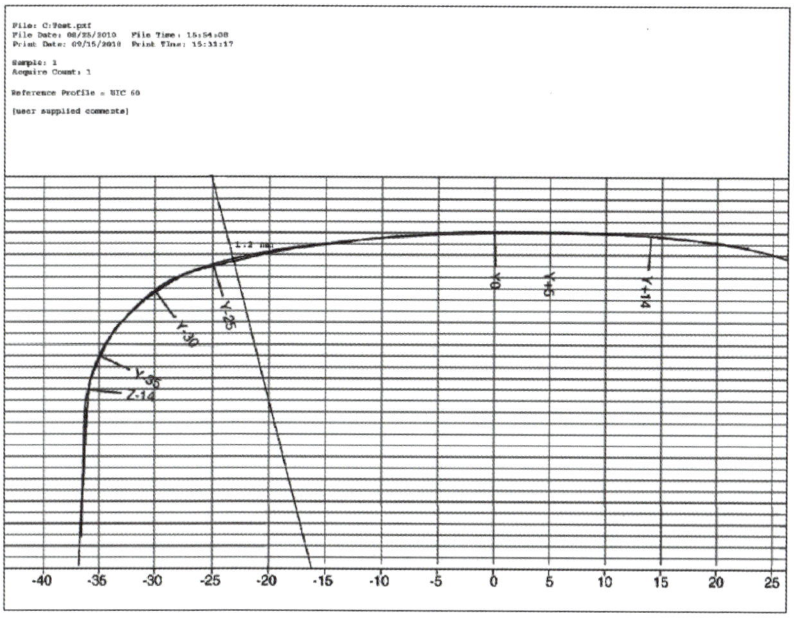

Figure 312: Enlarged Profile Deviation Plot

6.3.3. Overhead Contact Wire Reports

The following electronic data is provided per measuring run after post-processing for every section measured:

- Quantity of all recorded contact wire defects
- Electronic listing of recorded events
- Mast pole report containing contact wire height and stagger at each calculated change in contact wire stagger
- Number of defects per kilometer
- Electronic listing of defects exceeding the "deviations constituting an urgent repair" threshold value

6.3.4. Clearance and Ballast Profile Reports

The Tunnel and Clearance System sends one crosscut for every received data break to the Computer Measuring and Analyser System. The system stores the data at the appropriate location in the geometry file. Another software package evaluates the crosscuts. The report provided is an exceptions report providing the type and size of the exceedence at a specific location.

7. EVALUATING THE EFFECTIVENESS OF THE MAINTENANCE STRATEGY

The TQI values can be used to produce an accumulative distribution of track quality curve (also known as an s-curve, see Figure 313) to evaluate if the maintenance strategy is effective and if the maintenance expenditure is sufficient to maintain the track to set standards.

The s-curve shows the percentage of the track (in 200-metre break lengths) which is smaller than a specific TQI value. Multiple consecutive measuring runs can be compared on one curve and trends established. In Figure 313 six consecutive runs are compared with one another. This is a valuable tool for management since it indicates the improvement or deterioration of the track over time. This curve can be prepared per section, corridor or for the whole railway of several thousand kilometres, depending on the level of management for whom the report is intended.

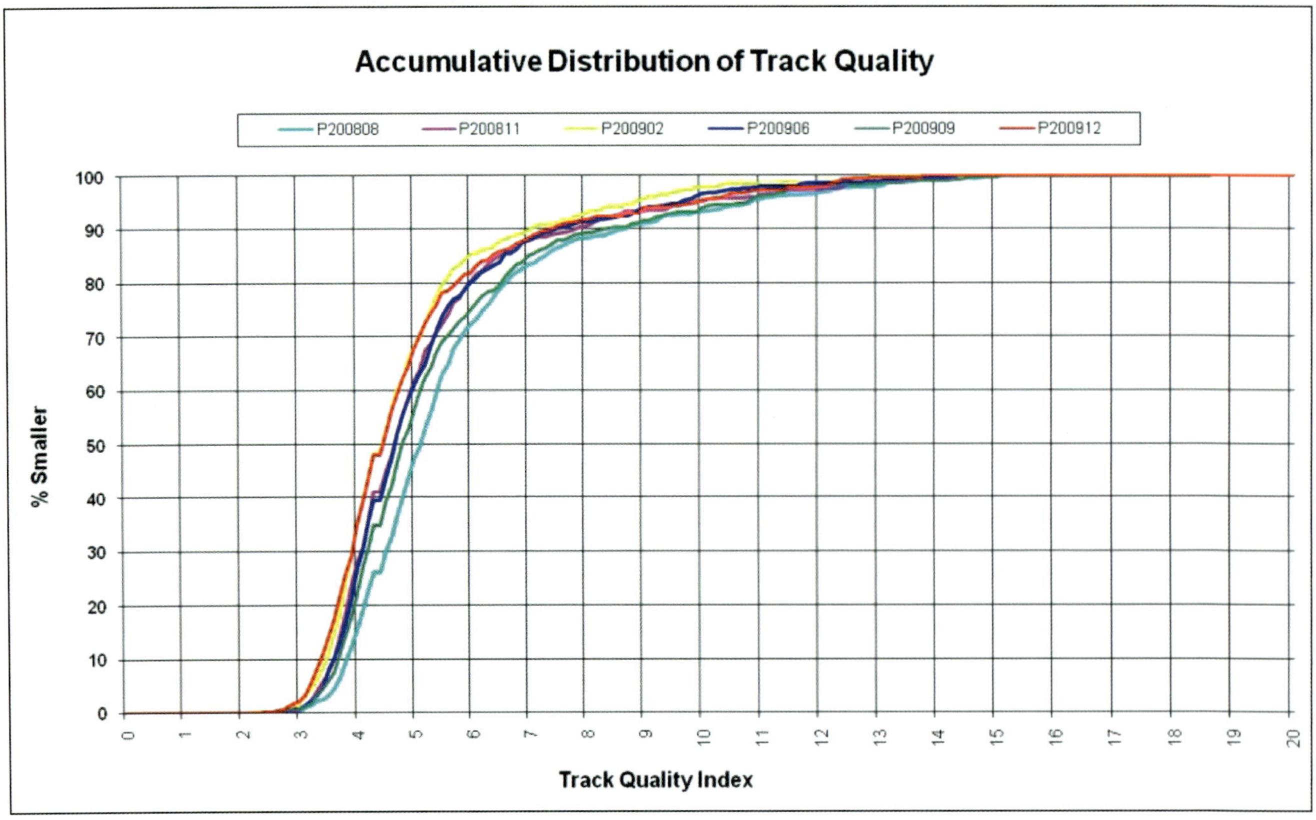

Figure 313: Example of the Accumulative Distribution of Track Quality Graph (S-Curve)

The following conclusion can be derived from this curve:

- The closer the curve starts to the y-axis (on the 0% line), the better the best part (the best 200-metre section) of the track section and vice versa.
- The closer the curve ends to the y-axis (on the 100% line), the better the worst part of the track section and vice versa.
- The total condition of the track section can be related to previous measuring runs by observing horizontal shifts in the s-curve with regard to previous measurements. A horizontal shift to the right indicates deterioration in condition and vice versa.
- The steeper the gradient of the s-curve, the more uniform the track condition. The ideal situation would be to have the curve as vertical upright as possible, which is a track section that will be easier to maintain, since the types of maintenance actions should be very similar over the entire section. Likewise, a very flatly shaped curve means that the overall track condition over the track section is very varied. Maintenance on such a section will be more difficult, since the types of maintenance actions that must be performed on this route will be varied.
- The worst 10% (90%–100% area) normally indicates the condition of track events like turnouts. Monitoring the trends of this area of the chart shows improvement or deterioration of these events.
- The best 10% is never maintained by hand, but kept in check with on-track machines like tampers. If this shows deterioration, the maintenance cycles are too long and indicate insufficient financial investment or a poor maintenance strategy.
- In areas where temperature changes can influence track quality, this will also be observed in the s-curve. Summer measurements will tend to be further to the right on the chart than winter measurements

8. CONCLUSION

Measuring the track infrastructure geometry by any means other than mechanised methods is practically impossible. It was once calculated that to measure only the track geometry over one kilometre would take a surveyor one month to complete. It can be argued that measurements are not always required at 250 mm sampling intervals but even at one metre sampling intervals, it would still take a week. The higher the track speed and axle loading, the shorter the required sampling intervals and the more frequent the measurement would have to be. Furthermore, the accuracy of measurements using, among other, a string and ruler, cannot be compared with the accuracy of lasers.

Mechanised track measuring and recording using measuring cars is a valuable tool if the reports are used to capacity for the following purposes:

- locating and prioritizing corrective maintenance spots (short-term maintenance planning and priority work);
- planning the mechanised maintenance program (routine preventative maintenance);
- track condition monitoring by superimposing the curves of several runs of the vehicle to determine condition changes;
- by imposing minimum and maximum thresholds for track condition, the results from the infrastructure measuring car will indicate to managers the need to intensify financial investment, and, in some cases, even to reduce financial investment; and
- auditing the effectiveness and quality of maintenance work performed.

Track condition monitoring and analysis aids effective track maintenance planning, which in turn supports cost effective and efficient track maintenance, which in turn produces a higher track quality, which again reduces the level of stress exerted on rolling stock and track, and which ultimately reduces the lifecycle cost of the track.

Chapter 10
Rail Flaw Detection

1. INTRODUCTION

Ever increasing demands for rail traffic with regards to speed, axle loading, train lengths, punctuality, comfort etc. places great demands on track safety. Rail defects such as rail breaks can have disastrous consequences in terms of loss of life if a train should derail, especially high speed trains at 300 km/h! On freight trains the loss of high value commodities of especially heavy axle and long trains is extreme in case of a derailment, not even to mention the cleaning up costs in potentially remote areas where access is difficult. In some cases the commodity such as manganese and other hazardous materials cannot be left for environmental reasons.

Other track defects that may hold potential disastrous consequences will normally give a measurable and visible warning that deterioration is taking place and may provide enough time to correct such deviations. The rail on the other hand can fail without warning due to internal defects which are not necessarily visible or due to head checks which are not easy to spot. See CHAPTER 3, paragraph 2.19 for more information on rail defects.

Detecting internal rail defects is therefore very important to avoid disasters. For this reason non-destructive evaluation or testing (NDE or NDT) methods using ultrasonic rail detection systems are common on most high capacity lines today.

Rail surface defects such as head checks also have to be detected, measured and categorised which is the area where eddy current methods are gaining acceptance. Ultrasonic signals are not reliable for detecting surface defects. Rail surface defects can also be detected using options available on a track measuring and recording vehicles using, among other, video monitoring systems.

Over and above the systems discussed in this chapter it should be noted that various other systems are being developed and researched such as electromagnetic acoustic transducers, alternating current field measurements sensors, advanced multi-frequency eddy current sensors, field gradient imaging, etc.

2. ULTRASONIC RAIL FLAW DETECTION

Internal rail defects can be as a result of many factors including:

- Rail fatigue due to rail age, long sleeper spacing, etc.
- Dynamic loading of the rail due to running surface defects or flat spots on wheels
- Built in during the casting of the rail
- Over stressing of the rail during handling (loading and unloading, CHAPTER 19)
- Over lifting of the rail during tamping thereby exceeding the maximum rail stress (CHAPTER 11, paragraph 4.7)

These internal rail defects must be detected using non-destructive examination methods and appropriate action taken before catastrophic rail failure. Ultrasonic rail flaw detection is such a NDE method carried out using multiple ultrasonic transducers mounted to either rail-bound, road/rail, hand push trolleys or hand held equipment. Ultrasonic examination of rails was used for the first time as long ago as 50 years.

Carrying out ultrasonic flaw detection and analysing the data is a specialist activity for which qualified, experienced and licenced staff is required.

2.1. Working Principle of Ultrasonic Flaw Detection

In simple terms, ultrasonic rail testing equipment consists of a pulser/receiver that produces high voltage electrical pulses and a transducer that converts the electrical pulse into ultrasound waves of between 2 and 4 MHz.

The transducer sends the sound signal (pressure waves) into the rail at close proximity using water as a couplant to displace the air between the transducer and the rail. The acoustic impedance mismatch between air and rail steel is large which means that without the couplant most of the sound wave will be reflected and very little transmitted into the rail.

When there is a discontinuity (such as a crack or air gap) in the wave path, part of the energy will be reflected back from the flaw surface in the form of a pulse echo due to the difference in density of the two materials. The sound wave will also be reflected as a back wall echo at the opposing wall of the rail being tested. See Figure 314.

The sound pressure (signal amplitude) reduces with the distance travelled into the material due to scattering (reflection and refraction of the sound in directions other than its original direction of propagation) and absorption of the wave. The speed of sound is however constant for a specific material density and can only be altered by a change in the mode of energy.

The return wave signals are picked up by the crystal transducer and transformed back into an electrical signal. Their time travelled between sending and return of the pulses and their respective amplitudes can now be used to determine the distance to, position, orientation and size of discontinuities (defects) in the rail.

The reflected electrical signal provides a large amount of data for which processing software is required to transform it to real time usable information. The information is then displayed graphically on a screen as a side view of the rail as well as for producing defect reports. The GPS location of defects is also recorded and displayed.

The most common formats to display ultrasonic data are using the A-, B- and C-scan formats, each providing a different way of looking at and evaluating the internal rail defects.

- A-Scan Presentation: This method displays the amount of received ultrasonic energy (signal amplitude) as a function of time (related to the travelled distance within the rail). From the A-scan the relative discontinuity size can be estimated.
- B-Scan Presentation: This method shows a cross-sectional profile view of the rail. The travel time of the sound waves is displayed on the vertical axis and the position of the discontinuity along the length of the rail is displayed on the horizontal axis. From the B-scan, the depth of the discontinuity and its dimensions in the scan direction can be estimated.
- C-Scan Presentation: This method displays the relative signal amplitude or the travel time of the sound waves as a shade of grey or a colour for each of the positions where data was recorded.

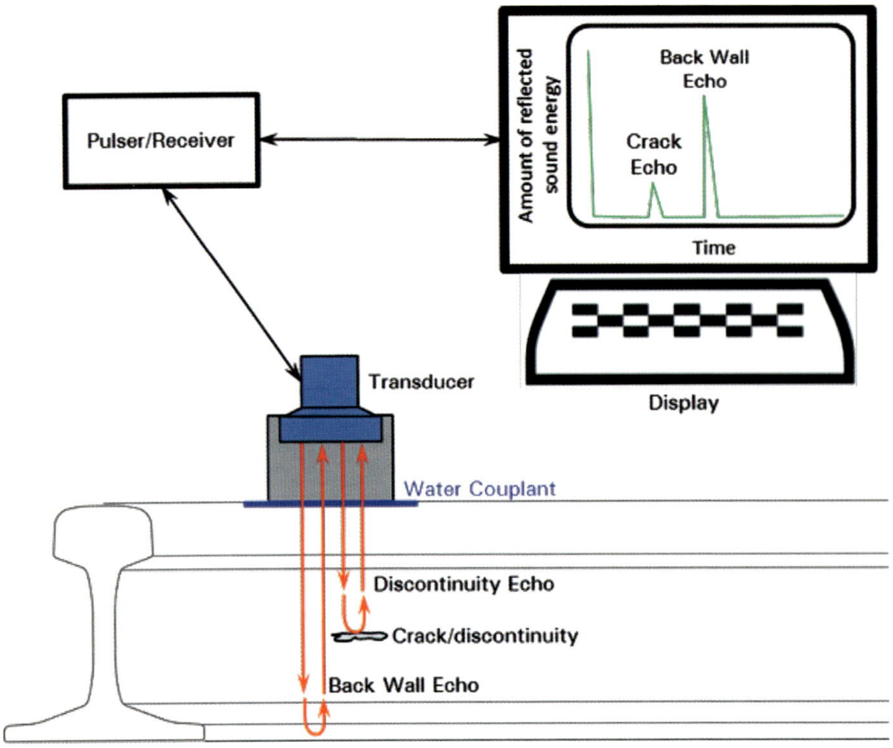

Figure 314: Working Principle of Ultrasonic Rail Flaw Detection (0° Probe)

2.2. Understanding Ultrasound for Effective Rail Flaw Detection

In pulse-echo testing, a transducer sends out a pulse of sound energy and the same or a second transducer listens for reflected echo. If one transducer sends and receives, very shallow defects cannot be detected due to the very short time between sending and having to receive a reflected echo.

Since the detection of discontinuities (defects or flaws) in the rail using ultra sound is based on reflected sound waves, it is important to understand the law of reflection (derived from Fermat's principle) which says "the angle of incidence (Θi) equals the angle of reflection (Θr)". See Figure 315.

If this law is applied to Figure 314 above but the discontinuity in that illustration is at an oblique angle as opposed to parallel to the measuring surface, it means that the reflected sound wave will not return to the transducer and no discontinuity will be detected (Figure 316).

Therefore, ultrasonic examination of rails cannot only use transducers that send sound beams 90° to the surface. These are called straight-beam probes which transmit and receive longitudinal waves (pressure waves).

It is therefore clear that detecting the above discontinuity will only be possible when the discontinuity is hit vertically by the sound beam meaning that the transducer must send the signal at angle to the surface (Figure 317).

Transducers whose beams enter the rail at an angle are called angle-beam probes. Standard angle-beam probes transmit and receive transverse waves, also called shear waves. Angle-beam transducers are mounted in a wedged sled to introduce a refracted shear wave into the rail.

Due to the law of reflection there will be no reflected back wall echo when using an angle beam transducer since the sound wave will be reflected away from the probe (the angle of incidence equals the angle of reflection, see Figure 315 again and also Figure 318).

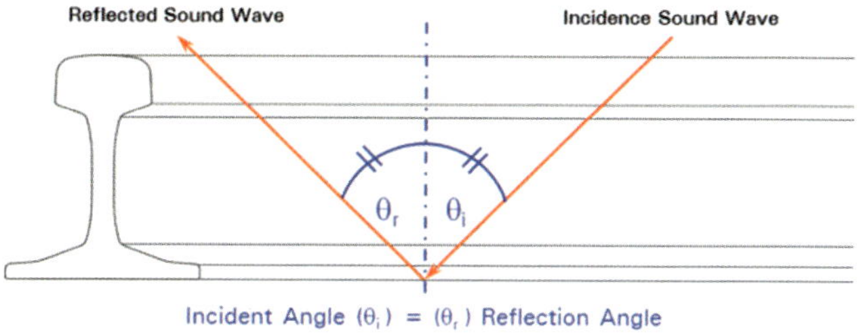

Figure 315: Law of Reflection

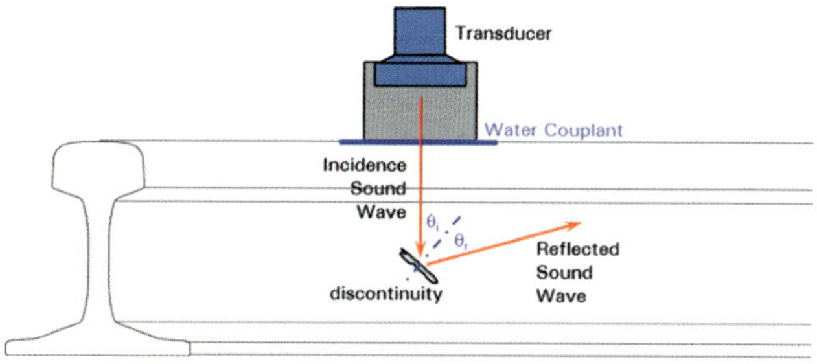

Figure 316: Law of Reflection and Oblique Angle Defects (0° Probe)

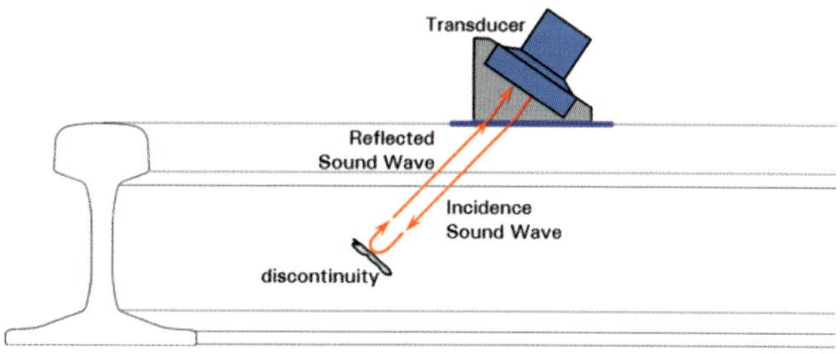

Figure 317: Angle-Beam Transducer for Detecting Oblique Angle Defects

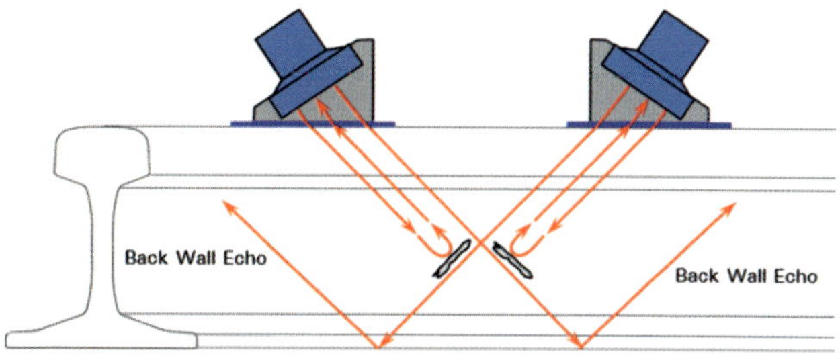

Figure 318: Probes of Different Angles and Directions are Required

It should also be clear that for effective detection of a wide range of defects in the rail of different shapes, sizes and direction of propagation; angle beam probes of more than one incidence angle have to be used and that probes of the same incidence angle would be required to face forward and backward towards the centre. Probes can also face towards the field and gauge side to provide wider coverage of defect detection on the railhead, web and rail foot.

Depending on the supplier of the equipment and whether it is handheld, push trolley or vehicle mounted equipment a combination of the following angle probes are being used:

- 0° (2 or 4 MHz): Defects detected by this probe include bolthole cracks in the web and horizontal longitudinal defects in the head web and foot.
- 30° (2 MHz): Detects vertical longitudinal cracks.
- 35° and 37° (2 MHz): This probe also detects bolthole cracks and defects in the web and foot.
- 45° (2 MHz): This probe is used for testing flash butt welded rail joints and half-moon defects in Thermite welds below the web-foot junction.
- 55° (2 MHz): This probe also detects vertical longitudinal cracks.
- 70° (2 MHz): This probe examines the railhead area and the top section of the web for transverse railhead cracks.

The ultrasonic probes are housed either in acrylic wedges or in fluid filled membrane wheels, with up to six probes per wheel. A complete configuration uses as many as 18 ultrasonic transducers per rail to increase the probability of detecting defects.

For accuracy the probes need to pulse the defect at least 4 times which limits the detection speed to around 60 km/h but up to 100 km/h is possible depending on the supplier and system being used. The detectability of defects decreases with the increase of speed.

The ability of ultrasonic probes to locate small discontinuities (sensitivity) as well as their ability to locate discontinuities that are close together (resolution) generally increases with higher frequencies (shorter wavelength) signals.

The ultrasonic probes used in rail flaw detection have a frequency of 4 MHz (longitudinal wave) and 2 MHz (transverse waves). Therefore, cracks smaller than 0.8 mm cannot be detected.

2.3. Typical Ultrasonic Flaw Detection Equipment and Vehicles

There are various suppliers of ultrasonic rail flaw detection equipment mounted to vehicles or push trolleys. The following is just a short description of some the equipment available.

2.3.1. Track-Bound Ultrasonic Rail Flaw Detection Vehicles

Figure 319 shows a telescopic axle with two membrane wheels per rail attached to a Plasser & Theurer machine. Each of these wheels contains six probes, making it twelve in total per rail.

Figure 319: Telescopic Axle Equipped with Ultrasonic Flaw Detection Wheels

The telescopic axle can be lifted when the flaw detection vehicle travels over switches and crossings while the axle also assists with guiding the probes on the centre of the rail.

The advantage of rail-bound vehicles is its size and space for work stations to view and analyse recorded data. It is also capable of carrying large volumes of water for spraying on the rail as the couplant.

The flaw detection vehicle is equipped with colour monitors, computer systems for analysis of the data, digital oscilloscopes and a video system for visual monitoring of the probe wheels.

The ultrasonic flaw detection vehicle records the GPS position together with every defect recorded. This specific system allows synchronisation of all ultrasonic measurements with the Plasser & Theurer geometry measuring system. The location of detected flaws is normally also paint marked against the rail to assist with finding the exact position of the flaw.

Push trolleys or hand measurements are very slow and only short distances can be measured per day. Though not as accurate or sensitive as hand held equipment, rail flaw detection vehicles can cover large distances in a day measuring at least 60 km/h. It is therefore common for flaw detection vehicle to detect flaws and hand held equipment to follow the rail-bound vehicle by road or rail and only examine the flaws that were detected for accurate size, position and risk.

2.3.2. Road/Rail Ultrasonic Rail Flaw Detection Vehicles

There is not much difference between the rail-bound and road/rail vehicle with regards to the type and number of ultrasonic probes it can carry. Road/rail vehicles have the advantage of mobility but may also have disadvantages with regards to maximum speed on track and even on road, availability and condition of service roads, availability of access to the track, less space for work stations etc. All the parameters must be carefully considered before the decision is made between rail-bound and road/rail.

2.3.3. Push Trolley Ultrasonic Rail Flaw Detection Equipment

There are many products available from many suppliers. Hand push trolleys can be supplied to measure one rail or both rails at the same time.

They can be equipped with a variety of probes and the defects displayed, recorded and analysed in real time.

The size and portability has its advantages. The lower speed also has a higher resolution and the operator can move back to make double sure. The downside would be that limited water can be carried on the equipment. However, since the vehicle must be taken to the track section water can be brought along.

This method is however too slow to be part of an ongoing preventive maintenance program. Its use is rather as part of a corrective maintenance programme where internal defects over relatively short distances is suspected and the equipment used to determine such.

2.3.4. Hand-Held Ultrasonic Rail Flaw Detection Equipment

As with push trolleys, hand held equipment is too slow to be part of a preventive maintenance plan. Because the probes can be moved in all directions around the rail, it is very accurate and sensitive with a high resolution. Hand held equipment will often be used in conjunction with high speed equipment

Hand held equipment is also more suitable for flaw detection in turnout crossings than hand push trolleys due to the crossing shape and the problem of maintaining acoustic coupling with a push trolley.

2.4. Actions to be taken after Detection of Defects

Once an internal defect has been detected and the size, type and severity been determined, one of the following actions may be a priority to avoid a catastrophic rail failure:

- Ideally the defect must be cut out immediately and replaced with a closure rail of minimum 4,2 metres in length. Some smaller defects may be cut out within the maximum gap of a wide-gap Thermite weld in which case a closure rail is not required. Cutting out of the defective section must be carried out using approved equipment for the purpose and within procedure.

- Where the above corrective action is not immediately possible the rail must be clamped to contain a break using joggled fish plates or Robel rail clamps.
- Some Robel rail clamps can be used at speeds up to 160 km/h (see Figure 32). If the track speed is higher than that or other joggle plates are used, a speed restriction of 30 kph or less must be imposed.

2.5. Actions to be taken after Rail Breaks

It is important from a rail management and statistical point of view to determine, classify and record the cause and position of all rail breaks.

Unless the fractured rail was because of known external causes such as a derailment, both sides of the fractured rail must be analysed by a metallurgical laboratory to determine the cause of the fracture.

The outcome of the analysis must be used to determine preventive actions especially where the fracture occurred within a relatively short period since installation of the rail or weld, there are repetitive breaks in the same location or where several breaks occur within close proximity of one another on the same line.

3. EDDY CURRENT RAIL SURFACE FLAW DETECTION

The following is a very brief introduction to eddy current rail examination as it is a very complex principle which requires experienced and qualified operators of the equipment.

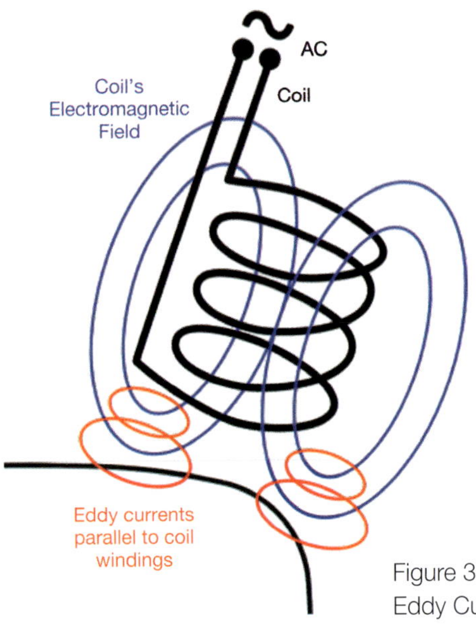

Figure 320: Eddy Currents

Rail cracks that propagate into the rail causing rail breaks often originate from the rolling contact fatigued (RCF) surface on the rail (CHAPTER 3 paragraph 2.19). The RCF area is very hard and prone to crack development if the RCF is not removed by preventive grinding. The reliability of conventional ultrasonic methods is poor for detecting surface or near-surface defects, especially if they are smaller than 4 mm.

Non-destructive eddy current (EC) examination is more suited for detecting nearly all surface or near-surface defects and is a method that is gaining reliability as the use and development of this method progresses. It is however not suited for internal rail defects resulting in simultaneous use of EC and ultrasonic examination on track inspection vehicles for more reliable rail flaw detection and better qualification of defect types and causes.

The eddy current method makes use of electromagnetic induction to detect rail flaws. An electrical coil of conductive wire (the test probe) is charged with alternating electrical current which produces a magnetic field in and around the coil. If the coil is brought close to another conductor, such as the rail, current will be induced into the rail. The induced current that flow in a circular path in planes perpendicular to the magnetic flux (i.e. parallel to the coil's winding) is called eddy currents.

The eddy currents will also produce its own magnetic fields that interact with the coil's magnetic field. The changes in the voltage caused by the electrical conductivity and magnetic permeability of the rail provide information regarding defects in the rail.

4. RAIL SURFACE MEASURING AND MONITORING SYSTEMS ON THE TRACK RECORDING VEHICLE

High end track recording vehicles such as those supplied by Plasser & Theurer can measure a vast number of rail parameters including rail wear and rail surface defects such as corrugations, head checks, small fissures and cracks. For more detail on these systems refer to CHAPTER 9 paragraphs 4.2, 4.3 and 4.4.

Chapter 11
Track Lifting, Levelling, Aligning And Tamping

1. INTRODUCTION

Under repeated loading from traffic the track progressively moves, causing deviations from the desired vertical and horizontal alignment (geometry, CHAPTER 8). Ballast tamping is the process used to re-arrange the ballast under the sleeper to restore the geometry and elasticity of the track structure. The result of this process is clearly illustrated by the "before" and "after" photos in Figure 321 and Figure 322.

However, over a period of time the track geometry will deteriorate again. Consideration of this period of deterioration is very important since this will determine the track tamping cycle. The longer the tamping cycle, the less the disruption to traffic, less the demand on railway resources and the cheaper the track will be over its lifecycle. The durability of the tamping process therefore plays an important role.

Durability can only be proven over time with traffic. This, however, poses a problem since many factors other than tamping can influence durability, e.g. fouled ballast, rail defects, joints and welds in the track, flat spots on wheels, etc. Choosing a tamping machine supplier will therefore have to be based on their processes used.

When having to choose a tamping machine for a specific railway or track section, all of the above must be considered. However, there are also various other network and infrastructure considerations, such as the length of the line, the traffic density, the number of turnouts on the line, etc.

Figure 321: Vertical Alignment Defect Clearly Visible before Tamping

Figure 322: Tamping Removed the Defect to Restore the Alignment

2. BASIC TAMPING PROCESS

Figure 323 shows the most important working components of a tamping machine, Figure 324 the working components of a tamping unit and Figure 325 provides a schematic illustration of the tamping process in four simplified steps.

Step 1 – A basic tamping machine indexes forward and comes to a standstill with the tamping tines of the tamping unit straddling the sleeper to be tamped on both sides.

Step 2 – The lifting and ligning unit works in conjunction with the measuring system and grips the rail under the crown, lifts the track to a predetermined height while correcting any vertical alignment defects in the track and at the same time slews the track to correct the horizontal alignment (simultaneous levelling and aligning).

Step 3 – The tamping units are lowered. The vibrating tines enter the ballast and stop at a pre-determined depth. The tines are vibrating in order to fluidise the ballast stone to permit it to re-arrange and settle in a dense matrix. Vibration also greatly reduces the force required to penetrate the tamping tines into the ballast.

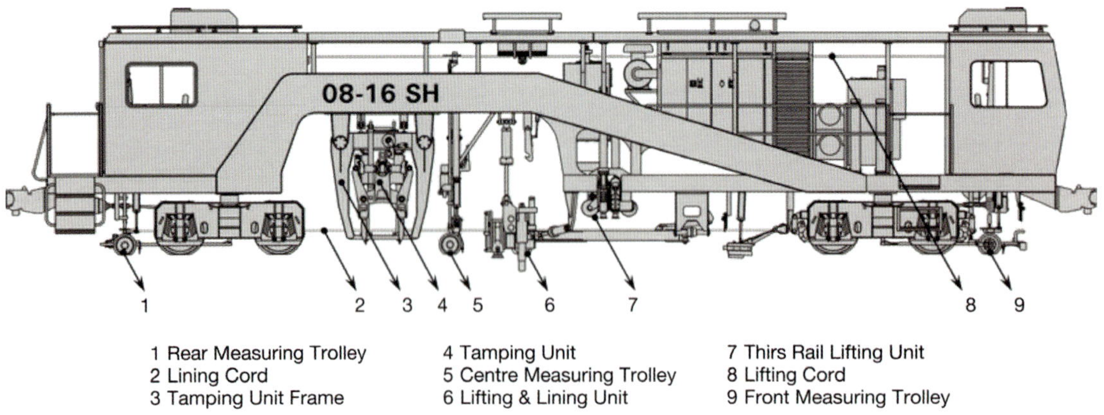

Figure 323: Working Components of a Universal Tamping Machine

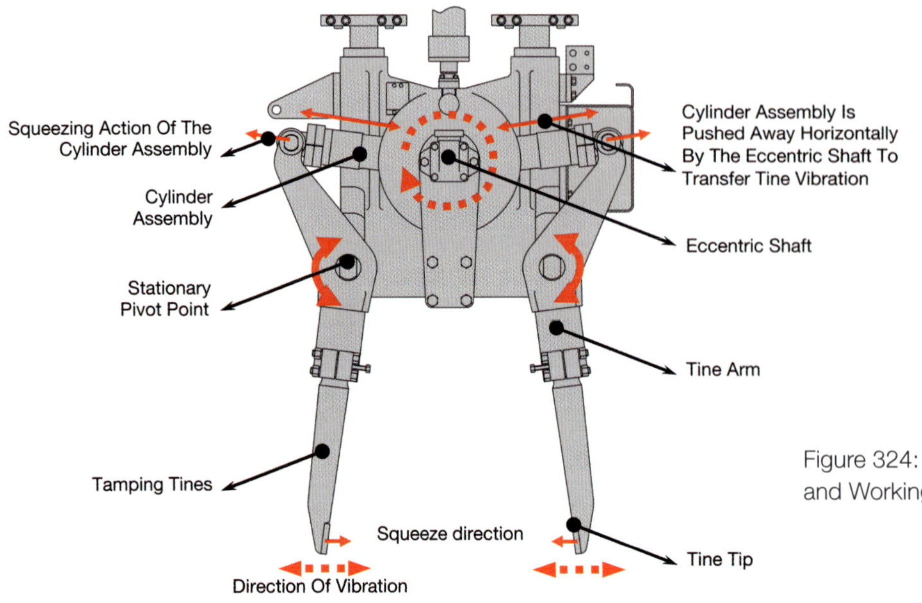

Figure 324: Tamping Unit Components and Working Principle

Step 4 – The cylinder assembly exerts a force on the tine arms which perform a squeezing motion of the tines. The tines compact ballast underneath the sleeper in the void created by the lifting process. The tamping machine indexes forward to the next sleeper and the process repeats itself.

Due to the automatic ending of the tamping process of each individual tine pair when reaching a squeeze pressure set to a certain size, this guarantees a uniform, homogeneous tamping of the track.

3. A SPECIALISED TAMPING MACHINE FOR EVERY APPLICATION

Many different ballast tamping machine designs are available to satisfy every possible tamping requirement, from tamping machines small enough to fit on a small truck, low-cost low-production machines tamping one sleeper at a time to high production mainline tamping machines using the continuous action tamping principle and tamping up to four sleepers per cycle. Other machines are universal and though more specialised for tamping turnouts, they are equally capable of tamping the plain track without having to make any changes to the machine, such as removing tines. Other machines crossed the boundary between high production plain track and high production turnout tamping. Each of these machines have specialised components which identify the function of the machine.

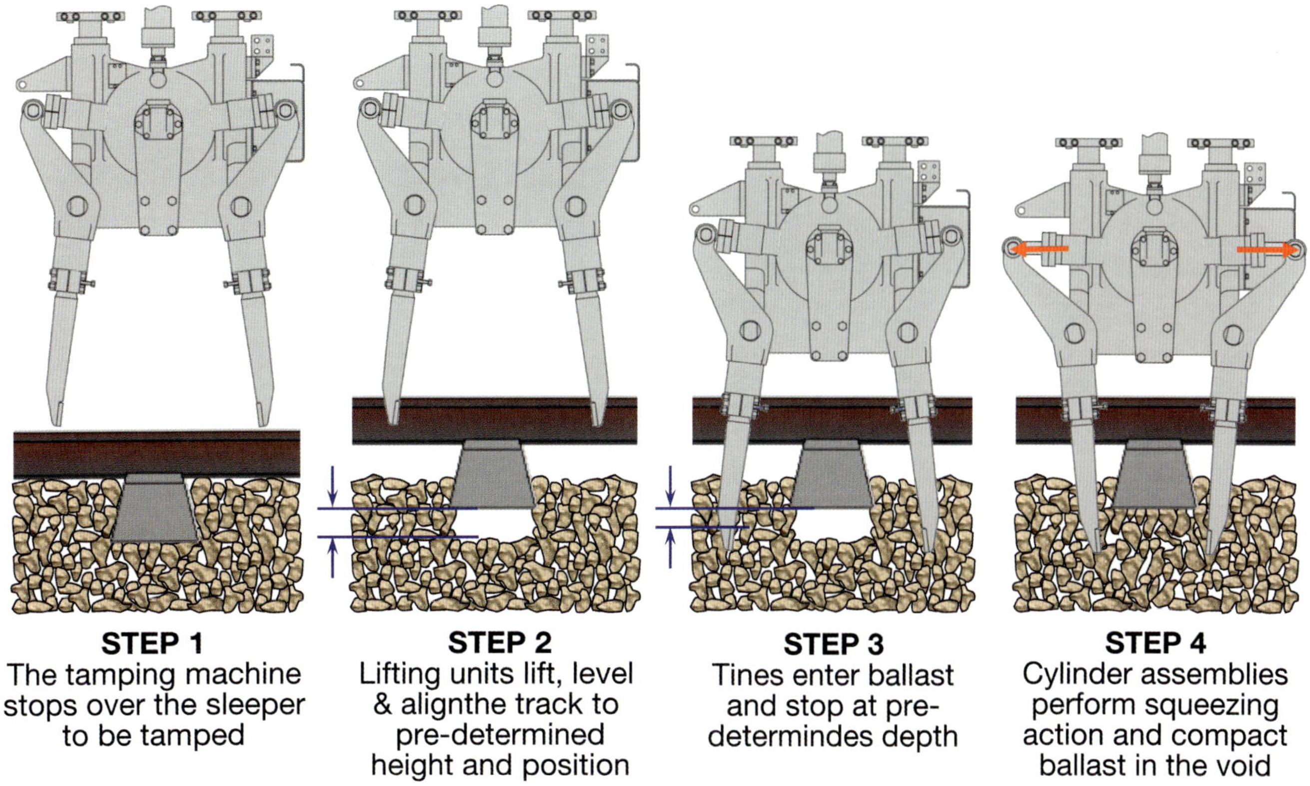

Figure 325: Tamping Process Description

3.1. Plain Track Tamping Machines

Plain track tamping machines are designed for high production tamping on the open track on very busy lines where the maximum number of sleepers must be tamped in short available maintenance windows. These machines cannot tamp the restricted areas in turnouts and neither were they designed to reach the turnout portion of a switch.

The production capability of mainline tamping machines has improved tremendously over the years, from tamping less than 4 sleepers per minute in 1953 (Plasser & Theurer's first prototype machine the HGL followed by the VKR 01 which was produced from 1955) to tamping in excess of 70 sleepers per minute today with the Plasser & Theurer 09-4X (Figure 327).

Figure 326: VKR01Tamping Machine 1955

Figure 327: 09-4X Continuous Action High Production Tamping Machine

There are, however, many different mainline tamping machine models and not all of them are focused on maximum production, since conditions vary from railway to railway and even from track section to track section within the same railway.

The production rate of tamping machines is determined by (i) the number of sleepers tamped per cycle (insertion) and whether (ii) the index or (iii) continuous action tamping principle is used.

(i) Modern tamping unit designs allow for one to four sleepers to be tamped with every cycle. The more sleepers tamped per cycle, the higher the production rate and the more homogeneous the track longitudinal level (top) will be.

(ii) With the index tamping principle, the entire machine will move from sleeper to sleeper to be tamped.
The production rate is limited due to the weight of the machine that must be accelerated and braked.

(iii) With the continuous action tamping principle the main frame of the machine will move forward at a constant speed while the cyclic braking and acceleration for the tamping action is performed by an auxiliary frame to which the tamping units are mounted.
If this principle is combined with multiple sleepers tamped per cycle, the maximum possible production is achieved.

3.2. Universal Tamping Machines

In turnouts where track is restricted, some tines may hit an obstruction such as the turnout rails, switch blade or crossing. Turnouts must then either be omitted or tamping machines with specialised features must be used to tamp the turnout effectively. In 1962 Plasser & Theurer developed the first turnout tamping machine using specialised features, the WE 75 with one tamping unit and the WE 275 with two tamping units (Figure 328). A significant feature was that each individual tine could be tilted sideways according to the needs.

Specialised features required for tamping turnouts include tamping units capable of reaching in-between restricted areas, specialised lifting units capable of lifting the track in restricted areas as well as third rail lifting devices to lift the long sleepers in a turnout, tamping unit frames that can slew sideways to reach the turnout portion of the turnout while standing on the straight portion of the turnout and rotating tamping unit frames to tamp the skew sleepers of the turnout at right angles. The Unimat 08-475 4S (Figure 329) is a good example of such a highly specialised machine.

Universal tamping machines are equally capable of tamping the mainline at tamping rates similar to that of other single sleeper mainline tamping machines.

Figure 328: First Turnout Tamping Machine with Tilting Tines 1962

Figure 329: Unimat 08-475 4S High Production Turnout Tamping Machine

3.3. Bridging the Gap Between Turnout and Plain Track Tamping

Universal tamping machines are, however, still biased towards specialised turnout tamping. To improve the machine's plain track production capacity, universal machines may combine high production continuous action tamping with specialised high speed turnout tamping, using two sleeper split head units, third rail lifting and integral dynamic track stabilising all into one machine (Figure 330).

Figure 330: DYNA-CAT Universal Tamping Machine

3.4. Spot Tamping Machines

Tamping machines are designed for tamping long track sections continuously. This will produce the lowest possible unit costs of tamping for the machine type. Spot tamping over very short distances is very costly due to the establishment costs of machinery and the low production that would be achieved. However, this may, at times, be necessary until the root cause of the need for spot tamping has been rectified, such as a failed formation, severe rail surface defects causing high impact loads, etc.

Small cost effective universal tamping machines such as the Unimat Sprinter produce the same quality and production as other similar single sleeper tamping machines but have specialised software for the elimination of spot faults and may have additional features such as shoulder ploughs and a sweeper unit and even loading space for materials or other tools.

Figure 331: Unimat Sprinter

3.5. Lightweight Tamping Machine

Some tamping requirements are too small, remote or unpractical due to traffic conditions to send out a railbound tamping machine. For this purpose very small tamping machines are available which can easily be loaded on a small road truck or track lorry for high levels of mobility yet with uniform compaction that only mechanised methods can achieve. However, these machines cannot lift, align or level which will require a labour gang with track jacks and a level ahead of the machine. They may be ideal for very short lines such as sidings as well.

3.6. Handheld Vibratory Tampers

Though handheld vibratory tampers cannot be considered "mechanised maintenance", they are still very often used and fulfil an important role in support of mechanised tamping machines to tamp in areas of a turnout where it is not possible for the tamping machine to reach. They also fulfil a limited application for addressing very short track profile problems over just a few sleepers in areas of a weak formation or where rail surface defects caused track profile irregularities between scheduled tamping cycles. Two different types of machines are shown in Figure 334 and Figure 335.

Figure 332: Small Tamping Machine

Figure 333: The Machine Fits on a Small Truck or Track Lorry

Figure 334: Electrical Horizontal Vibratory Hand-Held Tamping Units

Figure 335: Vertical Tamper Available in Petrol, Electrical or Hydraulic

4. TAMPING MACHINE COMPONENTS

In the previous paragraphs reference was made to various tamping machine components and features. In this paragraph those components, their unique features and benefits will be discussed.

4.1. Lifting and Aligning Unit

Tamping machines are equipped with a combined lifting and aligning unit (viii) mounted in front of the tamping units (iii) between the bogies (see Figure 336). There are various designs depending on the machine type but they can generally be divided between lifting and ligning units for universal tamping machines (Figure 336) and those for plain track tamping machines (Figure 337).

4.1.1. Universal Tamping Machines

Lifting and ligning units on universal tamping machines such as the Unimat illustrated in Figure 336, have special features such as:

- lifting hooks (vi) which grips the rail under the crown or base for lifting in restricted track in turnouts;
- roller clamps (v) for high speed lifting during plain track tamping;
- either one or two flanged rollers (vii) which runs on the rail crown to transfer the lateral force to the track for horizontal alignment.

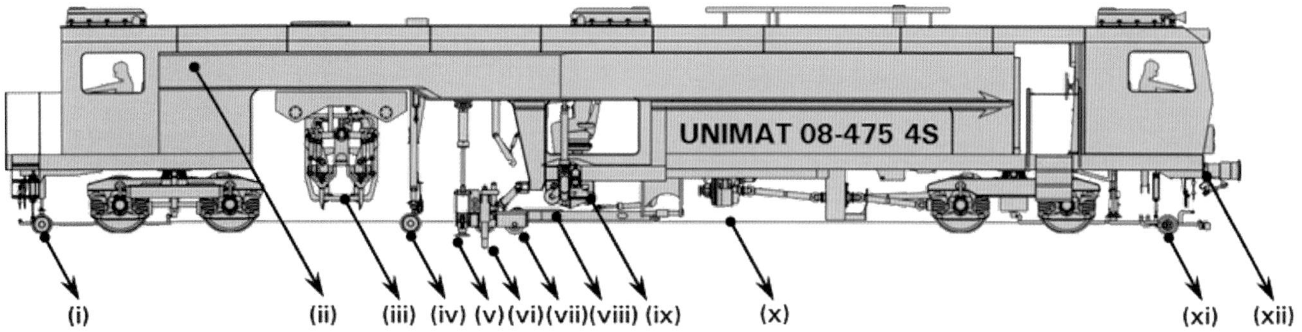

Figure 336: Location of Components (Universal Tamping Machine)

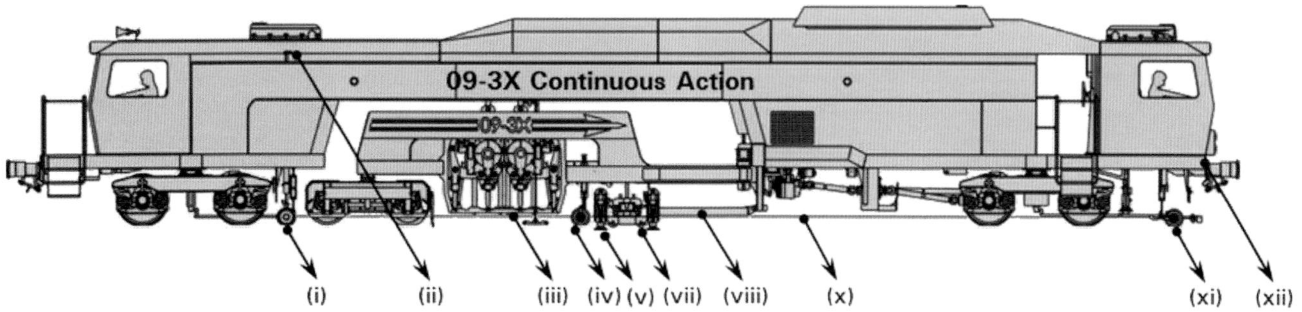

Figure 337: Location of Components (Plain Track Tamping Machine)

4.1.2. Plain Track Tamping Machines

High production plain track tamping machines such as the 09-3X continuous action 3 sleeper tamping machine (Figure 337) are equipped with a lifting and ligning unit with double roller clamps to grip the rail under the crown for high speed lifting on the open line and two flanged rollers which run on the rail crown to transfer the lateral force to the track for alignment. See Figure 339 and Figure 340.

The positioning of two roller clamps either side of the rail means that the application even on fish-plated joints is automatically secured. During forward travel the lifting and lining cylinders are switched to pressureless mode. The lifting and lining operation is controlled automatically by the measuring systems. The lining operation starts simultaneously with the lifting operation. As soon as the target values are reached, lining and lifting operations are automatically switched off.

Due to the roller clamps gripping the rail on both sides, transmission of the lifting force is guaranteed in the centre of the rail.

Figure 338: Specialised Lifting and Ligning Unit found on Universal Tamping Machines

4.2. Measuring System

Refer to (Figure 336 and Figure 337) for the location of the components referred to in this paragraph. A fully automatic measuring system determines the relative difference in the vertical and horizontal positions of the track at a centre measuring trolley (iv), which is the recording point, between two reference points, the front (xi) and rear (i) measuring trolleys. Using this system for correcting the level and line of the track is called the compensation method.

The relative vertical alignment is established with the use of a steel cord above each rail (ii), extended from the rear to the front measuring trolleys (see also Figure 341). The reference line for relative horizontal alignment is a steel cord (x) extended along the centre of the track (see also Figure 342).

The lifting and ligning unit works together with the measuring system to lift the track to a uniform height removing any vertical defects in the process and to simultaneously slew the track to correct any horizontal defects in the track.

To ensure a residual lift after tamping, research has shown that the minimum lift should be around 20 mm to provide enough space under the sleepers to rearrange the ballast stones. This will however depend on the ballast size; the larger the ballast stone, the higher the minimum lift will be. The machine will therefore uniformly lift the track by an approximate pre-set lift of 20 mm.

Figure 339: Lifting and Ligning Unit found on Plain Track Tamping Machines

Figure 340: Forces in the Centre of the Rail (Plain Track Lifting Unit Shown)

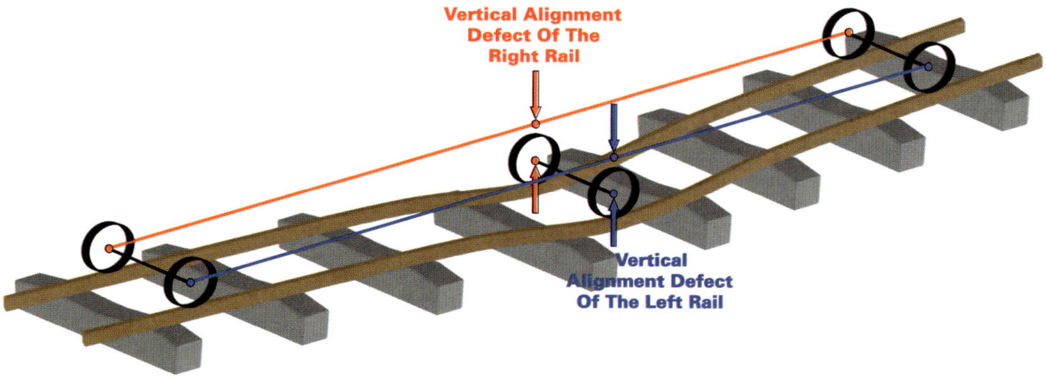

Figure 341: Illustration of a Vertical Alignment Defect

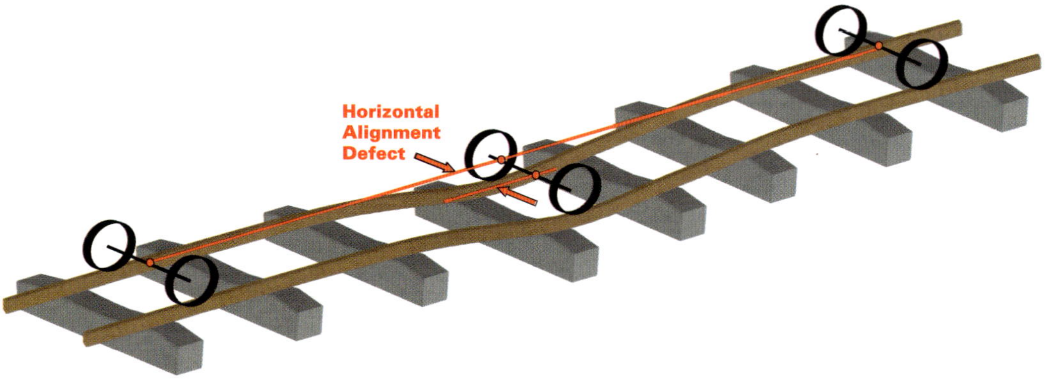

Figure 342: Illustration of a Horizontal Alignment Defect

Where a vertical alignment defect is encountered, the measuring system will detect the relative vertical difference in height and lift the track by the pre-set height as well as the depth of the vertical alignment defect (also referred to as a slack). Refer to Figure 343 and paragraph 5.1 below.

However, defects which are shorter than the chord length between the front and rear measuring trolleys of the tamping machine can be corrected successfully but if the defects are longer than the chord length, the machine will follow these defects without removing it.

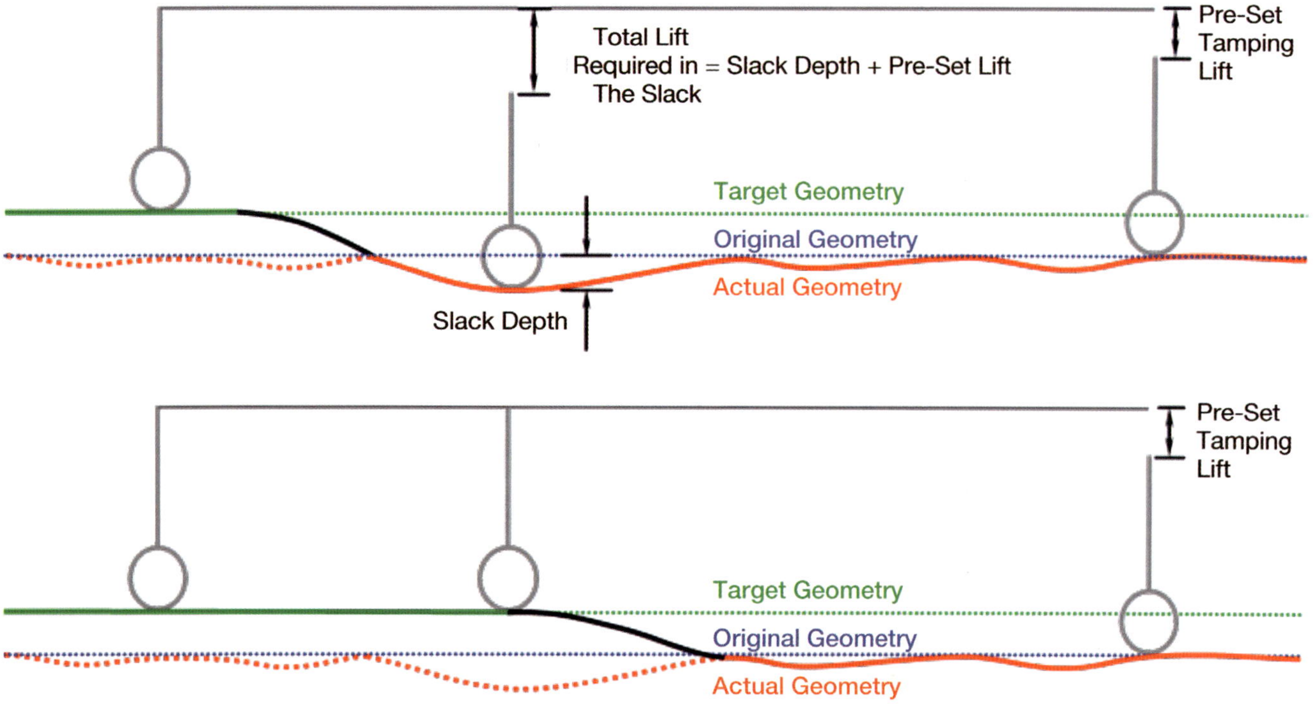

Figure 343: The Lifting Principle

Tamping machines of high quality therefore use optical (Figure 344) or laser equipment (Figure 345) mounted to a mobile trolley which is moved 100 to 150 metres ahead of the tamping machine, depending on geography and other conditions. This system is often referred to as design lifting and lining.

The tamping machine is then guided by a straight line which is aimed at a target board at the front of the machine (xii in Figure 336). When the machine moves forward the lifting and ligning wires are adjusted via remote control (or automatically with laser systems) so that the reticule of the viewfinder is lined up with the marks on the target board. This effectively lengthens the chord length to the distance between the rear measuring trolley (i) of the machine and the position of the mobile trolley. This provides utmost accuracy in the vertical and horizontal alignment of tangent (straight) track and is referred to as the precision method.
An exact (precise) longitudinal level of the track can also be produced when the correct longitudinal level has been measured beforehand using a separate vehicle such as the EMSAT. However, for this method, longitudinal level fixed points must be available. The amount of lift is then transferred electronically to the machine.

Figure 345: Laser Guiding System

The measuring system can be supplemented with a computer-based system (for example the WIN-ALC system) which can be used to measure the track and for automatic calculations and/or setting of the offsets in curves.

Especially in curves the ALC avoids the time consuming and potentially very inaccurate measuring of the curve by hand. It also avoids the resulting alignment errors associated with hand measuring of curves and manual adjustment of the correction values on the machine. See CHAPTER 8, paragraph 2.2 for more information on curve measuring and tamping.

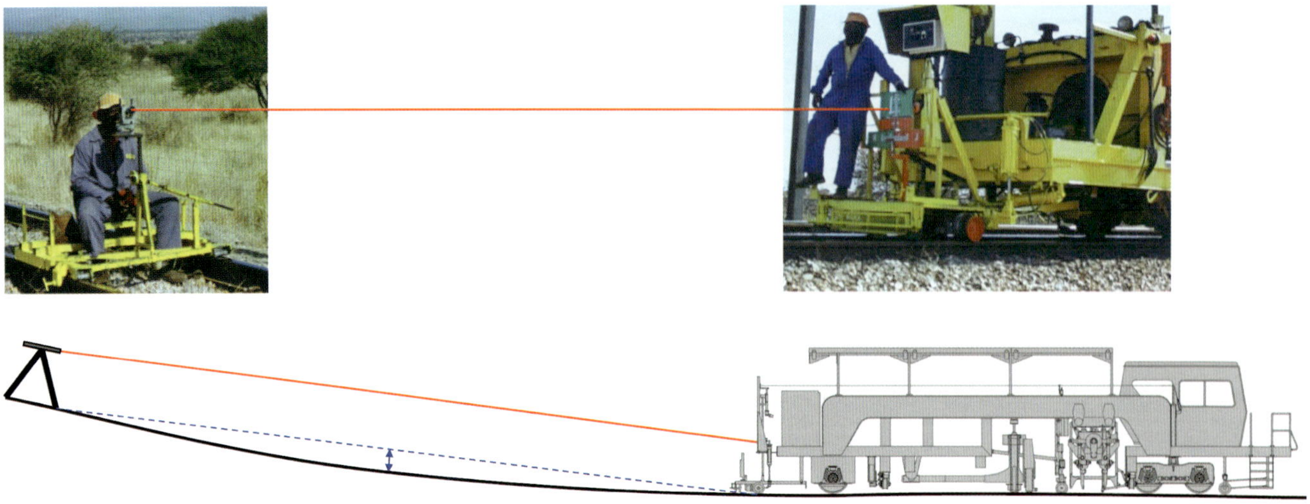

Figure 344 : The Use of Optical (as Illustrated) or Laser Equipment to Permit Removal of Long Wave Defects

4.3. Third-Rail Lifting Device

Modern universal tamping machines are fitted with a hydraulically operated, telescopic third-rail lifting system on either side of the machine [see Figure 336 (ix) and Figure 346], which is synchronised with the combined lifting and ligning unit. This clamp assembly provides controlled lift of the outside turnout rail (the curved closure rail) in the area of the long sleepers during turnout tamping operations and eliminates the need for cumbersome, manually-placed track jacks, hydraulic hose reels and the related labour and maintenance costs.

For switches with divided concrete sleepers the lifting of the third-rail is carried out by a laser control.

The standard two point lift on the long sleepers on concrete turnouts leads to overstressing and damage to the rail fastenings due to the weight of the long sleeper, the added weight of the turnout rails, as well as the turning moment caused by the long sleeper being lifted at one end. The entire load is carried by the fastenings of the two rails (see Figure 347).

An additional synchronised lift at the curved closure rail distributes the weight of the sleeper and rails across three lifting points and eliminates the turning moment which reduces the lifting force at the middle rail by almost half (see Figure 348). Third-rail lifting devices are therefore essential to tamp turnouts on concrete sleepers. On some railways the use of third-rail lifting devices is prescribed in its regulations.

Figure 346 : Third Rail Lifting Device

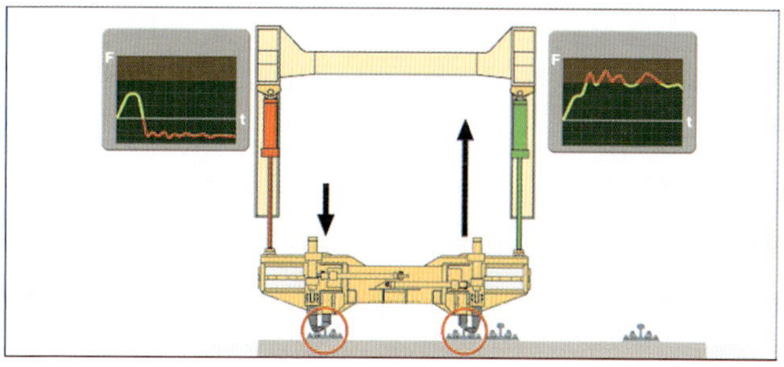
Figure 347: Without Three-Rail Lifting

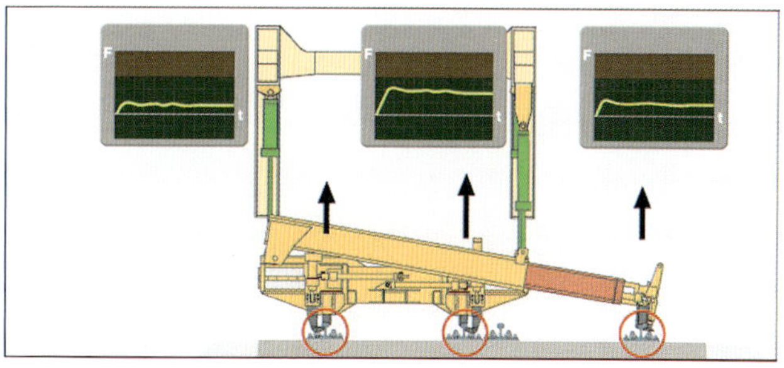
Figure 348: With Three Rail Lifting

4.4. Tamping Units

It is the tamping units that carry out the main function of the tamping machine. Refer again to Figure 324 in paragraph 2 above for the components of a simple tamping unit design.

Due to the variety of requirements for tamping, such as production capabilities, specialised units for tamping turnouts etc, a large variety of tamping unit designs are available to fulfil these specialised requirements. The following are features that set the different tamping units apart:

4.4.1. Number of Sleepers Tamped per Cycle

The number of sleepers that are tamped per cycle (insertion) determines the production capability of the machine. Machines tamping from one to four sleepers per cycle are available (Figure 349). Single sleeper tamping machines are still very common since high production is not always required and a lower production at a lower machine price can be preferred for some applications, such as tamping behind medium production ballast cleaning machines and short or low traffic density lines.

As the traffic density increases and the available maintenance windows become fewer and shorter, the production of the machines must increase in order to complete the tamping cycle. This is the sphere of multiple sleeper continuous action tamping machines.

The machine purchase price must, however, not be confused with the unit price of production since higher production machines may be more expensive than lower production machines to procure but their unit cost of production (cost per sleeper tamped) is generally much lower. On high capacity lines, the fewer number of occupations required to maintain the tamping cycle required for the traffic throughput, may bring about higher income to the railway when the opportunity cost of traffic is considered. In other words, if fewer occupations are required to tamp the track, more trains can be running resulting in more income for the railway.

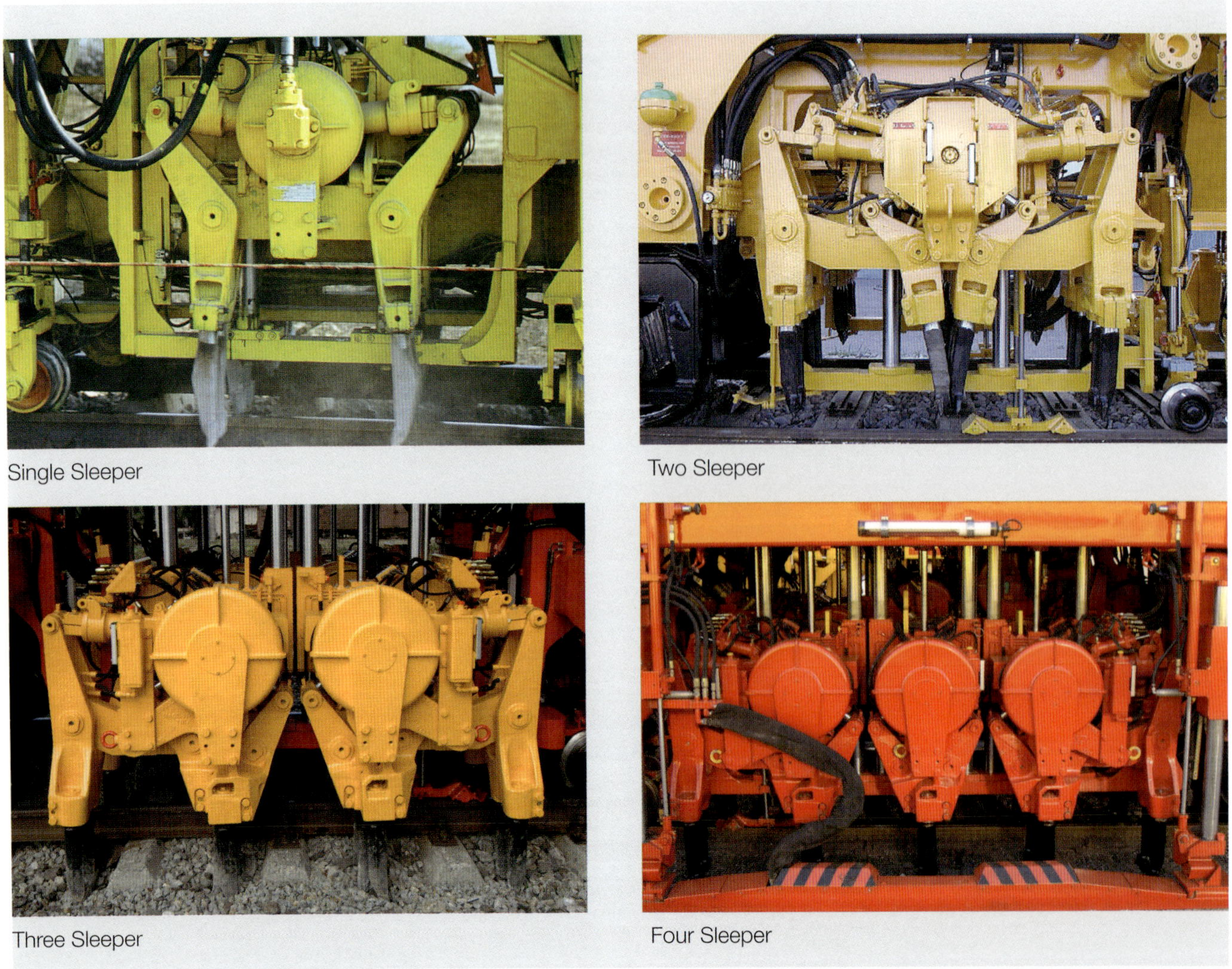

Single Sleeper

Two Sleeper

Three Sleeper

Four Sleeper

Figure 349: Examples of Tamping Unit Layouts

The 09-4X mainline continuous action tamping machine for example can tamp up to 4 sleepers per cycle and is the fastest machine available in the world, requiring the least time and number of occupations to complete the required tamping cycle.

If necessary, multiple sleeper units can be switched to single sleeper tamping where sleepers out of square or poor sleeper spacing or restricted track is encountered (Figure 350 and Figure 351).

Figure 350: Single Sleeper Tamping in Restricted Track with a Two Sleeper Tamping Unit

Figure 351: Single Sleeper Tamping in Restricted Track with a Three Sleeper Continuous Action Tamping Unit

4.4.2. Tine Configuration

The tamping tines are the wear components of the tamping unit (Figure 352 and Figure 324 in paragraph 2). The tamping tines are the components that enter the ballast to compact the ballast under the sleeper. The tine's wear surfaces may be clad with tungsten plates which are wear resistant to prolong tine life. Once the plates have worn past a specified percentage of area, the tine is replaced. The tine design may also be different for the different positions they are located in the unit.

Tamping units of single sleeper tamping machines are either equipped with 4 or 8 tines per rail per tamping unit as can be seen in Figure 353 and Figure 354.

The tine plates (tips) on 4 tine units are, however, larger to make up for the fewer tines and provide similar compaction characteristics as 8 tine units. The units are used for predominantly older generation turnout tamping machines.

Eight tine per rail tamping units are used on universal tamping machines making them equally suitable for turnout or open track tamping.

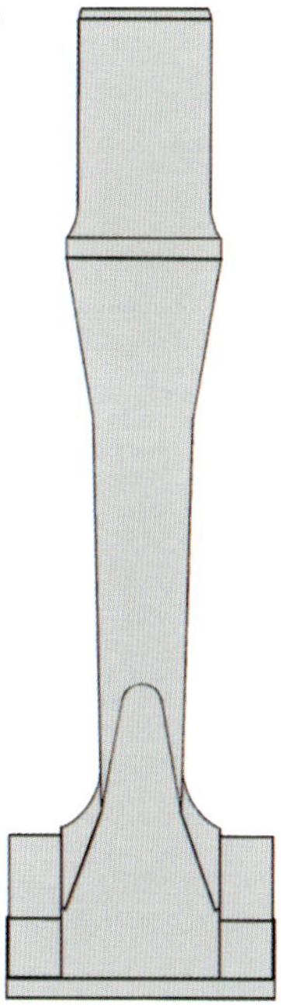

Figure 352: Tamping Tine

Figure 353: 4 Tine Units per Rail on Turnout Tamping Machines

Figure 354: 8 Tine Units per Rail on Universal Tamping Machines

Figure 356: Tilting up to 15° Inward or 85° Outward

Figure 357: Different Tine positions

4.4.3. Specialised Turnout Tamping Units

In turnouts where track is restricted, some of the tines may hit an obstruction such as the switch blade or curved closure rail of the turnout portion. True universal tamping machines will therefore have specialised tamping units to reach in-between restricted spaces. Two basic designs are used, either tilting tines or split units.

4.4.3.1. Tilting Tines

Tamping units equipped with tilting tines avoid hitting rails that may obstruct one set of tines by tilting those tines out of the way. See Figure 355. These units are available in 4 and 8 tine configurations but on narrow gauge lines 8 tine units may not fit in the structure gauge.

As with all universal tamping units, the units can be displaced laterally on horizontal guide columns (paragraph 4.5.1 below) so that the tamping tools are centred exactly over the area to be tamped. The tamping tines are tiltable in line with the sleepers and can thus be tilted to the centre of the rail or away from the rail at any angle up to 15 degrees inwards or 85 degrees outwards (Figure 356).

This means that any number of combinations of tine positions can be set. It is only necessary to tilt out the tines where there is actually an obstacle. Each sleeper is thus tamped with as many pairs of tines as possible.

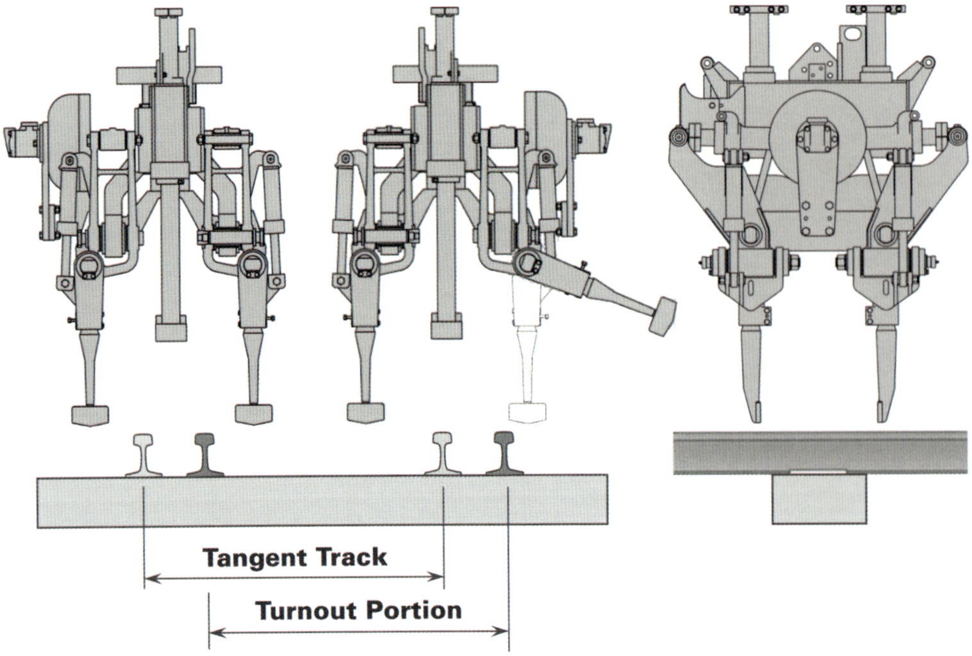

Figure 355 : The Use of Tilting Tines to Avoid Tines Hitting Turnout Rails

In the basic position (tines tilted inwards) the 16 laterally tiltable tamping tines have the same position as on plain track tamping machines, i.e. 8 pairs of tines tamp the track.

The combination of laterally displaceable tamping units and individually tiltable tines ensure that all sleepers are tamped, even in narrow sections between converging rails.

Due to the individually tiltable tines, even where tamping on both sides is not possible, i.e. around the switch power rod, sleepers can at least be tamped on one side.

4.4.3.2. Split Units

Split tamping units are in principle longitudinally divided in two (field side and gauge side), and can be raised and lowered individually for unparalleled versatility (vertical split). See Figure 358.

Each one of the total of four units can be lowered and put into action separately from the other. Furthermore, some split units can individually be displaced laterally as well (horizon split) to find the best area in the restricted track to enter the ballast. The split units can be locked together, in which case they will act like conventional plain track tamping units.

Figure 359: Tilting Tine and Split Technology Combined

4.4.3.3. Split Units with Tilting Tines

The two technologies can also be combined to provide the ultimate in reaching the restricted space in a turnout, i.e. split tamping units with tilting tines.

In Figure 359 the technologies are combined but due to this machine working on cape gauge, the units on the gauge side are not tiltable due to restricted space.

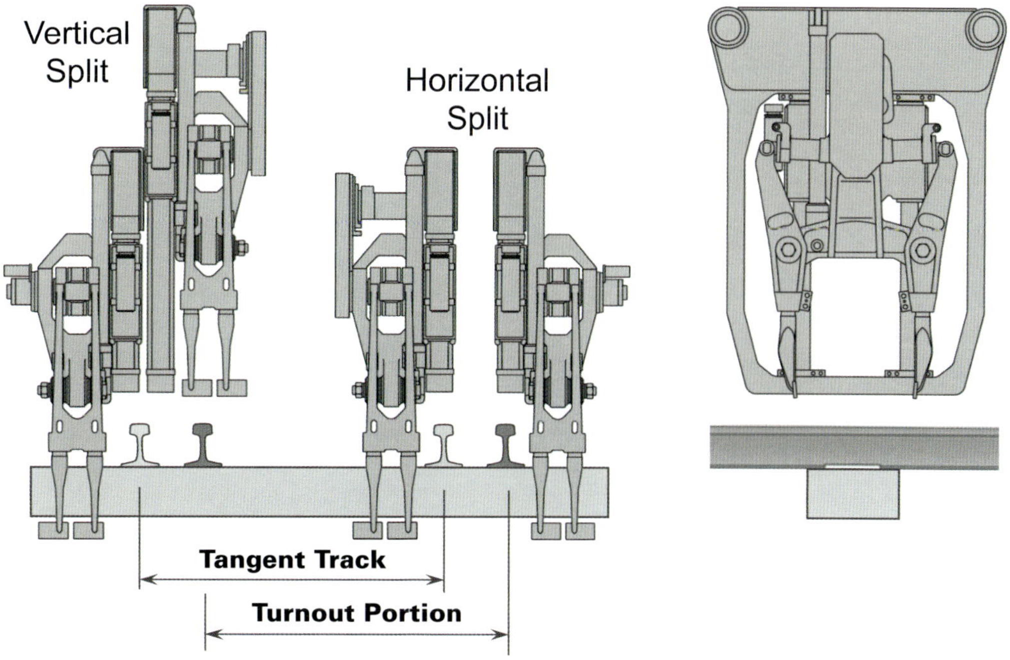

Figure 358: The Use of Split Units to Avoid Tines Hitting Turnout Rails

4.5. Specialised Tamping Unit Frames for Turnout Tamping

4.5.1. Slewing Tamping Unit Frames

Though tilting or split units prevent tines from hitting obstructions in turnouts, the areas between the obstructions must also be tamped for which universal tamping machines need slewing tamping unit frames.

Figure 360 : Double Slewing Tamping Unit Frame

Figure 361 : Single Slewing Tamping Unit Frame

Figure 362: Tamping Unit on Telescopic Arm

This allows the tamping units to slide laterally across the track to find the best place in the track to enter the ballast.

Combined with a third-rail lifting device, this concept is often referred to as synchronous three rail lifting and four rail tamping.

Depending on the machine model, the tamping unit frames slide on guide columns to allow either double slewing (Figure 360) or single slewing (Figure 361) reach. On double slewing systems the guide columns are mounted to an auxiliary frame which can be slewed beyond the frame of the machine. In single slewing systems the guide columns are attached to the main frame of the machine. Refer to Figure 363 for the typical maximum reach achieved with double (a) or single (b) slewing tamping unit frames.

Some machines don't use guide columns but the outer tamping unit can be slewed laterally to the track on telescopic arms making it possible to tamp more of the turnout section because of its greater reach (Figure 362).

For high production in turnouts, double slewing tamping unit frames are required to reach the turning out portion while the machine is standing on the straight portion of the turnout. This avoids unnecessary moving of the machine and may even avoid an occupation of both lines.

Figure 364: Maximum Distance Single Slew (Green) and Double Slew (Blue) Tamping Unit Frames can Reach From the Straight Section

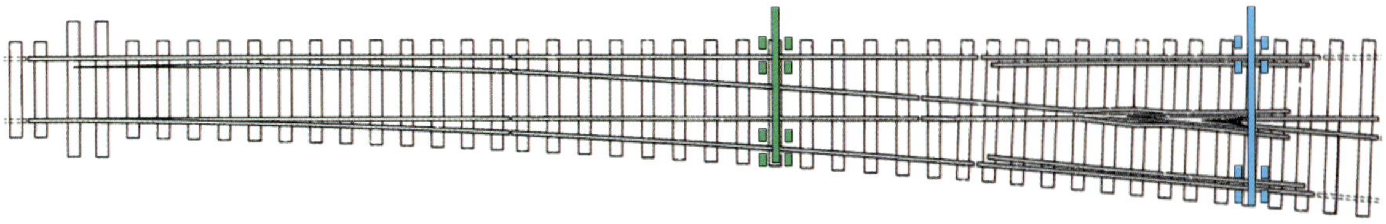

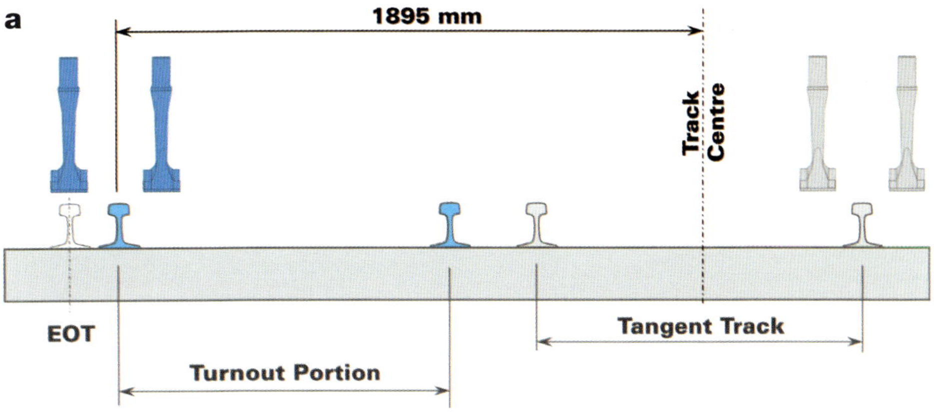

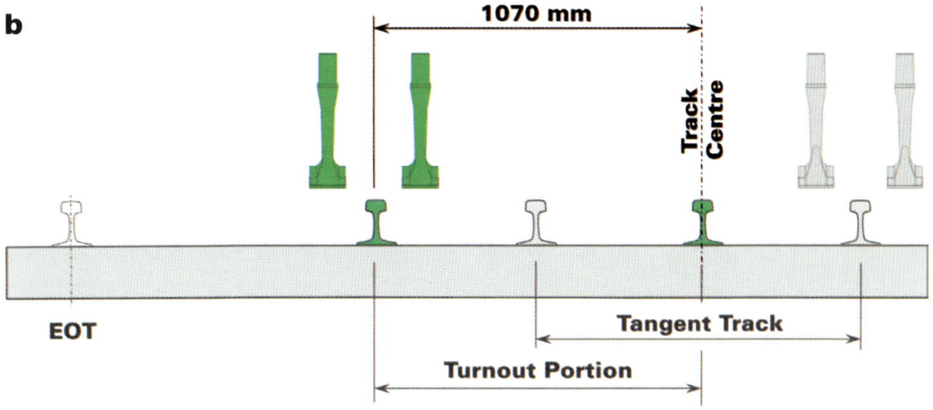

Figure 363: Typical Maximum Reach of Universal Tamping Machines to the Turnout Section while Standing on the Tangent Section
(a) Double Slewing Reach and
(b) Single Slewing Reach

Note: EOT denotes the end of the turnout. The illustration shows how close the 2 tamping unit frames can get to the end of the turnout while the machine is standing on the tangent (straight) section.

4.5.2. Rotating Tamping Unit Frames

An additional feature of modern universal tamping machines is that of rotating the tamping units through the angle of the skew sleepers of the turnout. These tamping units are mounted to a turntable that ensures right angles to the sleeper when the turnout portion is tamped. This avoids potential squaring of the skew sleepers and improves production times.

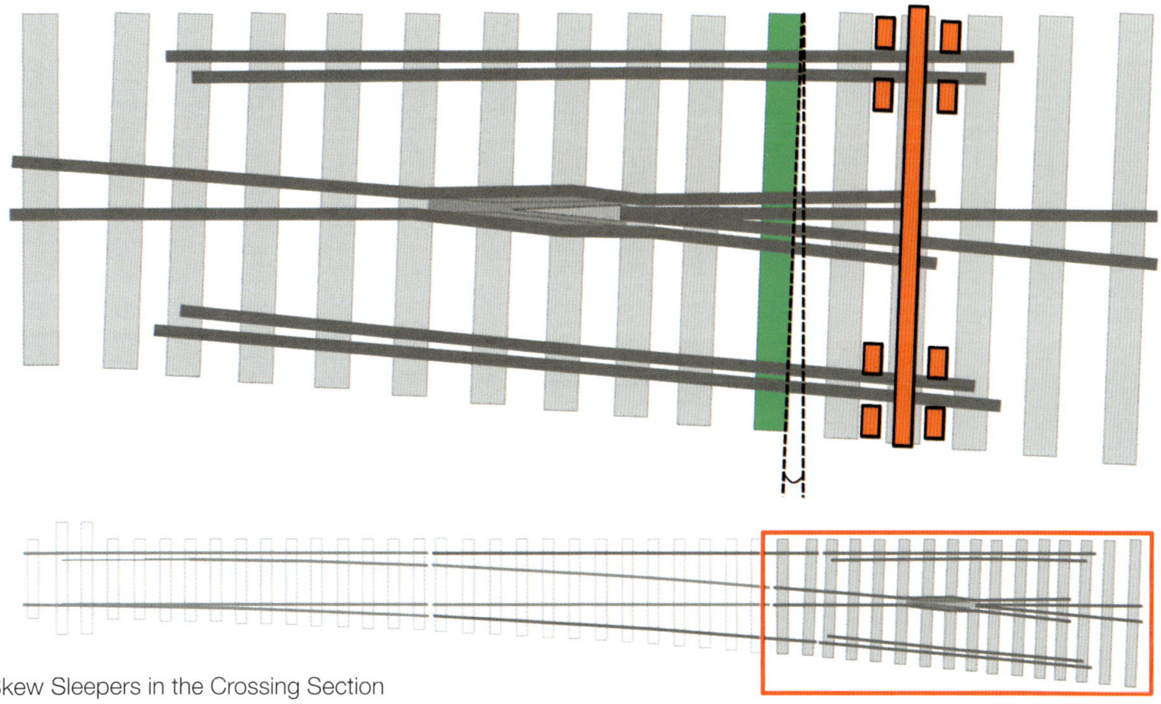

Figure 365: Skew Sleepers in the Crossing Section

4.6. Auxiliary Satellite Frame for Continuous Action Tamping

A conventional tamping machine must move from sleeper to sleeper for the tamping operation. The machine must therefore accelerate and brake again between sleepers and is referred to as index tamping. Though this principle is still used on many modern tamping machines, its production capability is limited due to the acceleration and braking limitations of heavy on-track machines using steel wheels on steel rail. The acceleration and braking is also very uncomfortable for the operator of the machine and causes fatigue to set in very quickly at higher tamping rates. The limit for index tamping using a 2-sleeper tamping machine is around 33 sleepers per minute. Therefore, only lower production, lower cost and specialised tamping machines use index tamping.

In 1983 Plasser & Theurer introduced the first continuous action tamping machine, the 09-32 CSM (see Figure 366) which tamped two sleepers per insertion and produced 30% more than the fastest machine available at the time. This was achieved by the separation of the main frame and an auxiliary satellite frame on which the tamping units where mounted. This allows continuous motion of the main frame while the cyclic braking and acceleration for the tamping action is performed by the auxiliary frame. Only around 20% of the machine mass must therefore be braked and accelerated.

machine tamps four sleepers per cycle and achieves a production rate of more than 70 sleepers per minute.

The continuous action principle was traditionally only used on plain track tamping machines. The plain track tamping speed of universal tamping machines were therefore always limited to that of single-sleeper index tamping machines. However, with the introduction of the DYNA-CAT (refer to Figure 330) and the 09-4S universal tamping machine series (Figure 367), continuous action tamping with two-sleeper split tamping units and integrated dynamic stabilisation were combined in one machine to provide the best possible production rates on turnouts while it achieves high production rates on plain track as well. The continuous action tamping principle has the following advantages:

- Increased output compared to indexing machines.
- Lower energy costs because a lower mass (just the satellite) must be accelerated from insertion to insertion.
- Reduced wear on the brakes.
- Reduced strain on the machine frame and drive system, which reduces maintenance costs and increases reliability.
- Ergonomic advantages for the operator/s of the machine.
- Other continuous-action work processes such as ballast regulating and dynamic stabilisation can be incorporated into the machine.

Figure 366: The 09-32 CSM Continuous Action Tamping Machine with the Tamping Units Mounted to a Separate Auxiliary Frame

Figure 367: 09-4S Universal Tamping Machine

4.7. Wheelbase

Wheelbase refers to the distance between the centre of the bogies or centre of two single axles of the tamping machine. During the tamping process, the track is lifted at the lifting unit by at least 20 mm to ensure a residual lift. Where a vertical alignment defect is encountered, the lift becomes progressively more by the depth of the defect as was illustrated in Figure 343.

The importance of wheelbase is illustrated in Figure 368, which shows the maximum rail stress vs the lift applied by the machine at the lifting unit and is the result of research done by Plasser & Theurer on tangent track with UIC60 rails and concrete sleepers.

The position of the EN Standard limit (45 kN/cm²) for the maximum rail stress in UIC60 rails is indicated on the graph. The UIC Standard limit which is much lower is also indicated. Tamping machines with different wheelbases were then used to lift the rail, the rail stress established and the results plotted.

The graph clearly shows that tamping machines with a 6 metre wheelbase cannot lift more than approximately 40 mm before the maximum rail stress is exceeded due to the bending radius of the rail. Considering that the minimum maintenance lift by a tamping machine is already around 20 mm, a 6 metre wheelbase machine will not be able to even lift out a vertical alignment defect exceeding 20 mm in one pass. This machine will therefore not be appropriate for heavy haul lines where a larger ballast stone is used. This research also highlights the danger of over stressing the rail during tamping when short wheelbase machines are used. Such over stressing could have been the reason of many rail breaks.

Machines with a wheelbase of at least 10 metres would be required to ensure that deep defects can be rectified in one pass, albeit with more than one insertion, to ensure high production.

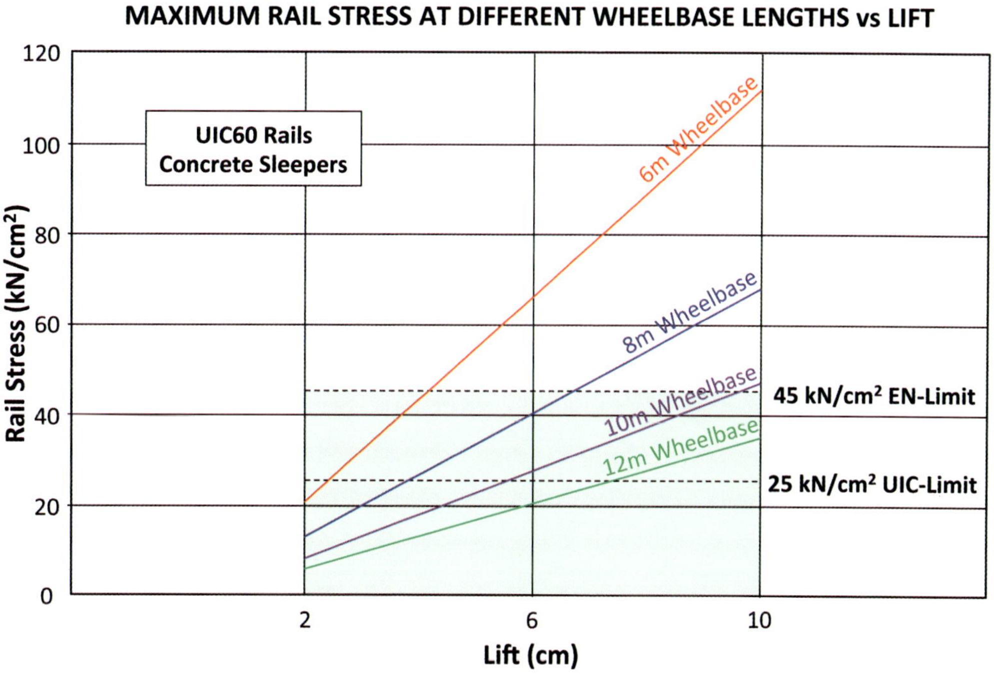

Figure 368: Maximum Rail Stress versus Lift Applied by Tamping Machine

5. HOW TO CHOOSE A TAMPING MACHINE SUPPLIER

What would be the selection criteria when choosing a tamping machine supplier? Before price should even be a consideration, there are many other considerations such as production capability, the quality of the product, the features available versus those required, the reputation of the supplier, the experience of the supplier, the support base of the supplier, availability of spares, etc., but at the end of the day it is the product that the machine provides that will have the greatest impact on the long term integrity, and lifecycle cost of the track.

The following operating parameters of a tamping process may influence track maintenance durability, reliability, availability and safety and must therefore be considered. The supplier must be able to show that their process has been thoroughly researched and that it will ensure optimisation of these parameters.

Figure 369 : Minimum and Maximum Lifting Height

5.1. Lifting Height

A predetermined minimum lifting height is instated in the track to ensure that the vertical level of the track is uniform behind the machine and to provide the necessary space for the ballast stones to rearrange in relation to one another. The value of this minimum lifting height has been thoroughly researched due to its effect on the durability of the tamping process. Refer to paragraph 4.2.

In the book "Track Geotechnology and Management" by Prof Ernie Selig and John Waters, the lift given by a tamping machine is plotted against the settlements that occurred in the subsequent 66 weeks under traffic. The objective was to determine the residual lift achieved after traffic as a measure of durability. See Figure 370 for the results.

Each point on Figure 370 corresponds to a sleeper end. A trend can clearly be identified and can be represented by the two red lines.

It can be seen that for relatively low lifts, the lift given by the tamping machine (for example 10 mm) is less than the settlement (approximately 13 mm) that occurred in the subsequent 66 weeks.

A possible explanation for the high settlement that occurred with a low tamping lift may be that by squeezing and vibrating the ballast, the ballast stone will expand upwards into the void but, due to the limited void space, the ballast cannot re-arrange completely and cannot interlock in a new stable position.

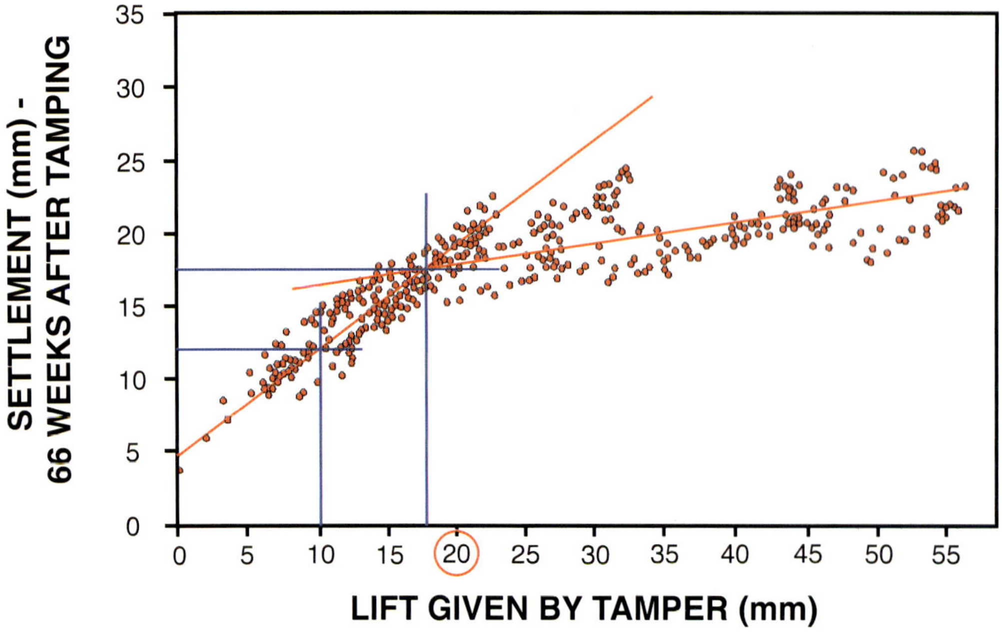

Figure 370 : Track Lift vs Settlement

Re-imposition of the vertical traffic loading will re-compact the ballast and the particles will adopt their original positions with respect to each other. The track will thus revert to its original irregular geometry, also referred to as the memory of the track.

Where the two red lines cross (at about 17 mm lift), the lift given by the tamper is more or less equal to the settlement, as measured. Therefore, in this experiment, a lift of 20 mm during maintenance tamping was required to ensure a residual lift that would achieve a lasting improvement in the track condition.

However, it is important to consider that in this experiment 50% of the ballast bed envelope had a size of approximately 20 mm (D50 = 20 mm). On certain lines such as heavy haul lines, the D50 can be 30 mm and the minimum lifting height will have to be adjusted accordingly to approximately 30 mm.

If a lift of more than 5 to 10 mm over the minimum is required a single tamp may not be sufficient anymore and double tamping at each sleeper is advisable to ensure durability.

5.2. The Frequency of Tine Vibrations

In practice it soon became very clear that by merely squeezing the ballast into position under the sleepers requires huge forces and does not provide even compaction. The reason is that the ballast stones are unable to re-arrange in relation to one another. Forcing the ballast stone will result in crushing.

Vibration of the tamping tines is necessary to liquefy the ballast stone to permit it to re-arrange and settle in a dense matrix. Vibration also greatly reduces the force required to penetrate the tamping tines into the ballast. The frequency of the vibration is, however, of utmost importance and will have to be:

- the highest possible frequency for ease of penetrating the ballast; and
- it should also be a frequency low enough to produce a lasting lift, since at higher frequencies the ballast stone can become so liquid that it will flow around the tine plates preventing adequate compaction.

The vibration of the Plasser & Theurer tamping unit tines is induced by an eccentric shaft within the tamping unit. See Figure 371. The cylinder assembly which

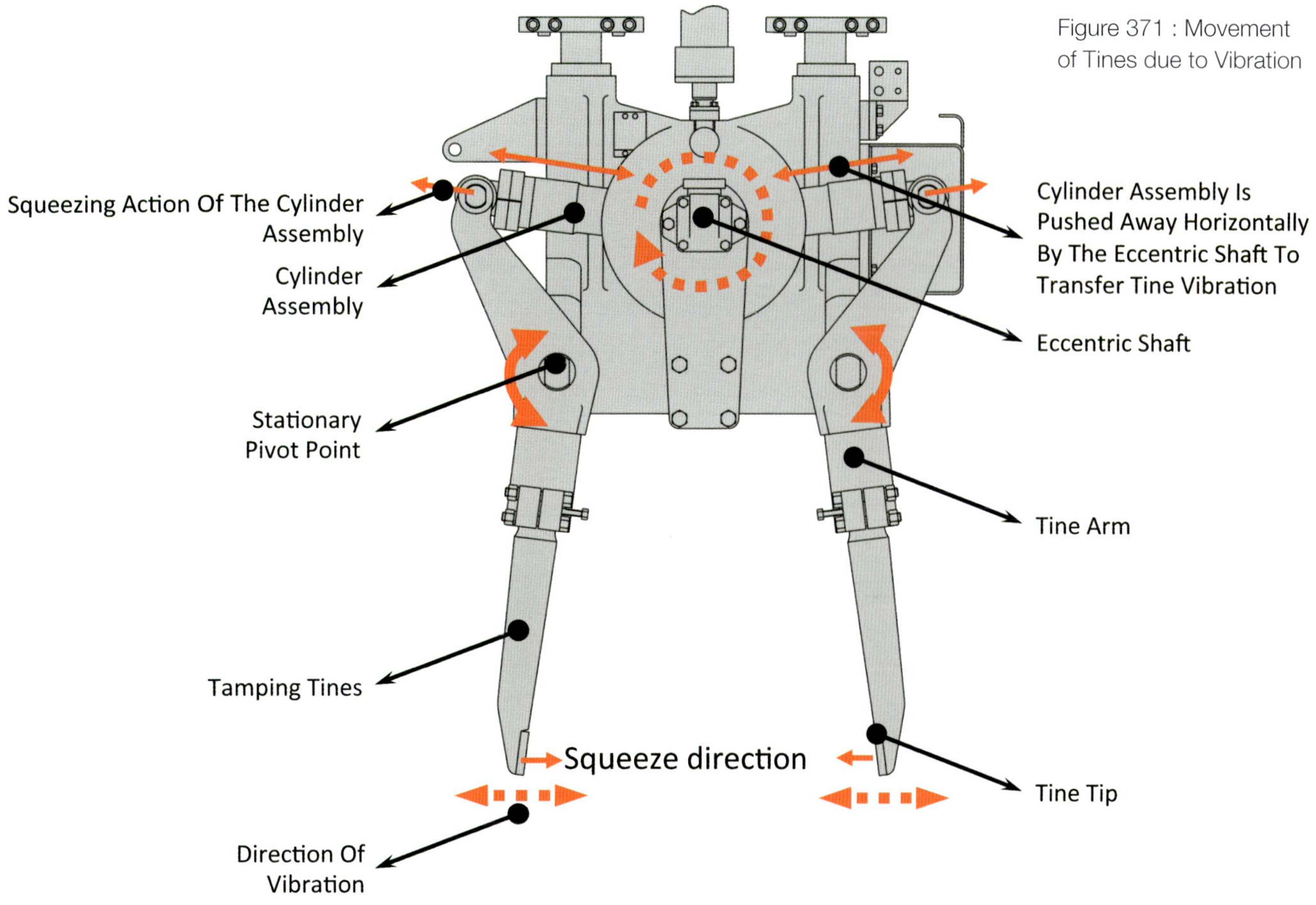

Figure 371 : Movement of Tines due to Vibration

produces the squeezing force of the tines also transfers the vibration caused by the off-centre cam of the eccentric shaft. This will cause the tines to move forward and backward around the stationary pivot point. The revolution speed of the eccentric shaft determines the frequency of the vibration and the size of the off-centre cam on the shaft determines the amplitude of the vibration.

To establish the ideal frequency, tests carried out by the Technical University of Graz have plotted different frequencies of tine vibration against the lift achieved (Figure 372).

As can be seen from Figure 372, frequencies in the range of 16 to 25 Hz produce a lasting lift, whereas frequencies exceeding 40 Hz result in a negative settlement of the sleeper. However, at a frequency of 16 to 25 Hz, the tines experience difficulty in penetrating the ballast. The graph in Figure 372 shows that the ideal frequency is between 25 and <40 Hz.

The negative settlement at higher frequencies can be attributed to a phenomenon referred to as the plasto-elastic deformation or also known as liquefaction of the ballast stone.

The Technical University of Graz study therefore also plotted the relative plastic deformation of ballast against different frequencies (Figure 373). The higher the plasto-elastic deformation, the more liquid the ballast stone becomes and the more difficult it is to pack the ballast in a dense matrix. It just flows away.

Figure 373 shows that at a frequency of 35 Hz, a minimum deformation of the ballast occurs, meaning that, at this frequency, the highest possible durability is achieved by tamping (based on clean and properly graded ballast).

Thus, the optimum tamping frequency and the frequency used by Plasser & Theurer tamping machines, for example, is 35 Hz.

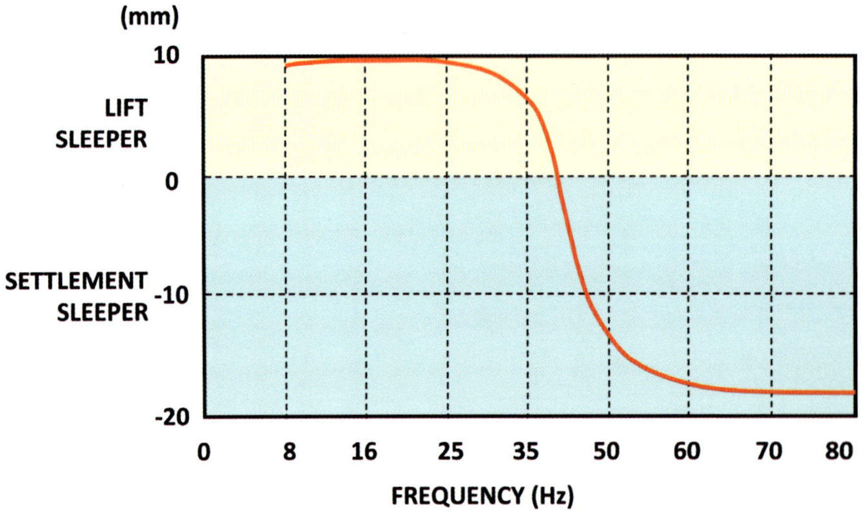

Figure 372 : Frequency vs Sleeper Settlement

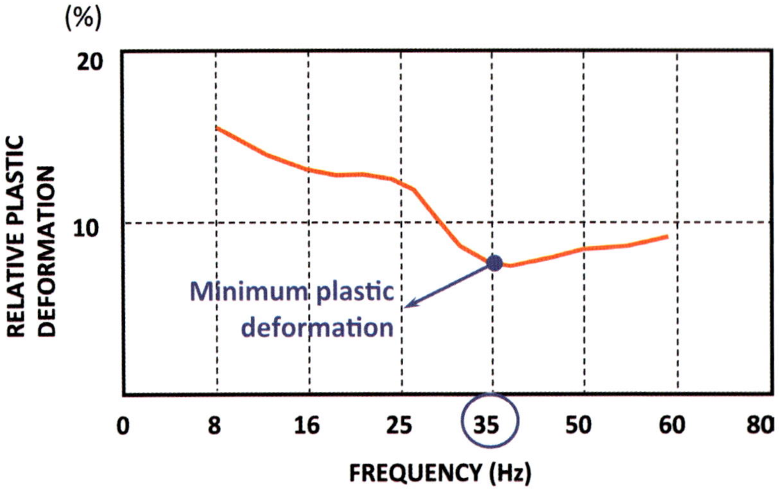

Figure 373 : Determining Minimum Plastic Deformation

5.3. The Amplitude of Tine Vibrations

The distance the entire cylinder assembly moves horizontally away from the centre of the eccentric shaft is determined by the size of the off-centre cam on the shaft. This distance travelled by the cylinder assembly determines the amplitude of the tine as the tine arm pivots around a stationary point. The tine will vibrate forward and backward, the total distance representing twice the amplitude. Refer to Figure 374 and Figure 375.

Practice has shown that amplitude greater than 5 mm causes crushing of the ballast whereas amplitude less than 5 mm fails to transfer the energy of the vibration to the ballast stone.

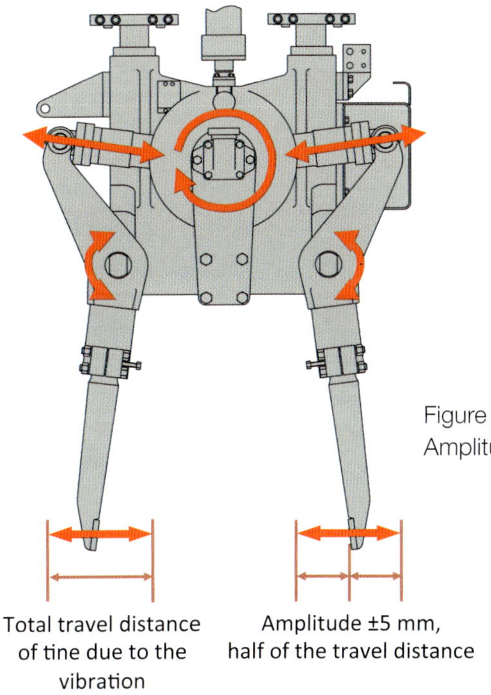

Figure 374: Amplitude

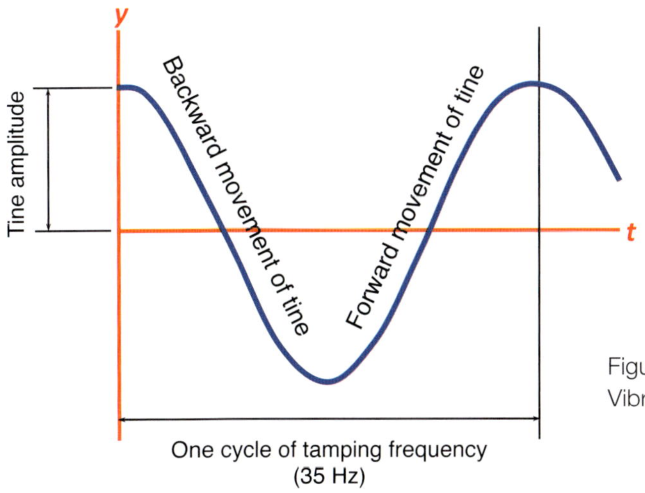

Figure 375 : The Tine Vibration Sine Wave

5.4. Tamping Depth

When the tamping tines enter the ballast, they come to rest at a pre-set depth. The tamping tine depth has great practical significance for durable packing of the ballast under the sleeper. When the depth is too shallow, the tines may hit the sleepers, and when the depth is too great, there is uneven and incomplete packing of the sleeper-bearing area.

This can be attributed to the liquefaction of the ballast and the ability of the stone to flow over the tamping tine plate.

The ideal setting range of the tamping depth is approximately 15 mm free space between upper edge of the tamping tine plate and the lower edge of the sleeper and should be electronically controlled (Figure 376).

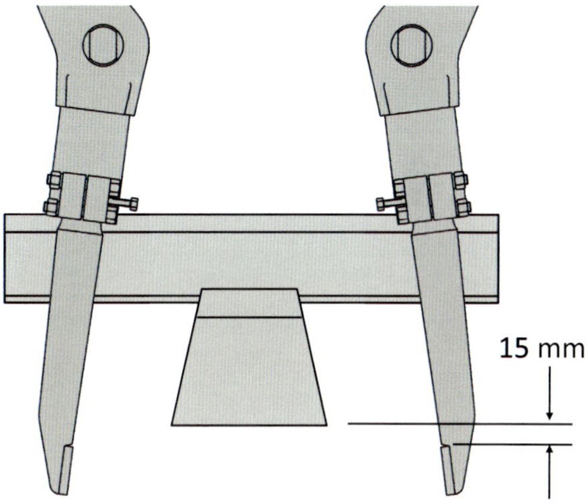

Figure 376 : Tamping Depth

5.5. Squeezing Speed and Time

The squeezing speed (closing speed) and duration of the tamping tines in the ballast also influence the durability of the tamping process.

To squeeze the ballast into the void, a certain kinetic energy is required to overcome the inertia of the ballast stone. A minimum length of time for contact between the tine and the ballast stones is therefore required to transmit the required energy.

At first one will conclude that energy transmission takes place during the entire period and squeezing distance of the tine. This is, however, incorrect. Due to the vibration induced forward and backward movement of the tine (Figure 378), which is much faster than the squeezing speed, energy can only be transmitted during a short impact period between the tine tip and the ballast face during the tine's forward motion of the oscillation.

During the oscillation the tine will create a void in the ballast of at least twice the size of the amplitude i.e. 5 mm forward and 5 mm backward of the zero position (Figure 378). During the forward motion of the vibration cycle impact with the ballast stone will only take place once the tine has travelled the total void distance, in other words, if the tine was not moving forward at a squeezing speed, impact would have taken place at the turning point of the vibrating cycle.

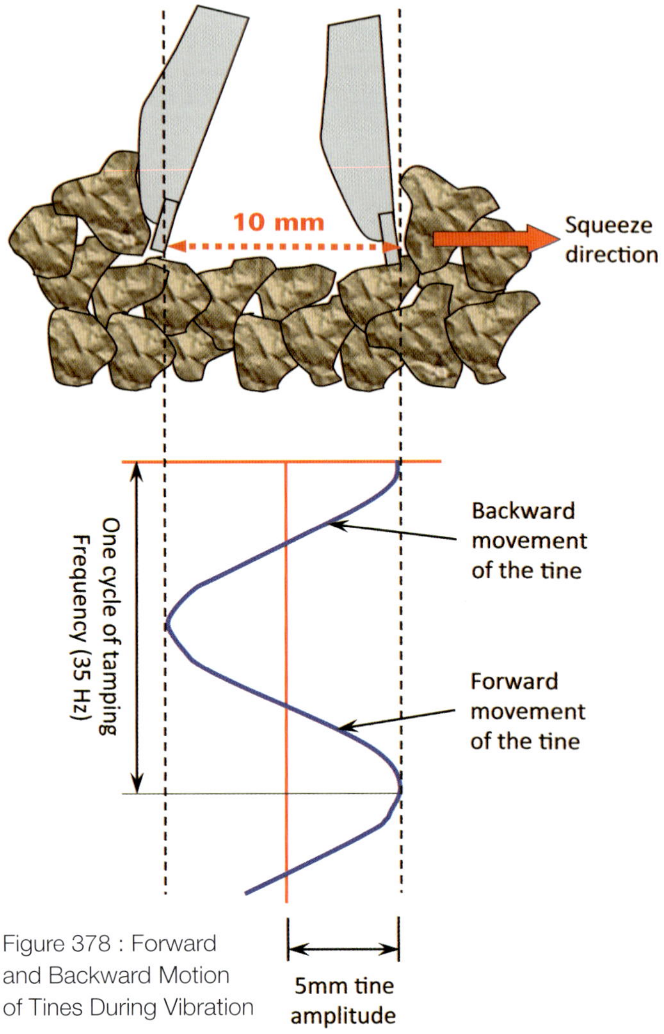

Figure 378 : Forward and Backward Motion of Tines During Vibration

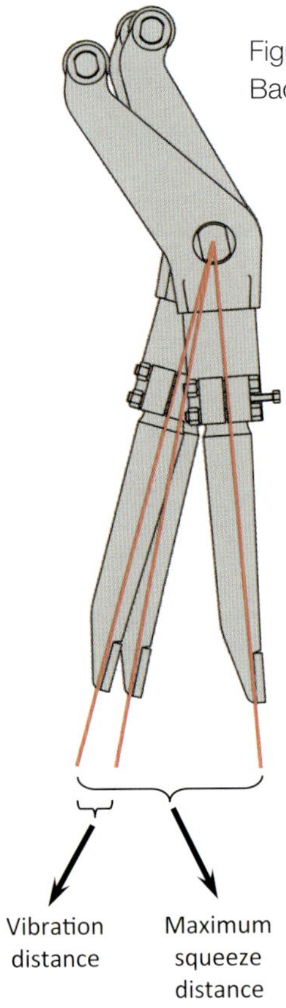

Figure 377: Forward and Backward Movement of Tines

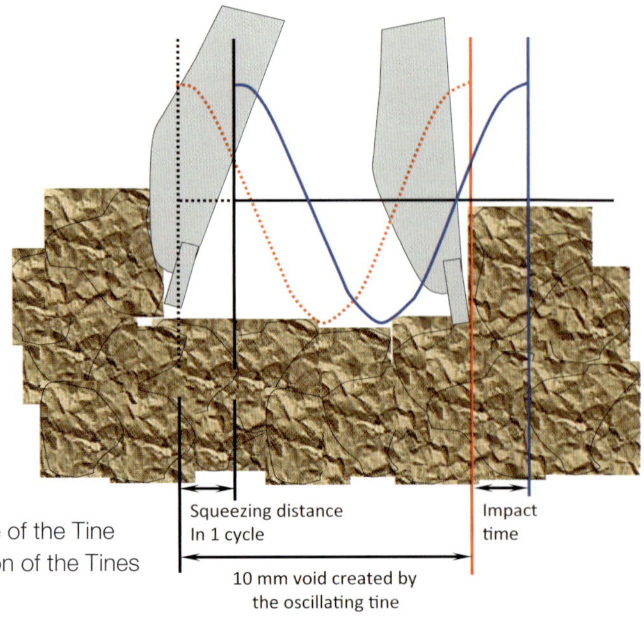

Figure 379: Impact Time of the Tine due the Squeezing Action of the Tines

However, impact will occur before the tine has completed its forward cycle of the vibration because the tine is at the same time moving forward at its squeezing speed (see Figure 379).

After impact the tine tip will remain in contact with the ballast until the oscillation starts its backward movement. This means that only during a short phase of the forward movement will the tine make contact with ballast stone. The impact duration is therefore a function of the squeeze speed.

Dr Johann Fischer in his Thesis, Influence of Frequency and Amplitude on the Stabilization of Track Ballast of 1983, plotted the squeeze movement over time on the sine graph. Figure 380 illustrates how a fast squeezing speed (short time – ti2) will have a longer impact duration than a slow squeezing speed (ti1).

The rationale is that the faster the squeezing speed, the earlier during the forward oscillation will the tine start making contact with the ballast and in so doing remain in contact with the ballast. The tine has travelled faster through the void space.

The question still is – what is the required impact duration to transfer the necessary energy to overcome the inertia of the ballast stone?

From the findings of a research report of G Kronenberg in 1961, a guide value for the minimum impact duration of 5 milliseconds was established. In this manual this will be taken as a given.

Figure 381 shows that the squeezing speed that will guarantee a 5 milliseconds impact duration at 35 Hz is 150 mm/s. It can also be seen from this graph that at higher tamping frequencies the ideal impact duration of 5 milliseconds cannot be achieved. The ideal squeezing time can now be derived. Normal squeezing distances are 120 mm.

$$\text{Ideal tamping time} = \frac{120\text{ mm}}{150\text{ mm/s}}$$

$$= 0.8\text{ seconds (minimum)}$$

To produce a good quality product, tamping machines should squeeze at approximately 150 mm/s permitting the tine to remain in the ballast bed for a minimum of 0.8 seconds and a maximum of 1,2 seconds, depending on ballast condition, which achieves the minimum required impact time between ballast and tine.

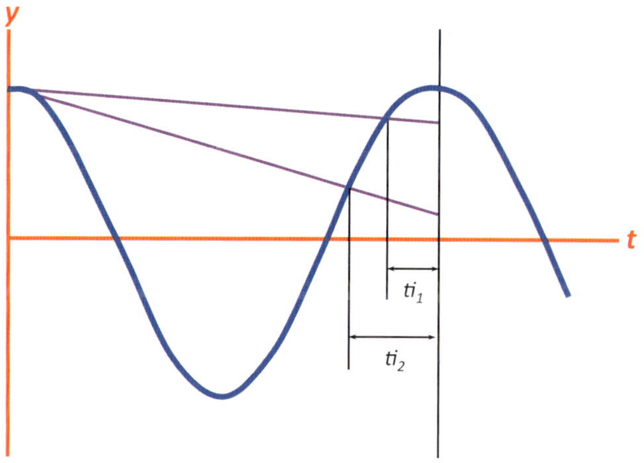

Figure 380: Impact Duration Between Tine and Ballast

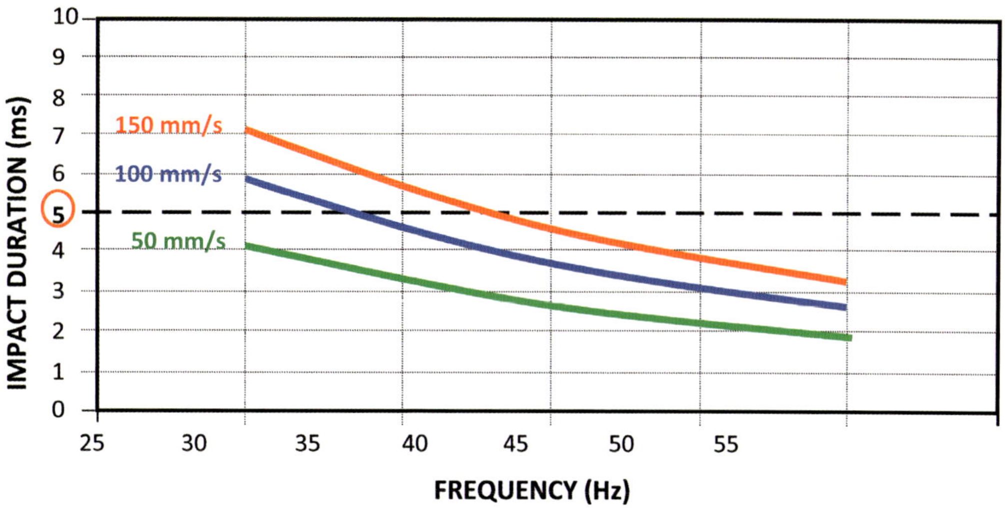

Figure 381: Required Squeezing Speed at 35Hz for a 5 ms Impact Duration

6. CHOOSING THE RIGHT TAMPING MACHINE FOR THE APPLICATION

When having to choose a tamping machine for the railway or track section, all of the above must be considered. However, there are also various other network and infrastructure considerations such as the length of the line, the traffic density, the number of turnouts on the line, etc. Following are some of the considerations:

6.1. Calculating the Tamping Cycle Based on Traffic Throughput

In CHAPTER 1 condition-based maintenance was suggested as the most efficient maintenance strategy. However, a railway must budget for the maintenance requirements in advance in order to ensure that the necessary machinery is available when required. The tamping cycle must therefore be determined. The following is an empirical formula that is often used to determine the expected tamping cycle on freight lines in South Africa:

Tamping cycle in months
$= 48 \div \sqrt{(\text{annual throughput in million gross ton})}$

If the annual throughput on a particular line is expected to be, for example, 5 million gross tons (the total tonnage including the weight of the rolling stock) the tamping cycle will be approximately 21 months if the above formula is used. In other words, the same spot on the line must be tamped again 21 months after the previous tamping cycle. However, one should consider that at lower traffic volumes, other external factors may influence the tamping cycle. If the expected traffic is 50 million gross tons, for example, the expected cycle will be approximately 6,8 months. This formula is used for heavy haul lines on cape gauge and may be too conservative for other gauges and traffic types.

6.2. Length of the Line

In practice, various lines/districts/sections may share maintenance machinery in which case the tamping cycle of all the lines that share machines must be considered. To illustrate the influence of the length of line on the type and number of machines required, a single line of 500 km and 1,000 km will be used as examples together with the above throughputs of 5 and 50 MGT.

Maintenance planners should always aim to limit the number of machines on the line due to their disruptive effect on traffic, especially on single lines. However, the choice will always be between one or more universal tamping machines which can tamp the turnouts as well as plain track, or a mix of universal and plain track tamping machines.

6.2.1. Scenario 1 – 5 MGT Throughput on 500 km Line

For the 5 million gross ton (MGT) throughput example, it means that a machine/s must tamp a total of 500 km in a 21 month period before it theoretically starts at the beginning again. One would expect a relatively low production machine would be adequate for such low throughput. A tamping machine with a nominal tamping rate of 16 sleepers per minute will tamp 960 sleepers in an hour. One can expect a maximum of about 4 working hours per day in a 240 working day year. The machine will therefore tamp 3,840 sleepers in a day. At a sleeper spacing of 700 mm, it equates to 2,7 km in a day or 54 kilometres in a 20 working day month.

This tamping machine would therefore complete one cycle of 500 km in 9,3 months which is much less than the calculated 21 months. Only one machine will therefore be required and a single sleeper index universal tamping machine will provide the flexibility for it can tamp plain track and turnouts. Turnouts take longer to tamp than plain track requiring consideration of the number of turnouts.

6.2.2. Scenario 2 – 50 MGT Throughput on 500 km Line

On a line with 50 MGT throughput, the 9,3 month cycle will be too long since the required tamping cycle is 6,8 months. A faster machine or more machines would therefore be required. Staying with the concept of using one machine, a high production continuous action universal tamping machine may produce up to 32 sleepers tamped per minute (nominal). The machine will produce 7,680 sleepers tamped per day (5,4 km/day or 108 km/month). This machine will therefore have a tamping cycle of 4,6 months which is less than the required 6,8 months which again leaves enough spare capacity for emergencies, turnouts, yards, etc.

6.2.3. 1,000 km Line and/or High Throughput

If it is a 1,000 km line with the same throughput as above (50 MGT) and the same universal tamping machine is used, the tamping cycle will still be 9,3 months and the required cycle will still be 6,8 months but a much longer line must now be tamped over this period. Even two machines will not have adequate spare capacity to maintain the tamping cycle of such a long line. A high production mainline tamping machine such as a 09-3X continuous action tamping machine combined with a lower production universal tamping machine for the turnouts and yards will have to be considered as one of various scenarios.

6.3. Number of Turnouts and Curves on the Line

With the length of the line there are also other practical considerations. Curves and turnouts may require a more frequent tamping cycle than tangent (straight) track. The more curves and turnouts there are on the line, the more the machine will have to be moving up and down the line to satisfy the tamping cycles of turnouts, curved and tangent track. As the line gets longer, one machine may be unpractical even if it is capable of keeping to the required tamping cycle. If emergencies and other sections that require more frequent tamping such as those on weak or failing formations are also considered, the time the machine will be traveling between working areas may diminish the machine's production beyond that which is required. More machines of lower production capabilities may be required, spread along the line.

6.4. Traffic and Maintenance Windows

The above scenarios have not considered the number of trains required to increase the throughput from 5 MGT to 50 MGT. Even if longer trains are used to carry the freight, it will still require more frequent trains. The higher the traffic density, the less time will be available for maintenance and the shorter the maintenance windows will become.

Tamping machines will be required to produce more sleepers tamped in the shorter windows to keep to the required tamping cycle. Therefore, high production machines would be required. If it is a single line as well, using more tamping machines cannot always be accommodated since train slots must be occupied to create a maintenance window and having to create more than one maintenance window will have a severely negative effect on income-producing traffic.

6.5. Number of Crossing Loops and Double or Single Lines

Double lines are more flexible due to the fact that traffic can pass on the adjacent line. On single lines a more careful consideration of maintenance is required, due to the effect it has on traffic.

On high traffic density lines tamping is often required to take place in-between trains which mean that the machine must vacate the line for oncoming traffic. This requires crossing loops where the train can cross the machine. If these crossing areas are too far apart, the machine may not be able to work for very long before it needs to start vacating the line. Once again higher production machines may be required and the travel speed of the machine will also become important.

7. CONCLUSION

Tamping is the most frequent of mechanised maintenance activities on the track and therefore warrants a great deal of understanding regarding the variety of machines and features available, together with their application, to ensure that the most cost-effective and efficient machine is obtained.

It should, however, be noted that various other factors not mentioned yet will also influence track durability. This includes the skill and experience of the machine operator, the accuracy of tamping unit settings, size and condition of the tine tips, etc. These parameters do not have a scientific basis but should also be considered to ensure that production targets are met and the required durability is achieved.

Chapter 12
Dynamic Track Stabilising

1. INTRODUCTION

There is scarcely another track laying and track maintenance action that has been as intensively investigated by railways and research institutes as dynamic stabilisation.

The findings from these extensive tests have contributed towards the further optimisation of the technique. By drawing on these tests this chapter will discuss the theory of dynamic track stabilisation.

Figure 382: DGS 62 Dynamic Track Stabilising Machine

2. UNEVEN TRACK SETTLEMENT AFTER MAINTENANCE

After months of ballast consolidation by passing trains the track panel is anchored in the ballast bed by high resistance to displacement.

However, horizontal and vertical deviations from the desired geometry develop over a period of time which requires cyclic maintenance tamping or ballast cleaning to restore the designed track geometry and the ballast resilience.

During the ballast cleaning process the fines are removed from the ballast and, if required, new ballast is added. This process dramatically increases the void spaces between the ballast stones through the entire depth of the ballast bed.

During the tamping process the sleeper is lifted and aligned. This creates a void below the sleeper and at the sleeper head. The tamping tines compact the ballast in the void below the sleeper, whereas the void at the sleeper head can only be filled by ballast, under gravity. This tamping process disturbs the ballast bed and increases the total volume of void spaces between the ballast stones below the sleeper and at the sleeper head.

Some of the ballast stones will now lie in an unstable position with only their corners and edges touching due to the greater void space.

Therefore, the ability of the track to resist forces exerted upon it directly after tamping and or ballast cleaning, is greatly reduced. So, for example, the track's lateral resistance is reduced to approximately half of its value before tamping.

The track's initial subjection to traffic at normal speed can therefore have a detrimental effect. To overcome this risk, the customary practice has been to let trains compact the track bed with their weight when passing at restricted speed for a period of time, after maintenance. In other words, speed restrictions are imposed.

The load from trains will cause very high pressure on the ballast edges and corners, resulting in high initial crushing and abrasion of the ballast stone and an uncontrolled re-arrangement of the ballast stones.

Figure 383 illustrates that the ballast structure has reduced volume but the gridlines have not changed their angle. The ballast stones have not rearranged themselves against one another. They were simply crushed into a new configuration, characterised by an uneven longitudinal and cross level. Graphically, this high initial settlement can be seen in Figure 384 which shows the track settlement in mm on the vertical axis against the traffic in million gross tons on the horizontal axis.

This period of high initial settlement is clearly visible during the first 100,000 tons of traffic. The track has now reached 70 to 80% of its original resistance. During this period speed restrictions are imposed due to the reduced lateral resistance of the track.

This process of compacting the track by the weight of passing trains at a reduced speed has a considerable cost impact on train operations, labour input, wear and tear on braking systems and train energy consumption, due to the braking and acceleration of the train before and after speed restrictions.

In addition, within 72 hours of ballast cleaning, the tamping machine must return to the ballast cleaned section to re-tamp it to correct the high and uneven settlement that would have occurred. This increases the total cost of maintenance and is disruptive to train operations.

In 1974/5 Plasser & Theurer developed the dynamic track stabilising machine with the aim of avoiding these problems associated with ballast maintenance.

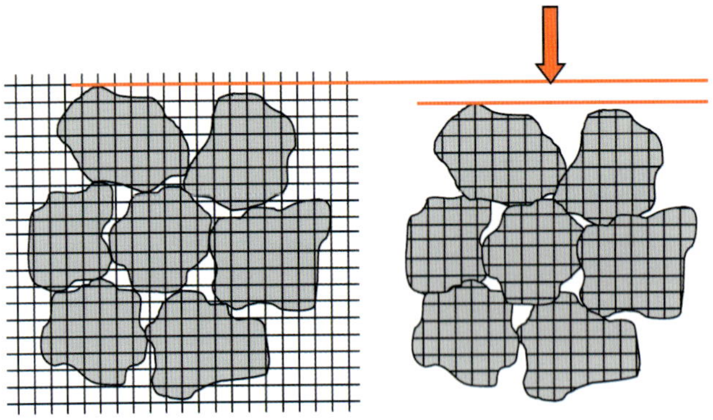

Figure 383: High Initial Crushing of the Ballast after Ballast Maintenance Caused by Traffic

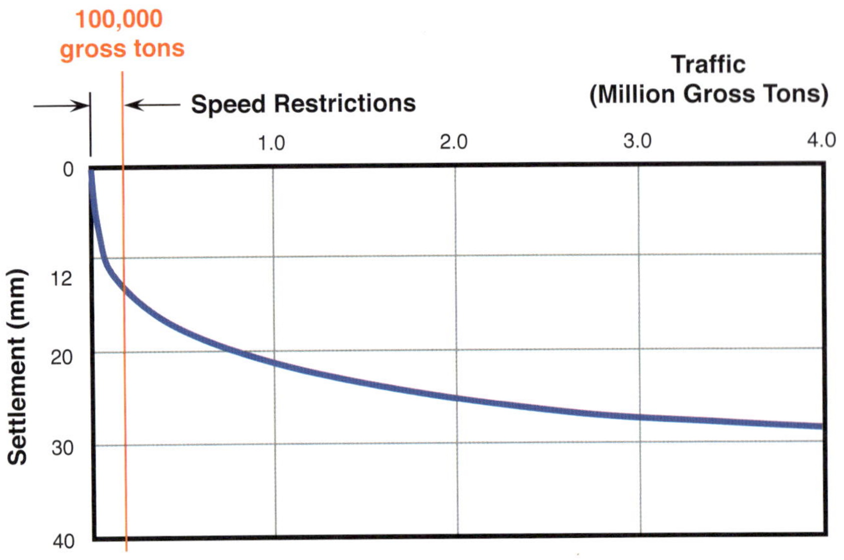

Figure 384: Track Settlement in Relation to Train Loads

*Source - Plasser & Theurer, ORE D117 PKP-test series

3. CONTROLLED SETTLEMENT WITH DYNAMIC TRACK STABILISING

The dynamic track stabilising machine is introduced directly behind the tamping machine after ballast maintenance such as tamping, ballast cleaning or track renewal.

It is an on-track machine which consists of two hydraulically driven, vibratory stabilising units located under the stabilising vehicle between the axles. It sets the track in horizontal oscillation directed crosswise/ laterally to the track whilst at the same time applying a controlled static vertical load. See Figure 385.

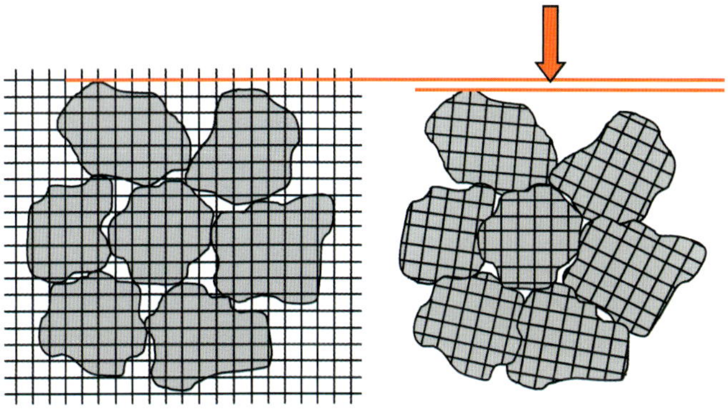

Figure 386: Homogeneous Settlement Made Possible by the Fluidisation Induced by Vibration

The liquefaction (fluidisation) induced by the vibration, together with the static vertical load, causes the ballast stones to settle closer together in a more homogeneous and compact grouping in the entire ballast bed.

This is illustrated once again by the ballast arrangement with a grid drawn over it, in Figure 386. The figure shows how the ballast stones changed their orientation to one another by the change in the angle of the gridlines on the stones.

The controlled settlement with dynamic stabilisation achieves a relative force-free re-arrangement of the ballast stones with only around 30% of the forces occurring due to traffic loading being transmitted. The ballast is therefore not unduly stressed, which reduces the abrasion and crushing of ballast during the initial high settlement under traffic, which in return prevents unnecessary ballast fouling.

The voids between the ballast stones are now reduced and instead of individual contact points the ballast stones have a larger number of contact surfaces. The settlement is homogeneous over the length of the dynamically stabilised track, which would otherwise have occurred unevenly under traffic loading.

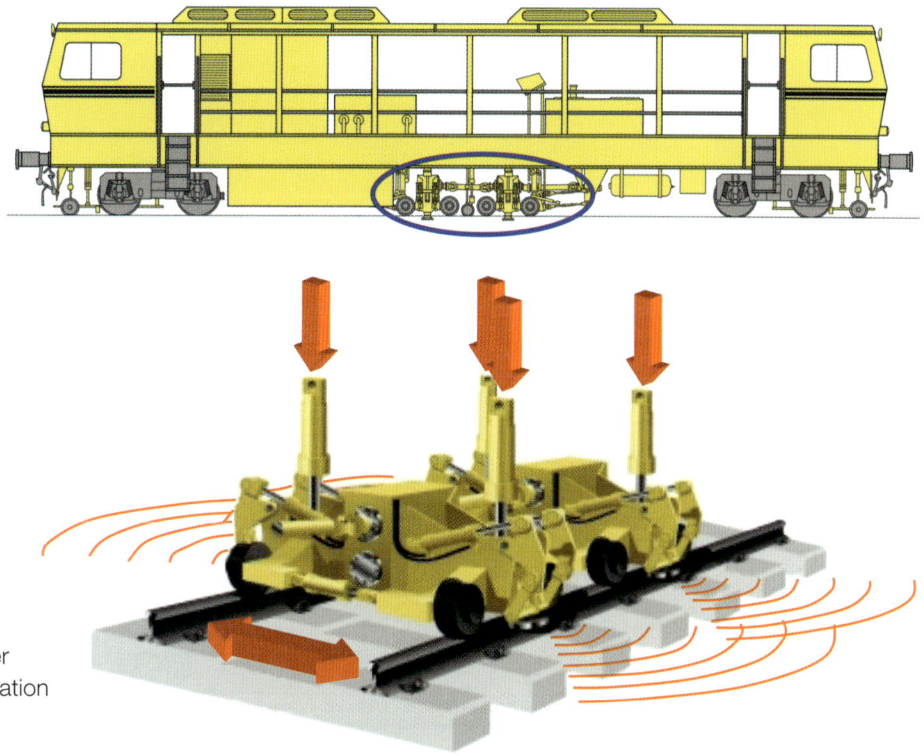

Figure 385: The Stabiliser Applies a Horizontal Vibration with Static Vertical Load

4. THE EFFECT OF DYNAMIC STABILISATION

4.1. The Effect of Stabilisation on Track Settlement

The dynamic track stabiliser effects within seconds the same settlement which the train loads achieve in days.

When a dynamically stabilised track is compared to a non-stabilised track on the same graph, (Figure 387), it shows how the high initial settlement has been removed by stabilisation.

In this example a settlement of 12 mm was achieved with dynamic stabilisation which is equivalent to approximately 140,000 gross ton traffic. However, a specific "amount of settlement" cannot be insisted upon since the amount of settlement depends upon a number of factors including:

- Height of the preceding lift by the tamping machine. During normal maintenance work the settlement introduced by the stabiliser is usually around 10–15 mm after a maintenance lift of 25 mm. The maximum settlement that can be achieved on new lines or ballast cleaned lines is around 25 mm. To achieve this, tamping and dynamic stabilising is done in layers of approximately 100 mm.

- The vertical load applied by the stabilising machine: Though increasing the vertical load results in greater settlement, theoretical work and practical experience has shown that the amount of settlement is not in direct proportion to load; it rather becomes less after a particular point has been reached. The amount of settlement is due to the law of plastic deformation under dynamic loads.

- The thickness of the incompletely consolidated ballast bed: Smaller settlements can be expected on thinner ballast profiles.

- Size and grading of the ballast stones.

- Degree of contamination of the ballast bed: Very contaminated ballast will have reduced void spaces, resulting in lower settlements. Very fine contaminants and moisture, such as silt in the ballast, can, however, act as a lubricant to enhance the movement and re-orientation of the ballast, resulting in higher settlements.

- Condition of the other permanent way material and components: Experience has shown that the effectiveness of this operation is markedly reduced on track with loose fastenings or worn base plates. Loose fastening prevents the transmission of vibrations from the rail to the sleeper, so preventing a state of liquefaction.

The repeated runs of the dynamic track stabiliser immediately following one another brings no noticeable improved consolidation compared to a single pass only. This will, however, lower the track, which is not desirable.

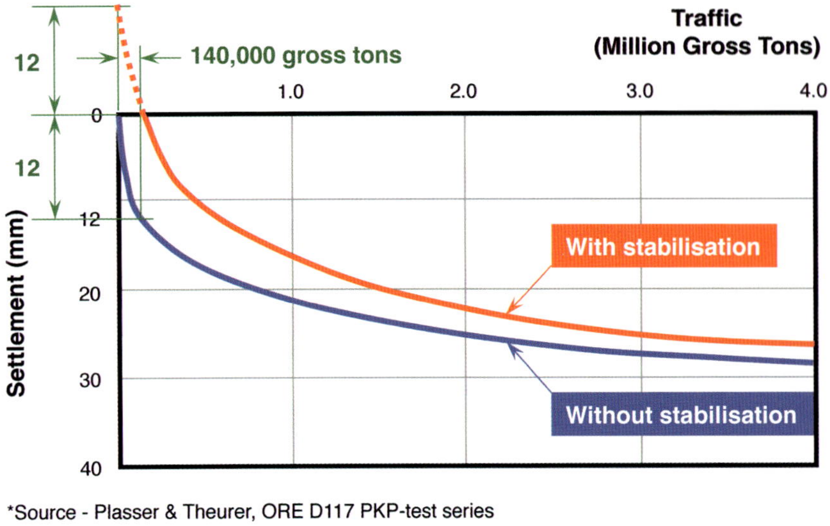

Figure 387: Track Settlement in Relation to Train Loads

4.2. The Effect of Stabilisation on Tamping Cycles

Transnet Track Testing Centre of South Africa did extensive testing and evaluation on the effect of dynamic stabilisation on the settlement and cant retention of curved track on the heavy haul coal line.

For comparison purposes of tamping alone versus tamping and stabilising of curves, 8 curves were selected for the field tests. All curves were grouped in pairs of which one was stabilised at the start of the test and the other not.

Each of the mentioned groups of test curves is characterised by its homogeneity with respect to ballast thickness and fouling, geology, track stiffness and geometric features.

One of the observations of the testing was the effect of stabilisation on the cant deterioration of the curve. The high percentage of cant loss can clearly be seen from the table in Figure 388 on the non-stabilised curves.

After 50 MGT the stabilised curves have lost only 30% on average of the given cant whereas the unstabilised test curves have lost nearly all of the cant (90% on average) given to them by the tamping operation.

From the field tests it was noted that none of the test curves required tamping after 50 MGT, which is the normal tamping cycle on the coal line. These conclusions were drawn based on the track roughness as well as the cant of each curve at the time.

Due to the higher initial quality of using dynamic stabilisation, the required tamping intervention interval improved from 50 MGT to 115 MGT, confirmed by the development of a ballast tamping model (Figure 389).

Figure 388: Cant Loss Comparison

Test no.	Curve no.	Cant before maintenance (mm)	Cant after maintenance (mm)	Cant after 50 MGT (mm)	Loss of cant (%)
N1*	75	18.6	20.6	19.7	45%
N2	79	14.7	18.6	16.1	64%
N3*	77	18.8	21.1	20.5	28%
N4	76	17.5	21	17.9	89%
S1*	188	8.3	20.6	15.9	38%
S2	159	8.8	9.9	8.7	110%
S3*	195	28.2	37.8	36.2	17%
S4	193	9.8	11.2	9.9	96%

***Stabilised curve**

Source - RTR 4 (1997), P J Gräbe, J S Maree

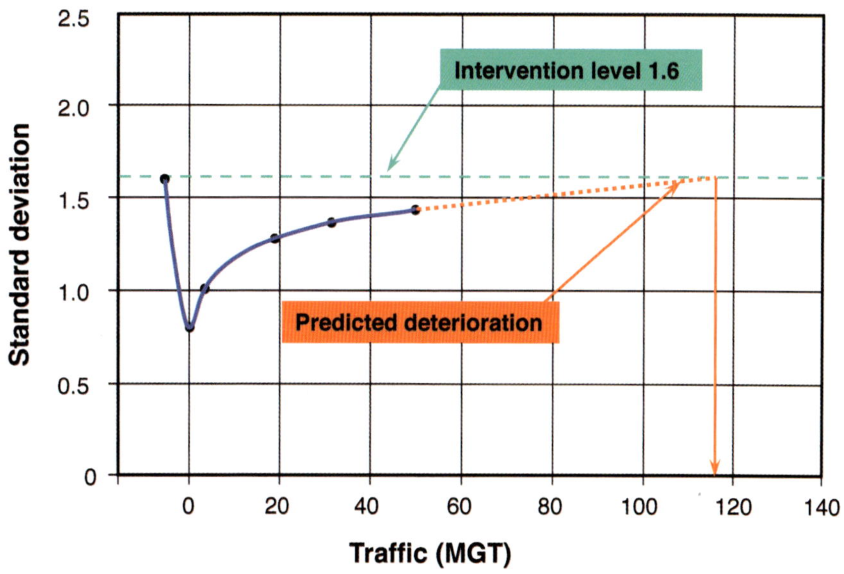

Figure 389: Ballast Tamping Model

Source - RTR 4 (1997), P J Gräbe, J S Maree

This model shows, on a graph, the track roughness or average standard deviation of the parameter "top" (vertical alignment) as measured by the track recording vehicle on the vertical axis and the traffic in million gross tons (MGT) on the horizontal axis. Refer to CHAPTER 9, paragraph 4.1 for a detailed description of the parameter "vertical alignment" as measured by the track recording vehicle.

By combining engineering inputs with track experience, an average standard deviation for vertical alignment of approximately 1,6 was found to be the maximum allowable roughness before tamping intervention must take place.

Consecutive standard deviations for a homogeneous section of track are plotted against increasing traffic (MGT). A mathematical trend is made through the available data to determine the deterioration function. The function is computed for future traffic and the expected time for maintenance is calculated where the predicted deterioration intersects the intervention level.

4.3. The Effect of Stabilisation on Track Stability

Track stability refers to the track's resistance to lateral displacement. If the lateral force caused by trains or thermal stress exceeds the track's resistance to lateral displacement, a speed restriction or line closure must be imposed so that operating safety is guaranteed at all times.

Several methods of measuring track strength have been developed by different railway administrations. These include:

- the single-sleeper displacement method, and
- the "07" method

Figure 390: Force Required Displacing the Sleeper By 2 mm

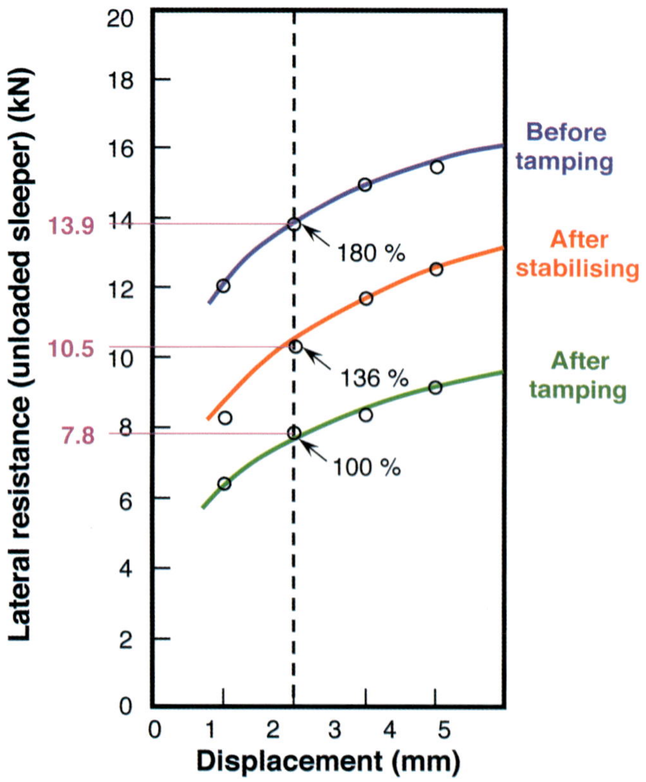

Source : Railway Gazette March 1977 - Prof.Dipl.-Ing.Dr.techn. Klaus Riessberger

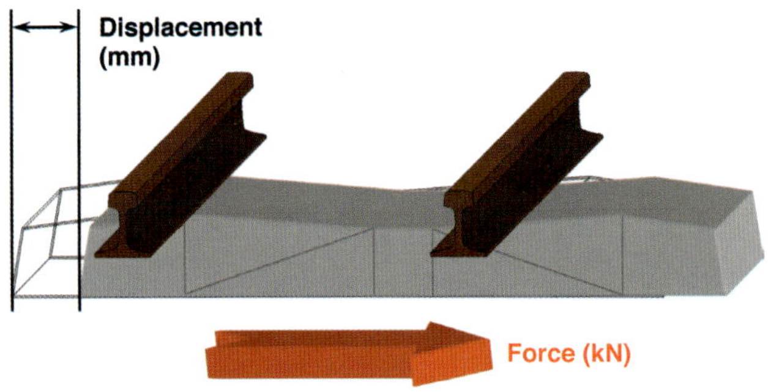

Figure 391: Single Sleeper Displacement Method

4.3.1. The Single Sleeper Displacement Method

This method was used by Franz Plasser research department in collaboration with the German Railways. Single sleepers are disconnected from the rails and laterally displaced using a hydraulic force. The force required, i.e the track's resistance to lateral displacement, is recorded and plotted on a graph (Figure 390). An average value is established from 20 individual tests.

For the purpose of comparison it has proven to be useful to measure the resistance force at a displacement of 2 mm of the sleeper, as a standard value.

For an equal shift of 2 mm after tamping a force of 7,8 kN per single sleeper was necessary (100%), while before tamping a force of 13,9 kN (180%) had to be applied. After stabilising with a dynamic stabiliser a force of 10,5 kN (136%) was required. This is an improvement of 36% on tamping without stabilisation.

4.3.2. The "07" Method

Netherlands Railways measured lateral resistance using a tamping machine on wooden sleepers. By using a special hydraulic circuit the track was loaded with a vertical load of 12 ton. The force necessary to slew the track was measured at the lining cylinders and the displacement against the lining chord.

Similar to the single sleeper displacement method the graph shows approximately half of the lateral resistance to be reinstated after stabilisation (Figure 393).

Another interesting result from these tests was that, even with the smallest lifts, there were no centre-bound sleepers after the use of the stabiliser.

This improvement of track stability after stabilisation can be attributed to the rise in the sum of contact surfaces between sleepers and ballast stones. Wider areas of the ballast bed take over the transmission of the forces. A new bearing support is produced for the sleeper.

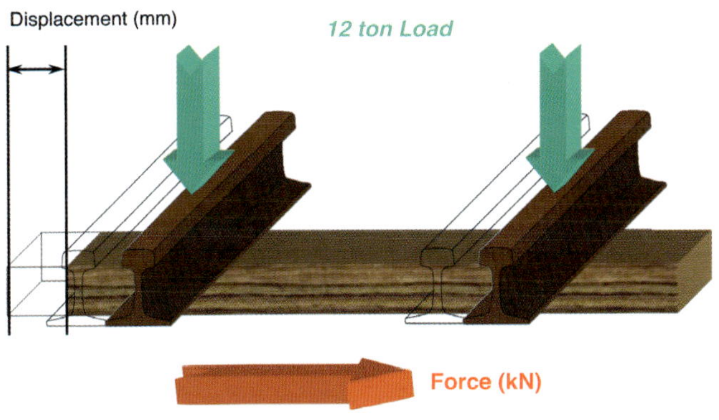

Figure 392: The "07" Method

Figure 393: Force Required to Displace the Sleeper

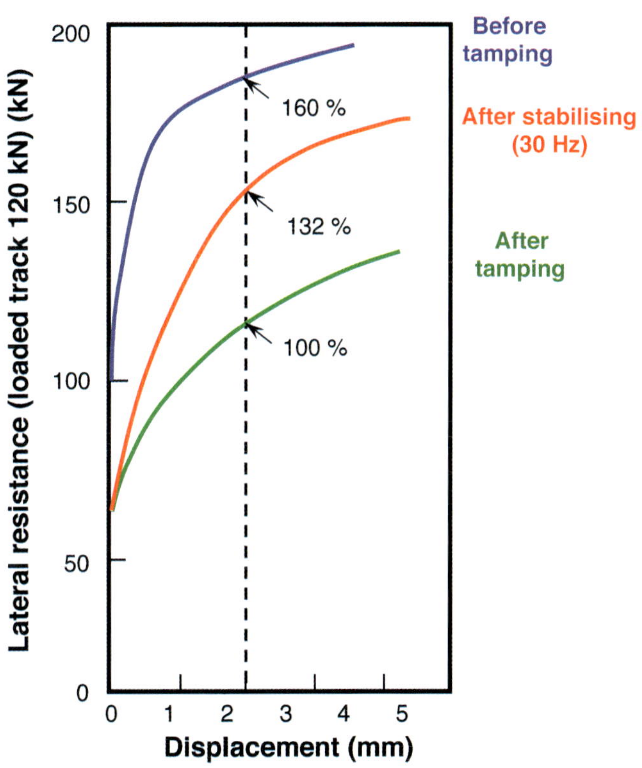

Source : Railway Gazette March 1977 -
Prof.Dipl.-Ing. Dr.techn. Klaus Riessberger

5. CONCLUSION

The aim of all track maintenance must be to place the track in the geometrically correct position at the highest initial quality and to conserve this state for the longest possible period of time.

Stabilising does not in itself improve the geometry of the track. It consolidates the ballast to improve the retention of the track geometry thus extending the period between maintenance cycles. It also increases the stability of the track which has the following advantages:

- The safety to traffic is considerably increased.
- The higher stability of the ballast bed will allow the maximum line speed immediately after maintenance.
- The need for follow-up tamping after consolidation by trains is avoided.
- By avoiding temporary speed restrictions the reliability of the rail service is improved.

Chapter 13
Ballast Distribution and Regulating

1. INTRODUCTION

One of the functions of ballast is to resist vertical, lateral and longitudinal forces applied to the sleeper to retain the track in its required position. The ability of the ballast to perform this function is controlled by the cross-sectional ballast profile which should conform to specified minimum requirements.

However, in due course deviations from the ideal profile will result, due to a lack of maintenance, people walking across the track, crushing of the ballast and the tamping process which lifts the sleeper out of the ballast, resulting in no or a too little ballast in the cribs or at the shoulder. The end result is the loss of vertical, lateral and longitudinal stability of the track, which causes alignment defects, wear and damage to the rails and rolling stock, loss of passenger ride comfort and derailments. The cross-sectional profile must therefore be restored on a regular basis.

The safety aspect of the ballast is, however, not the only reason why the management of ballast is a vital track maintenance activity. Ballast material is expensive due to its sheer quantity and the economical distribution of ballast along the track is a great source of potential savings. The ballast volume could be adequate but, spread over a wide area, excess ballast could exist on one section of the line and a deficit further down the line.

Ballast regulating machines are designed to bring about savings in the management of ballast through the distribution of ballast across and along the line, ploughing in the ballast that is spread over a wide area and correcting the cross sectional profile.

2. FUNCTIONS OF THE BALLAST BED

The ballast has the following functions (refer to CHAPTER 3, paragraph 5):

- The depth of the ballast under the sleepers provides the elasticity of the track and reduces pressures from the sleeper-bearing area to acceptable stress levels for the underlying weaker subgrade.
- The ballast provides voids for the movement of particles through the ballast and allows effective drainage of water entering the ballast bed.
- The ballast allows correction of the track geometry through lifting, levelling, ligning and tamping.
- The ballast bed resists vertical, lateral and longitudinal forces applied to the sleeper to retain the track in its required position.

3. REQUIRED BALLAST BED CROSS-SECTIONAL PROFILE

In order for the ballast bed to fulfil its functions, the ballast bed cross-sectional profile must conform to specified minimum requirements. The following is an example of a typical specification (Figure 394):

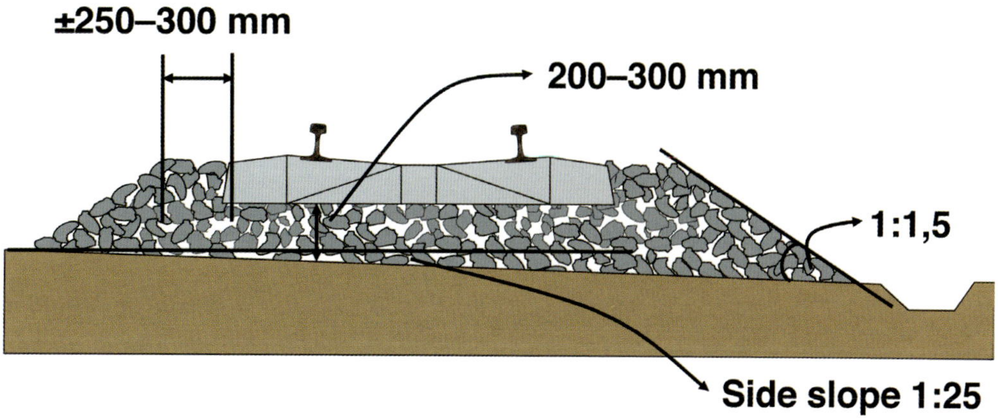

Figure 394 : Typical Specified Ballast Bed Cross-Sectional Profile

- The sleeper should rest inside the ballast with the cribs filled with ballast. A shoulder of approximately 250 to 300 mm is required to resist the lateral movement of the track. On sharp curves the shoulder can increase to 600 mm. Figure 395 shows a section of track with a very good cross-sectional profile.

- The ballast depth should be kept at between 200 to 300 mm under the sleeper, depending on the class of line.

- The formation should have a cross fall towards the drain of 1:25. This is required to direct the water which drains through the ballast towards the side drain.

Figure 395: Track Section with a Good Ballast Profile

Figure 396: Track Section with no Shoulder and too Little Ballast in the Cribs

4. CAUSES OF A POOR CROSS-SECTIONAL BALLAST PROFILE

There are various causes of a poor cross-sectional ballast profile (see Figure 396 for an example) including the following:

- A lack of ballast profile maintenance: Maintenance managers must monitor the ballast profile on their sections and plan for the application of regulating machines to correct any deviations speedily before the deviation causes further damage. Track recording vehicles may be equipped with ballast profile measuring equipment which is of great assistance to monitor the ballast profile and to plan for the effective utilisation of the ballast regulating machine. The ballast profile measuring equipment may also be fitted to larger ballast regulating machines which has the capacity to pick-up and offload large volumes of ballast. See CHAPTER 9 paragraph 4.7.

- People walking across the track: In densely populated areas, especially in poorer communities, which are often located close to the railway line, people crossing the railway line in several places and on a regular basis will spread the ballast over an uneconomical wide area.

- Ballast crushing: There are many reasons for the crushing of ballast, such as normal traffic loading and dynamic loading due to rail surface irregularities, poorly maintained joints and turnouts, poor track geometry, rolling stock defects, etc. The crushed ballast stones will settle closer together and reduce the void spaces.

- The tamping process: Over time and several tamping cycles, the tamping process may lift the sleeper out of the ballast resulting in no or too little ballast in the cribs.

5. EFFECT OF A POOR CROSS SECTIONAL BALLAST PROFILE

The ballast shoulder and ballast in the cribs provide the track's resistance to vertical, lateral and longitudinal displacement. This is especially important on gradients, curves and continuous welded rail (thermal stress) where great forces are at play.

If the ballast profile is inadequate to resist the force, the track may move causing horizontal or vertical alignment defects and, in severe cases, temperature kick-outs or buckling.

These defects will cause derailments in worst cases, and wear to all the track components and rolling stock due to increased dynamic loading and lateral acceleration. A loss of passenger ride comfort will also result and cargo may shift and/or get damaged in transit. Furthermore, ballast spread over a wide area is an economical loss unless such material can be recovered.

The ballast cross-sectional profile is reinstated and or maintained using ballast regulating machines.

Figure 397: PBR 400 Ballast Regulating Machine

6. BALLAST REGULATING MACHINES

A wide range of ballast regulating machines is available to fulfil a multitude of ballast management requirements. The most basic of requirements is to rectify the ballast profile for reasons mentioned above and, under certain circumstances, the ability for the machine to do just that is all that is required. The PBR 400 (Figure 397), for example, is by no means a simple machine but represents the one end of the spectrum and is suitable for small rail networks and following small tamping machines.

Ballast regulating machines may be equipped with specialised features to transfer ballast from the one side of the track to the other and may also be equipped to pick up excess ballast on one section of the track and transfer it by the use of a hopper on the machine to areas with a deficit, thereby ensuring the most economic use of ballast material.

The BDS 2000 represents the other end of the spectrum which features a large hopper in which to transport the excess ballast and can even accommodate an MFS material conveying hopper within the consist (Figure 398) for the ultimate in ballast management flexibility.

Figure 398: BDS 2000 Ballast Management System

7. BALLAST REGULATING MACHINE COMPONENTS

7.1. Shoulder Ploughs

Ballast regulating machines can either be equipped with articulating bi-directional (Figure 399) or fixed wing one-directional (Figure 400) shoulder ploughs. The articulated shoulder ploughs are more versatile than the fixed wing ploughs. Shoulder ploughs are mounted on either side of the machine and are hydraulically controlled from the machine's cab. They can be used together or independently of one another, variable up to shoulder angles of 45°.

Figure 399 : Articulating Shoulder Ploughs

Figure 400: Fixed Wing Shoulder Plough

Each plough can also be slewed vertically and laterally to the track axle (refer again to Figure 397). Therefore, the machine can produce any required ballast width and is capable of reaching ballast that is widespread.

The articulated shoulder plough has hydraulically adjustable blades on either side which may form a box that can store up to approximately 1 m^3 of ballast to deposit it when needed (see Figure 401).

In instances where there are obstacles such as mast poles, signals, etc. on the shoulder, the shoulder plough does not have to be lifted, but the operating span of the plough can be adjusted so that it does not touch the obstacle. In this way the piling up of ballast is therefore avoided. The adjustment is made in the shoulder plane, i.e. the shoulder angle remains constant during the by-pass.

Figure 401 : Possible Movements of Ballast by the Articulated Shoulder Ploughs

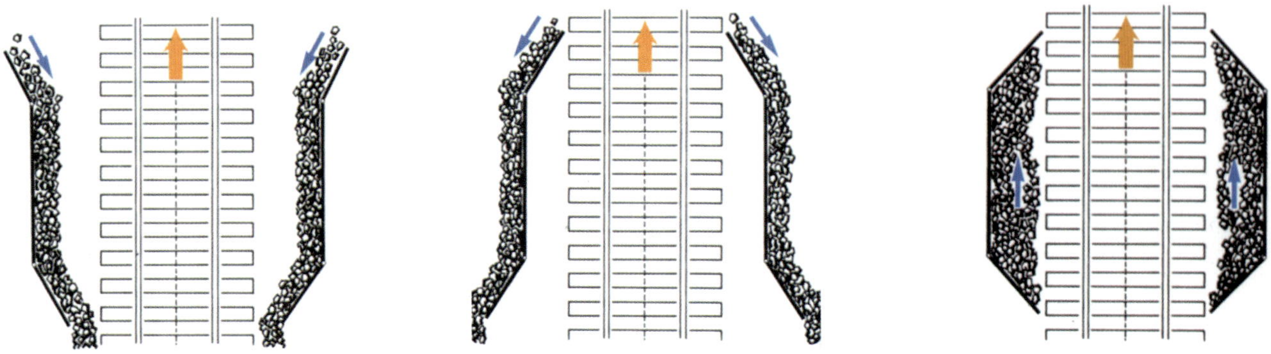

7.2. Transfer Plough

Similar to shoulder ploughs, two different types of transfer ploughs are available, one-directional located at the front of the machine (Figure 402) and bi-directional transfer ploughs located in the centre of the machine (Figure 403).

Lifting and lowering of the transfer plough along a parallel guide is hydraulically operated and controlled from the operator's seat. This movement allows the height to be adjusted with precision relative to the position of the sleepers.

Both transfer plough types consist of fixed and adjustable guide sheets. The fixed sheets are provided with welded-on tunnels to protect the rail fastenings from damage. The adjustable guide sheets in turn are adjusted either manually or hydraulically in the desired positions to allow the following ballast movements in a single pass (Figure 404).

Figure 402: One-Directional Transfer Plough Located at the Front

Figure 403: Bi-Directional Transfer Plough Located in the Centre of the Machine

7.3. Grading Plough

The grading plough (Figure 405) is used directly behind ballast wagons during ballast offloading or behind a ballast cleaning machine to distribute and even out the ballast. This is however only available on machines of heavy design and is equipped with a transfer plough in the centre of the machine.

Figure 405: Grading Plough

Figure 404 : Possible Movements of Ballast by the Transfer Plough Located at the Centre of the Machine

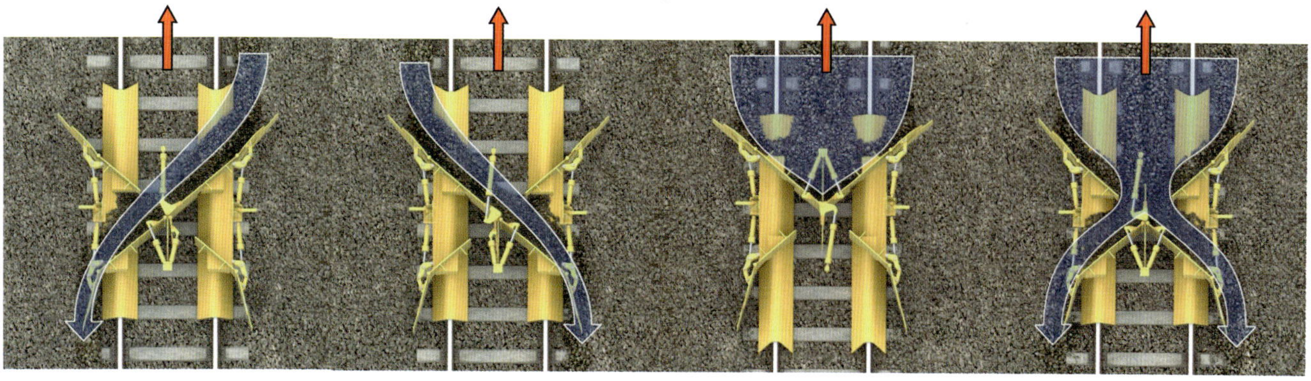

7.4. Sweeper Device

The sweeper device (also referred to as a broom box, see Figure 406 and Figure 407) is generally at the one end of the machine in a steel housing. The sweeper consists of a rotating brush of rubber hoses with which the ballast is swept from the sleepers and can be lowered and lifted hydraulically. Most of the units are equipped with a cross conveyor belt. Its purpose is twofold:

- It brooms excess ballast from the top of the sleepers and around the fastenings into the sleeper cribs or to the shoulder. This is important for ballast tamping for the machine operator to see where the sleepers are.

Figure 406 : Sweeper Device for Plain Track

Figure 407: Sweeper Device for Plain Track and Turnouts

Figure 408 : Rail Fastening Brush

- Two sweeper device designs are available; a smaller unit suitable for open line and working behind smaller tamping machines (Figure 406) and larger units of a specialised design for use in turnouts and behind high production tamping machines on the open line (Figure 407).

- Where the larger specialised sweeper device is fitted, it is also a surplus ballast pick-up device by sweeping the ballast onto a conveyor belt system and into a ballast hopper or into MFS units coupled to the machine (the BDS 2000 system).

7.5. Hopper

The capacity of the hoppers varies between 4 and 15 m3 depending on the machine model. The purpose of the hopper is to transfer excess ballast from one area to another. It works in conjunction with the sweeper device as explained above. The ballast is unloaded from the hopper via distributing chutes.

7.6. Rail Fastening Brush

One rail fastening brush is mounted over each rail. They are lowered pneumatically. Six special cable pieces per rail are pressed against the rail web to remove ballast stones lodged between the rail and fastening.

Some machines may be equipped with rotating rail fastening brushes.

8. CONCLUSION

The importance of maintaining the required ballast cross-sectional profile should not be underestimated, since neglect thereof will have many undesirable results.

Though often overlooked as a maintenance machine on its own, the ballast regulating machine fulfils the very important function of maintaining the specified ballast cross-sectional profile.

The large quantity of ineffective ballast that is recovered by the regulating machine during maintenance brings about huge savings in ballast material and labour costs, making this machine a highly cost-effective maintenance tool.

Chapter 14
Ballast Cleaning

1. INTRODUCTION

In order for the ballast bed to perform to requirements and fulfil its functions, it must be free of fine material. Refer to CHAPTER 3, paragraph 5. Once the ballast bed has reached a certain level of fouling, the fines must be removed using a ballast cleaning machine of high quality and production.

Figure 409: RM 62 Ballast Cleaning Machine (1964)

Plasser & Theurer developed and constructed the first ballast cleaning machine in 1959 with improved models following in quick succession. The first fully hydraulic on-track ballast cleaning machine was the RM 61 introduced in 1963. This was followed by the RM 62 in 1964 mounted on bogies and with spoil discharge to the front and was the first ballast cleaning machine in Africa (South Africa) on narrow gauge.

Ballast cleaning machines were a huge step forward in ballast bed maintenance since they were able to clean under the sleepers where it was most important, while reinstating the required formation cross fall. Furthermore, the production rate of mechanised ballast cleaning would be impossible to match with manual methods, even with an extremely large labour gang and there would be no comparison in quality either.

Figure 410: The RM 74 Ballast Cleaning Machine on Cape Gauge in South Africa

The next important step in ballast cleaning was the development of the RM 74 (Figure 410, narrow or cape gauge) and RM 76 (standard gauge) ballast cleaning machines in 1975 which was able to clean under turnouts and crossings as well as plain track. This model was so successful that it is still available today. Plasser & Theurer still remains the only supplier of ballast cleaning machines designed to clean under turnouts and crossings today.

The RM 74 was the highest production ballast cleaning machine on narrow gauge until the introduction of the RM 900 ballast cleaning machine (Figure 411). The RM 900 is a large machine which is only suitable for high traffic density lines with short maintenance windows.

Figure 411: The RM 900 Ballast Cleaning Machine

Between the production capacities of the RM 74 and the RM 900 is the RM 80 ballast cleaning machine. This machine can also be made to fit on narrow gauge, albeit with a different design, and is the most used ballast cleaning machine on standard gauge worldwide. On standard gauge there is however a large variety of models available depending on requirements.

2. BALLAST CLEANING MACHINE COMPONENTS

All Plasser & Theurer ballast cleaning machines work on the same principles with a few differences that bring about their difference in production capabilities. Figure 413 illustrates the working components of the RM 74/RM 76/RM 80 ballast cleaning machines (RM74 illustrated).

Figure 414: The Cutting Chain Being Connected Inside the Cutter Bar

2.1. Cutter Bar

The cutter bar is a casting which guides the excavating chain underneath the track. The cutter bar is inserted under the track before the occupation starts and before the ballast cleaning machine arrives on site.
In Figure 414 the cutterbar and the excavating chain can be seen during the process of connecting.

Figure 412: RM80 on Cape Gauge in New Zealand

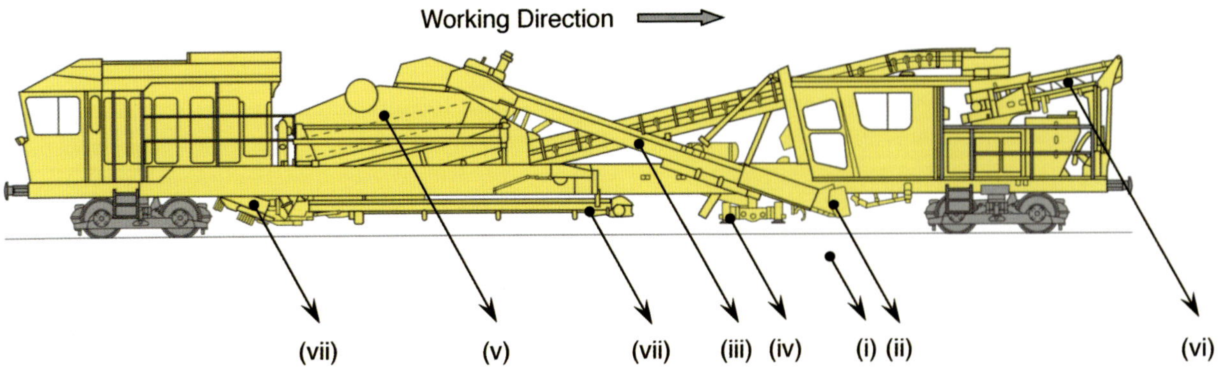

Legend:
(i) Cutter Bar
(ii) Excavating Chain
(iii) Guide Chute
(iv) Lifting Unit
(v) Screen Box
(vi) Spoil Conveyor
(vii) Distributor Conveyor

Figure 413: The Components of the RM 74/RM 76/RM 80 Ballast Cleaning Machines

The cutting depth can be controlled electronically to ensure a smooth longitudinal formation surface as can be seen in Figure 415. This is an important feature that prevents water from collecting on an uneven formation and to prevent the rails from copying the uneven formation during settlement.

The cutterbar also reinstates the formation cross fall to a side drain as illustrated by Figure 416. This cross fall is a prerequisite for good drainage.

The ballast cleaner can be fitted with a wide cutterbar (up to 7.7 metres) for cleaning turnouts and a narrow cutterbar for cleaning ballast in tunnels, bridges, next to platforms and other restricted areas.

However, for the cleaning of turnouts, more than just a wider cutterbar would be required. The chutes also need to be adapted for the longer reach. This is, however, a very important feature of ballast cleaning machines fitted with this option.

Specialised ballast cleaners for screening under turnouts have also been developed which works on a completely different concept than conventional ballast cleaners. Refer to paragraph 3 below.

Figure 415: A Level Formation Visible Behind the Cutter Bar

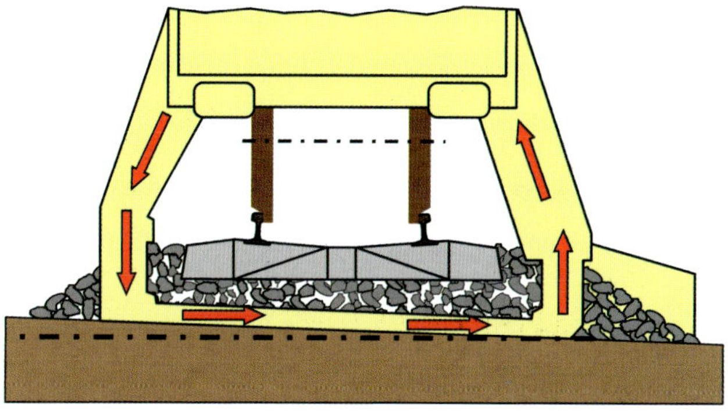

Figure 416: Cross Fall Reinstated by the Cutter Bar

2.2. Excavating Chain

When the occupation starts, an endless excavating chain (Figure 414) is connected inside the cutter bar. It is the hydraulically driven rotating excavating chain that excavates the entire ballast bed down to formation level. The chain consists of fingers (Figure 417) which break up the encrusted ballast and scraper plates (shovels) which transport the excavated material to the screens for separation.

Depending on the machine model, the chain is between 200 mm and 350 mm high. A special design chain of 150 mm is also available for excavating low ballast depth in, for example, tunnels. The chain scraper plates are wear-resistant castings, the fingers hardened and the chain links are connected by locking pins.

Figure 417: The Excavating Chain Consisting of Scraper Plates and Fingers

2.3. Guide Chutes

The excavating chain is guided up a guide chute [the up chute, (iii) in Figure 413] to a sprocket that drives the chain and back down the down chute.

At the beginning of the work shift, the cutter bar is connected to the up and down guide chutes using quick couplers. The excavating chain, which remains inside the chutes during travel and staging, is connected inside the cutter bar. For tensioning the chain, the down chute is hydraulically adjustable.

The top ends of the up and down guide chutes are pivoted to the machine frame (Figure 418) and allows the chutes to be displaced laterally without changing the formation cross-fall or the excavation depth. In so doing the chutes can be adjusted to reach the required vertical and lateral position to avoid obstacles or to position the excavating unit in the middle of the track during work in curves.

When the undercutting of the ballast bed starts, the chain conveys the excavated ballast in the up chute directly into the screen box for machines equipped with single screen boxes or onto a conveyor belt for machines equipped with two screen boxes such as the RM 900.

For protection against dust, noise and for safety reasons, as well as maintenance purposes, the chutes are fitted with hinged access doors on top.

Hydraulically operated plough flaps (see Figure 414) on both the up and down chutes make it possible to increase the excavation width thereby optimising the retention of available ballast.

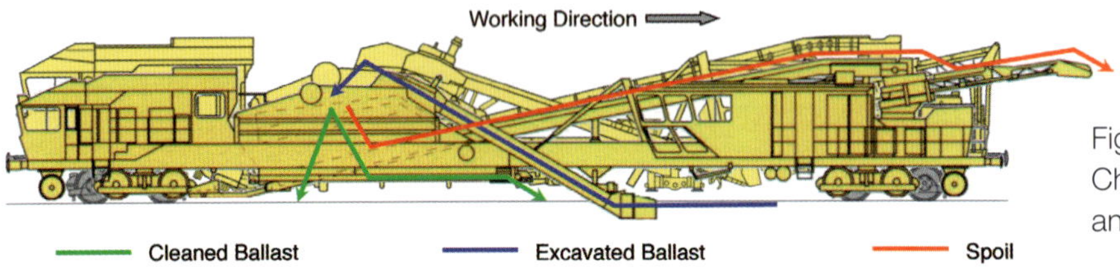

Figure 418: The Excavating Chain is Guided by the Up and Down Chute

Figure 419: The Flow of Excavated and Screened Ballast (RM 80)

Figure 420: The RM 900 Achieves Higher Production Due to the Greater Screen Area of Two Screen Boxes

2.4. Lifting Unit

The lifting and slewing unit of all ballast cleaning machines is located at the cutter bar (see Figure 417) and serves to support the track over the excavation working gap, to reduce the cutting depth by lifting the track where there is inadequate ballast and to avoid obstacles by slewing the track when required.

High production ballast cleaning machines, for example, the RM 900 are equipped with a second lifting unit located in front of the rear bogie which is integrated with the ballast ploughing and levelling unit. This reduces the lifting required by the tamping machine after ballast cleaning since the unit will lift the track to the required height to allow ballast returned to the track to flow underneath the track, saving time and increasing overall production.

2.5. The Screen Box/es

On the smaller RM 74/RM 76 and RM 80, the excavated material is transported by the excavating chain via the guide chutes up to, and dropped directly into a vibrating screen box (see Figure 419). The screen box is fitted with screens of different mesh sizes.

The RM 900 has a higher production rate due to, among other, the larger excavating chain and the greater screen area of two screen boxes. The excavated ballast is therefore conveyed on a central conveyor belt onto a longitudinally adjustable conveyor belt (see Figure 421), which equally distributes the ballast onto the 1st screening unit and via another conveyor belt onto the 2nd screening unit. Each screening unit works on the same principle, as explained for the smaller machines.

2.6. Spoil Conveyor

The spoil rejected from the screen box/es can either be discharged next to the track (Figure 422) or onto spoil conveying systems, such as MFS's (Figure 423), in front of the machine. Follow the red line on Figure 419 and Figure 421 to see the flow of the rejected spoil.

When spoiling material to the side, the spoil conveyors can be slewed to avoid obstacles such as overhead mast poles.

Some ballast cleaners have the added feature that when total rejection takes place, no material passes through the screening unit ensuring that the screens will not clog and slow down the process.

Figure 422: The RM 900 Using One of Two Spoil Conveyors to Spoil to the Side

Figure 423: RM 900 Discharging Spoil Into MFS Wagons

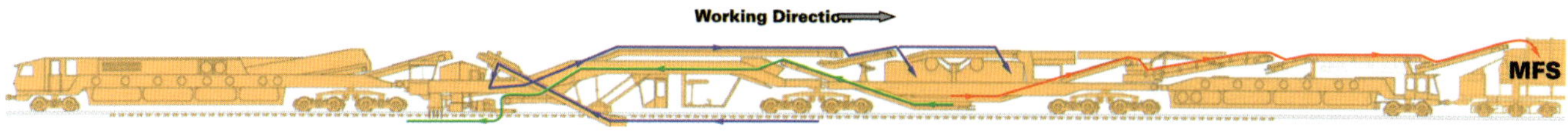

Figure 421: The Flow of Excavated and Screened Ballast (RM 900)

2.7. Distributor Conveyor

The re-usable screened ballast is returned to the track by the distribution conveyor (Figure 424). Follow the green line on Figure 419 and Figure 421.

The distributor conveyor discharges the ballast in front of a plough which, in the case of the RM 900, is equipped with roller lifting clamps and hydraulically adjustable flaps.

Figure 424: Distributor Conveyor on the RM 900 Discharging Re-Usable Ballast in Front of a Plough

Figure 425: RM 74 Working under Excessively Dry and Dusty Conditions

Figure 426: HGR Track Lifting Machine

On the smaller machines the ballast is discharged behind the cutter bar and also directly below the screen boxes to prevent too much ballast being placed under the sleepers. The plough assembly controls the height of the track and the adjustable flaps ensure that the ballast is distributed uniformly under the whole cross-section of the track and around the sleeper ends. The condition of the track behind the RM 900 ballast cleaner is therefore already in a good standard (provided that sufficient ballast is available), minimising tamping and the amount of lift required by the tamping machine.

Some high production ballast cleaners can also be fitted with conveyor belts to receive ballast from MFS wagons behind the machine. The new ballast can then be discharged in front of the ballast cleaner's plough to assist with a more even distribution of the ballast under and around the sleepers, once again saving regulating and tamping time behind the machine.

2.8. Dust Suppression Systems

The ballast cleaning machine may at times work in excessively dry and dusty conditions (see Figure 425). A dust suppression system consisting of spray bars may then be used which are located above the cutter bar, at the end of the main conveyor just before the screens and at the end of the 1st spoil conveyor.

Filtered water is pumped from a water tanker to the spray bar nozzles when it is required to suppress dust.

The system will require a large water tanker (40,000 litres) to provide water and this may need filling every day when in use. The water tanker would form part of the machine consist and would be located behind the ballast cleaner.

2.9. Track lifting Ahead of the Machine

Though not a component of the ballast cleaning machine, the HGR track lifting machine is an indispensable piece of equipment to lift the track before screening commences if there is a possibility that the cutting chain might cut into the formation.

3. BALLAST CLEANING OF TURNOUTS, SWITCHES AND CROSSINGS

Turnouts are at the receiving end of much higher dynamic loading than any other part of the track therefore also requiring more specialised and frequent maintenance intervention. The problem is however that turnouts, especially the latest modern designs on concrete sleepers are very heavy, long and wide requiring specialised machinery for the maintenance thereof.

Some ballast cleaning machines can be fitted with wider cutter bars adapted for turnout screening (see Figure 167). These machines may however experience difficulty when screening crossings or slips, especially on multiple lines such as metro lines where various different sets can be right on top of one another.

These were then mostly omitted or screened by hand, both options equally disastrous to the turnout in the long run.

Now continuous ballast screening in turnouts can be achieved with the development of the Plasser & Theurer URM 700 universal ballast cleaning machine (Figure 428) which allows screening any turnout, crossing or slip with infinitely variable cutting widths without obstructing the adjacent line, even on the multiple lines so often found on metro lines.

As can be concluded from the description "universal ballast cleaning machine', the machine can also be used for ballast cleaning in plain tracks or for shoulder cleaning alone. Due to the fast and independent installation and removal of the excavating guide bar, the URM 700 can also work cost-effectively on short sections of track (spot cleaning) which has not been possible before.

The working principle of the URM 700 is the same as conventional ballast cleaning machines. It consists of cutting chains to remove the ballast from underneath the sleepers down to formation level and transfer the material via a conveyor belt system to a screen box where the fine material is separated from the re-usable ballast. The fine material is conveyed again to material conveying systems ahead of the machine and the re-usable ballast is returned to the track.

The design of the excavating unit incorporates a horizontally circuiting excavating chain in a sword shaped cutterbar (see Figure 429). The excavating unit can be slewed either to the right or to the left under the skeleton track. Due to this pivoting suspension it is possible to carry out an infinite variation of the excavating width up to 6,100 mm. The excavating depth is variable and the formation cross fall can be adjusted.

Figure 427: Screening of a 1:20 Turnout Using the RM74 Ballast Cleaning Machine

Figure 428: URM 700 Universal Ballast Cleaning Machine with MFS Wagons Behind Supplying New Ballast and MFS Wagons Ahead Removing Waste Material

Figure 429: Sword Shaped Cutterbar

When slewed in, the excavating cutterbar is stabilised by a cylinder at the other end (see Figure 430). This provides for a higher working capacity. The excavating unit works without support only in short transition areas.

Shoulder excavating units on either side of the machine excavates the installation trench for the cutterbar at the start of work and pick up the excavated ballast during work (Figure 431).

Lifting units lift and support the track during the screening process. On turnouts the diverging track is also supported by the lifting unit.

4. THE BALLAST CLEANING PACKAGE

The Ballast cleaning machine does not work on its own. It needs a number of other machines to support it and collectively they are referred to as the ballast cleaning package. These machines are:

- MFS material conveying wagons – These wagons are used where spoiling of the fine material next to the track is not allowed or possible (CHAPTER 15).
- Ballast wagons – Due to the removal of the fine material, which to a large extent consists of crushed ballast, the ballast bed must be replenished with additional ballast to ensure the required ballast depth.

Figure 430: Stabilising of the Cutterbar at the Other End

- Ballast regulating machines – The returned ballast behind the ballast cleaning machine and the new ballast offloaded by ballast wagons must be regulated to the required ballast bed cross sectional profile. For this ballast regulating machines are used (CHAPTER 13).
- Tamping machines – These machines will lift, level, align and tamp the track to the required geometry (CHAPTER 11).
- Stabilising machines – The track behind ballast cleaning is not consolidated and will show reduced resistance to lateral displacement. Stabilising machines are used to consolidate the ballast which allows the track to be opened at normal section speeds directly behind the ballast cleaning operation (CHAPTER 12).

5. CONCLUSION

Ballast cleaning machines are vital tools in any railway today. Rapid deterioration and failure of track sections can normally be directly linked to inadequate ballast cleaning and the resultant intervention costs far exceed preventive maintenance using modern ballast cleaning machines.

Figure 431: Picking Up the Excavated Ballast

Figure 432: Ballast Cleaning Package Consisting of RM 74 Ballast Cleaning Machine, Ballast Tamping Machine, Ballast Regulator and Dynamic Stabilising Machine

Chapter 15
Spoil and Material Conveying

1. INTRODUCTION

In the absence of advanced spoil removal systems in the past, it was normal for ballast cleaning machines to spoil the fines screened from the ballast bed alongside the track. This had numerous disadvantages, including:

- The spoil from ballast, which has been treated with toxic herbicides for weed control, can and has been known to cause environmental pollution. The spoil is also contaminated with oil, grease and other contaminants that fell from or was spilled from the railway wagons and locomotives.
- Spoiling to the side can result in the blocking of drains (Figure 433). The contaminants inside the spoil material will also be dissolved in the drain water and spread over a wide area where the drain empties at the end of a cutting, especially if it runs off an embankment.
- Spoiling down an embankment may cause a landslide due to the added weight of the spoil, especially in the rain, which gives rise to erosion and structural problems for the track. See Figure 434.
- Ballast cleaning between tracks of multiple lines would necessitate spoiling between the tracks which will prevent adequate drainage, with disastrous consequences.
- The spoil may also cover other services such as cables and signal boxes causing additional maintenance restrictions and problems.
- Areas such as cuttings, tunnels, bridges and stations were often not screened due to unavailable space for spoiling. Some of the worst substructure problems today are in these restricted areas where screening had been omitted in the past.

The above problems caused by spoiling to the side as well as environmental awareness put pressure on railway owners to remove the spoil to approved and controlled spoil sites.

Figure 433: Ballast Cleaner Spoiling to the Side of the Track

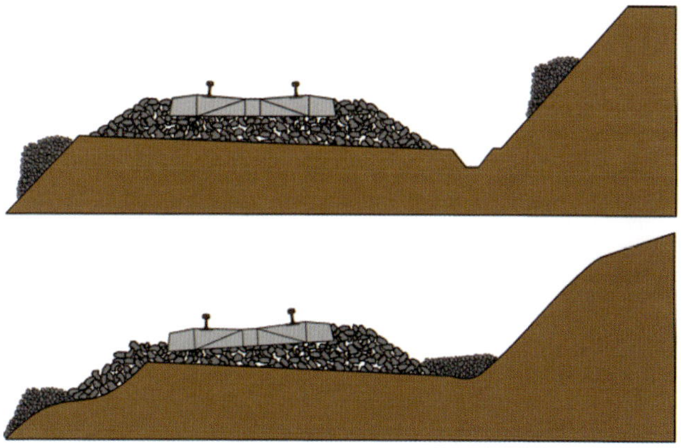

Figure 434: Problems associated with Spoiling to on Cuttings and Embankments

In the early years this led to the development of spoil conveying system consisting of wagons fitted with overhead conveyors (Figure 435). Mining type hoppers were also used (Figure 436) and are still being used today.

These systems permitted screening to be done in those previously omitted areas by providing storage of the spoil material for removal at the end of the shift.

Even though these systems allowed ballast cleaning of restricted areas such as multiple lines, tunnels etc, they are notorious for their disadvantages, among which:

- The mining hoppers have no way of transferring the spoil, which means that when a hopper is full, the ballast cleaning operation comes to a standstill at great cost while the full hopper is shunted out.

- Mining hoppers can also only be emptied from the bottom which requires the hoppers to go to a pit area where bottom discharge is possible. If this pit is far from the screening site it would either require many hoppers to cope with the production of the ballast cleaning machine or two locomotives to shunt the hoppers to the pit for emptying and returning them to the ballast cleaning site.

- High capacity cleaning, particularly in areas where the ballast is severely fouled, produces more spoil material than an overhead conveyor system of even 12 wagons can hold and therefore causes premature stoppage of the ballast cleaning operation. This increases the unit costs of ballast cleaning.

- An analysis of a screening contract in South Africa established that on average 22% of all working time is devoted to total rejection where poor ballast conditions are encountered. The total rejection capability of ballast cleaning machines far exceeds the overhead conveyor system loading rate of approximately 80 m^3 per hour. Thus for 22% of working time, the overhead conveyor system will restrict the ballast cleaner production and due to the rapid filling of the wagons, ballast cleaning will be aborted prematurely at great cost.

- After the shift each overhead conveyor wagon consisting of drop side rail wagons has to be individually emptied using a backhoe. On average this takes 25 minutes per wagon. This is usually only possible after hours and at night. Opening the wagon doors under load is also difficult and dangerous and serious accidents have occurred.

Figure 435: Spoil Removal Wagon with Overhead Conveyor System

Figure 436: Spoil Removal with Mining Hoppers

- The RM 74 ballast cleaner is limited to pushing a maximum of 6 full wagons. To maximize the ballast cleaner production, at least 12 wagons will be required. This necessitates changing 6 full wagons with 6 empty wagons since emptying a wagon during the shift is not possible. On average up to 2 hours per shift can be lost to change the two sets, depending on the distance to where the second set is staged.

- The overhead conveyor system proved to be very unreliable. Due to its design, the breaking down of one wagon paralyses the rest of the system.

- The large number of wagons causes staging problems.

A system was required without the disadvantages associated with the overhead conveyor system and mining hoppers which lead to the development of the MFS range of spoil conveying systems with hopper sizes up to 68 m^3.

The MFS concept has evolved with different specialised designs becoming available and more and more uses for the MFS conveyor systems being investigated for their productivity, safety and cost effectiveness.

2. MFS CONVEYOR SYSTEMS

MFS wagons for spoil removal are not self-propelled though self-propelled MFS wagons with crawler tracks are available for specialised off-track work at formation rehabilitation sites. Following are some of the MFS range that is available:

Figure 437: RM 900 Ballast Cleaning Machine Spoiling into MFS 38 Spoil Conveyor Systems

2.1. Conventional MFS Spoil Conveyor Wagons

The MFS is an open, high-sided hopper wagon, with a floor-mounted main conveyor belt which covers the entire width of the hopper and wear resistant sheets on the inner sides of the hopper. Each MFS wagon is equipped with its own diesel engine to drive the conveyor belts. The rotation speed of the conveyor belt can be controlled. In addition, the hopper wagon is equipped with a slewing transfer conveyor belt at the one end which operates independently from the floor conveyor belt to either transfer its load forward, or discharge its load to the side.

The transferring, slewing and discharging ability is the strength of the MFS as a spoil removal system when working with the ballast cleaner. In the following process description, three wagons are used though six wagons are generally used during ballast cleaning operations. However, the number of wagons used depends on the distance to the spoil disposal site and the tractive limit of the locomotive.

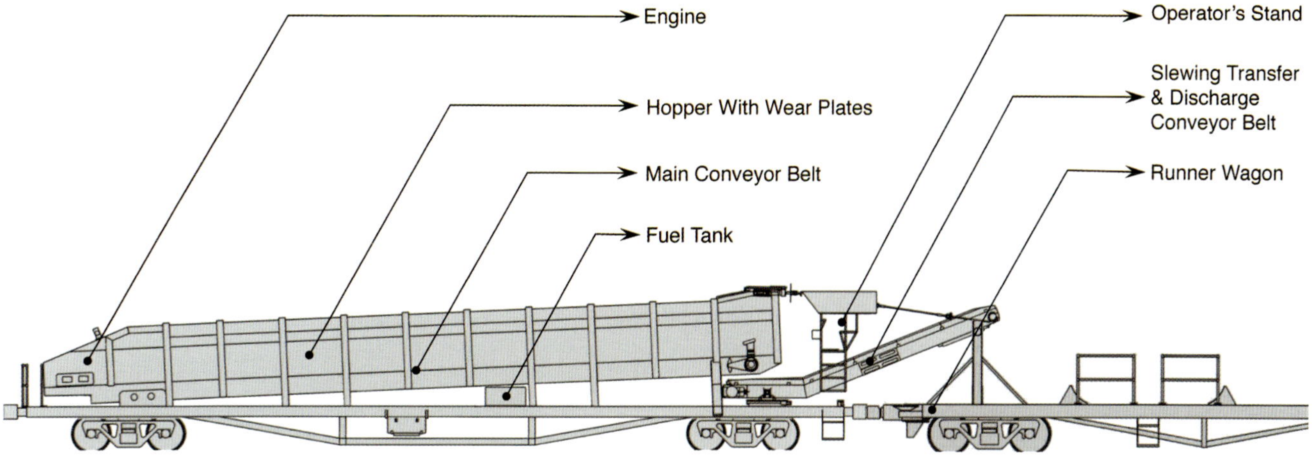

Figure 438: Layout of the MFS 40 Conveyor Systems

Figure 442: The Conveyor is Set to a High Speed to Transfer the Spoil to the Next Hopper

Figure 443: Filling of the Last Hopper to Capacity by Slowing Down the Conveyor Belt Speed

- The MFS wagons arrive on site in train formation together with the ballast cleaner, ballast wagons and a tractor loader on the runner wagon. The runner wagon is required due to the transfer and discharge conveyor belt protruding over the buffer (refer to Figure 438) and also to transport the tractor loader which is used at the spoil disposal site to level the spoiled material.

- The ballast cleaner excavates the ballast down to formation level and separates the ballast which fits within ballast stone envelope (refer back to Figure 107) from the fine material (spoil). The fines are spoiled into the hopper closest to the ballast cleaner (MFS 1 in Figure 441). The main floor mounted conveyor belt is set to a high speed to transfer the spoil to the next hopper as fast as it is received using the transfer conveyor belt (see also Figure 442.

- The spoil is transferred from hopper to hopper to the last hopper in the train (MFS 3 of Figure 441). The speed of the floor mounted conveyor belt of the last hopper is controlled by the operator to fill it to capacity (see also Figure 443).

- When the last hopper is full, the slewing transfer conveyor belt of the second last unit is switched off and the rotation speed of that floor conveyor belt controlled to allow it to also fill to capacity. When the second last unit is full, the next hopper will be filled to capacity. This process of filling the hoppers will continue until a predetermined number of hoppers are full. This predetermined number depends on the

Figure 439: The Ballast Cleaner and MFS Wagons Arrive from Staging Site in Train Formation

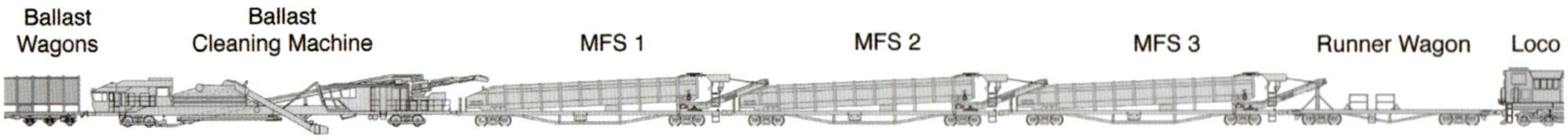

Figure 440: The Spoil is Transferred from Hopper to Hopper to the Last Hopper

distance to a suitable spoiling site where the hoppers can be emptied. The further the spoiling site, the more hoppers need to stay behind with the ballast cleaner so as not to prematurely stop the ballast cleaning process.

- The full units are uncoupled and moved to the prepared and environmentally approved spoiling site while the ballast cleaner continues to screen and discharge spoil into the remaining units of which the floor conveyor belt of the last unit is controlled to fill it to capacity.

- At the spoiling site the MFS wagons slew their transfer conveyor belts to the side and discharge their loads of approximately 40 m³ to a stock pile at 4,400 mm from track centre (Figure 445). All the wagons can be emptied simultaneously within approximately 6 minutes and is the single biggest advantage of the system.

  A tractor loader can be transported with the MFS train consist on the runner wagon if required. At the spoil site the tractor loader can be offloaded to level the spoil or load the spoil into trucks, depending on the spoil site requirements. It is also possible for the MFS to offload the spoil straight into trucks though this will delay the offloading time.

Figure 445: MFS Wagon Contents are Disposed by Slewing the Transfer Conveyor Belt to the Side

- When the empty wagons return to the working site, they can be coupled to the wagons that remained behind.

- The content of the units that remained behind with the ballast cleaner is transferred again to the last unit. This process of filling and emptying can continue uninterruptedly without affecting the ballast cleaning operation.

Figure 441: The Last MFS Wagons is Filled to Capacity

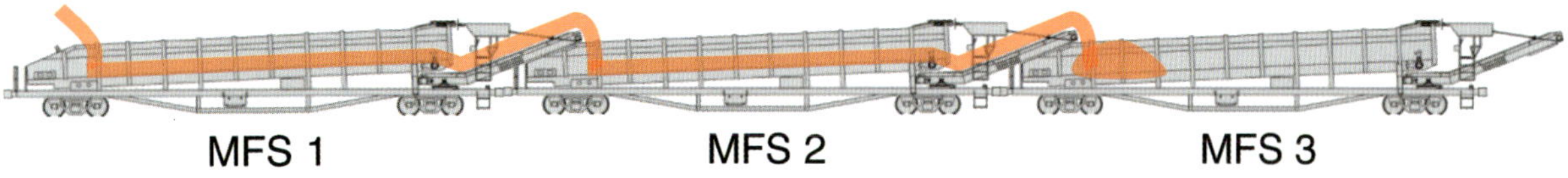

Figure 444: Full Wagons are Uncoupled and Hauled to the Spoil Site

Figure 446: Empty Wagons Return to Ballast Cleaning Site

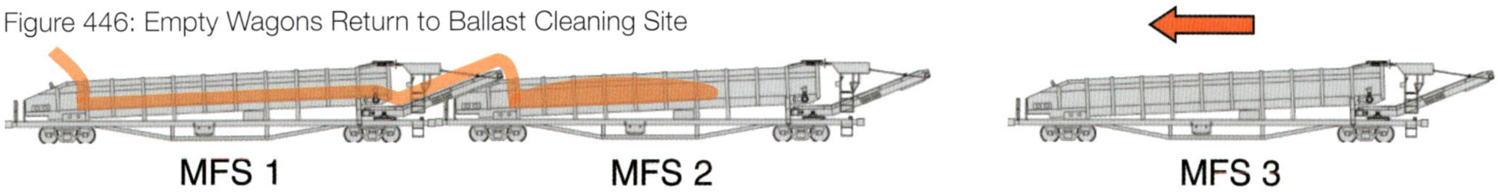

Figure 447: The Content of the Units that Remained Behind is Transferred to the Last Unit

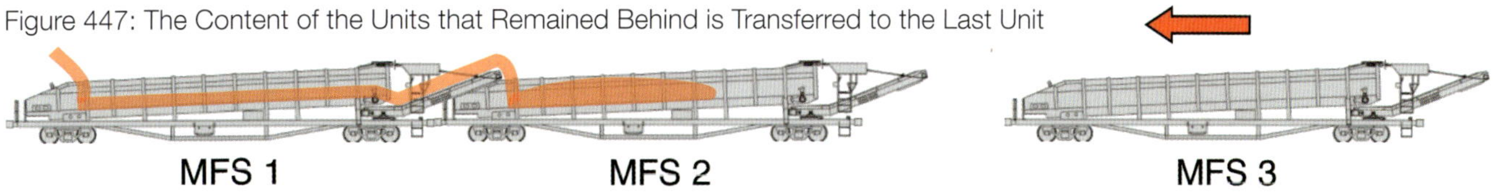

2.2. MFS Wagons with Crawler Tracks

Standard railbound MFS wagons, as described in paragraph 2.1, have been fitted with crawler tracks in Europe to move off track under its own power. There it is known as the MFS 40 ZW.

This allows the MFS wagon to travel to a major track rehabilitation site on rail bogies as part of the train consists and at the working site to lower the crawler tracks and travel into the excavation self-propelled on the crawler tracks. This allows efficient and uninterrupted removal of large volumes of excavated material (Figure 448) following the same working principle as described above. Similarly, the MFS 40 ZW wagon can also be used to offload new backfill material or ballast.

Figure 448: A MFS 40 ZW Wagon Travelling on Crawler Tracks to the end of the Excavation

3. OTHER APPLICATIONS OF MFS WAGONS

The MFS range of conveyor systems is known for its superior spoil-conveying capabilities from ballast cleaning sites. However, the system is very versatile and also ideally suited for the conveying and offloading of new backfill material and ballast at rehabilitation sites. Not only is it ideally suited to do so but would be a large improvement on some conventional methods in use.

3.1. Offloading Backfill Material

During formation rehabilitation very large volumes of backfill material may have to be offloaded, especially if extensive damage is experienced with the in-situ soil and deep excavation and backfilling is required. In such cases up to 3.2 m^3 of layered backfill material can be offloaded per metre. On a typical 10 kilometre rehabilitation section 32,000 m^3, or 3,200 tipper truckloads of backfill material alone can be offloaded.

When geo-cells (Figure 449) are used to strengthen the formation, the fact that vehicles cannot drive on the geo-cells adds another element to the problem. Back tipping (Figure 450) is required which slows down the formation rehabilitation process substantially and delays the re-opening of the line for traffic.

Figure 449: The Use of Geo-Cells for Formation Strength

Figure 450: Back Tipping Required if Geo-Cells are Used

MFS wagons can be used to bring in the backfill material where double lines are available at much improved production rates. Even on single lines production can be improved by using MFS 40 ZW wagons (Figure 448).

Under the protection of a short occupation on the adjacent line, MFS wagons filled with graded backfill material will be shunted in. The transfer conveyor belts of the MFS wagons will be slewed over the construction area and the backfill material offloaded simultaneously from all the MFS wagons straight into the excavation and/or straight into the geo-cells (where used) while the MFS wagons are shunted along the excavation gap, spreading the material evenly along the length to the required layer thickness (Figure 451 and Figure 452).

The number of required MFS wagons depends on availability and the length of the construction. If, for example, twenty MFS wagons are utilised at 40 m^3 per wagon, approximately one kilometre of a 200 mm layer can be offloaded within 30 to 60 minutes. This would have required approximately 80 tipper truck loads over a 2 day period to achieve the same.

The added value is the evenness of the offloading process, which reduces the levelling time and also prevents the tipper trucks travelling on the compacted subgrade (where geo-cells are not used).

Using the MFS system for offloading material would require a safe method for loading the MFS wagons. The use of a loading station is advisable to avoid damage to the MFS wagons (Figure 453).

Figure 452: MFS Wagons Used to Offload Backfill Material

Figure 453: Loading Station for MFS Wagons

Figure 451: Using MFS Wagons to Import Backfilling Material from the Adjacent Line

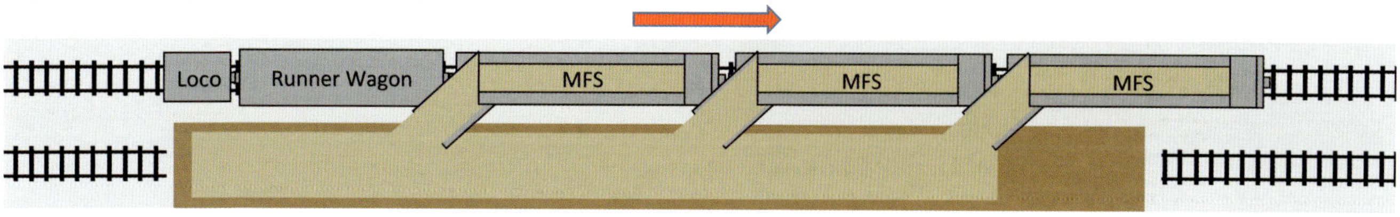

3.2. Offloading Ballast

Offloading the first layer of ballast poses a similar problem as with backfill material. The layer works would have been perfectly levelled and compacted to specification. Travelling with tipper trucks on the formation would damage the layer works. The ballast is therefore often offloaded using ballast wagons from the adjacent line (mostly for turnout formation rehabilitation though). Due to the method of offloading with ballast wagons, various passes are required (see Figure 454). It still poses a real problem for derailing of the ballast wagons.

Figure 454: Offloading Ballast from the Adjacent Line Using Ballast Wagons

Figure 455: Skid Steer Loader Used to Level the Ballast

The next problem is to level the ballast. For this, skid steer loaders are sometimes used, which cause less damage to the layer works but are still not ideal. The process is also very slow and during this time the line is closed.

Similar to the offloading of backfill material, the MFS wagons can be filled with ballast stone and the same offloading process from the adjacent line followed to quickly and evenly spread the 1st ballast stone layer, causing no damage to the formation layer and in a time that cannot be equalled.

4. CONCLUSION

The MFS range of conveyor systems have moved beyond being the ultimate in spoil removal from ballast cleaning operations. With the addition of self-propelled and MFS wagons with crawler tracks and using these wagons for material offloading as well, the MFS range is a very versatile piece of machinery which benefits the environment, track downtime and costs.

Chapter 16
Rail Surface Profiling

1. INTRODUCTION

Rail replacement is a very expensive exercise due to the cost, transportation, installation and welding of the rail. It is also disruptive to traffic and may have a large opportunity cost as well. It is therefore of utmost importance that the service life of the rail be extended for as long as possible. Maintenance to the rail and its complex crown profile is therefore one of the most important track maintenance activities to reduce the overall track lifecycle cost.

There are several elements to rail maintenance, most of which are discussed in CHAPTER 3 paragraph 2 and other parts of this manual. This chapter will focus only on on-track mobile rail grinding as a mechanised rail maintenance (preventive or corrective) activity and rail milling and rail planing as mostly corrective activities.

2. THE MAGIC WEAR RATE

Excessive forces created by the high traction capabilities of modern locomotives, train speed, long train length, high axle loading, poor rail/wheel interface design or irregular wear, poor maintenance, lack of greasing on curves, etc. lead to contact stresses exceeding the limits of the steel, resulting in rail surface fatigue (rolling contact fatigue) manifested in corrugations, head checks, spalling, shelling, indentations in the rail steel, cracks, etc. Refer to CHAPTER 4, Fundamentals of Rail and Wheel Interaction and CHAPTER 3 paragraphs 2.17 to 2.22 for detail regarding rail defects, their causes and remedies.

These surface irregularities cause higher dynamic loading and damage to all the track components, poor steering, rolling noise and the cracks may propagate into the rail resulting in rail breaks. If left untreated it will cause uneconomically high lifecycle costs of the entire track structure.

The rail running surface must be maintained by means of rail profiling to remove the work-hardened rolling contact fatigue (RCF) layer and defects and to reinstate the required rail crown profile at set intervention intervals (profiling cycle) or based on measured condition obtained from a track measuring vehicle (CHAPTER 9 paragraph 4.2 and also paragraph 6.3.2).

The rail profiling strategy and programme should however aim to limit metal removal since profiling is an artificial form of wear. The strategy should consider the targeted rail profile, wear development, crack growth, traffic type, speed and axle loading, different wear rates between tangent track and curves, etc. A too long profiling cycle (advanced rail wear, poor rail profile, heavy RCF and advanced crack growth before intervention) will require aggressive grinding with coarser stones and multiple passes which will increase rolling noise and wear to rolling stock wheels due to the rougher surface of the rail crown. If the profiling cycle is so long that the volume of material to be removed exceeds the capabilities of the grinding machine, corrective methods of milling or planing will be required which removes very large volumes of material. On the other hand a too short profiling cycle (more preventive in nature with one or two grinding passes) may cause accelerated artificial wear and can potentially be more disruptive to traffic.

Somewhere in-between is the "magic wear-rate (MWR)" which is the balance between maximum rail life (measured in million gross tons) and the total of both natural and artificial wear required to manage rolling contact fatigue (RCF). The MWR recognises that if there is no wear the rail will fail due to contact fatigue but excessive wear reduces rail life. Artificial wear is required since natural wear is too low to control RCF. This concept is illustrated by Figure 456 (International Heavy Haul Association's "Guidelines to Best Practice for Heavy Haul Railway Operations – Management of the Wheel and Rail Interface").

From Figure 456 the following can be observed:

- Point (ii) illustrates a very low wear rate of the rail as would typically be observed on tangent track purely due to natural wear from wheel/rail interaction which would result in a very high wear life. However, the natural wear will be slower than the development of RCF and crack growth, resulting in rail breaks which require premature rail replacement at a very short rail life at point (i).

- In a maintenance regime where rail profiling takes place to remove the RCF before excessively large cracks propagate, the total wear (artificial due to grinding together with natural wear) will be higher, illustrated by the curve moving towards point (v) from point (ii) as the grinding cycle shortens resulting in a lower rail life due to increasing material removal.

- The curve along points (iv) to (i) illustrates the rate at which rail life will reduce due to RCF as the total wear rate reduces.

- Where the two curves cross at point (iii) the highest rail life will be achieved for the lowest wear. This point is called the "magic wear rate" and is the point that maintenance managers are striving to achieve to reduce lifecycle costs and increase rail life to the maximum.

Detailed studies for heavy haul best practice for grinding suggests removal of the approximately 0.2 mm rolling contact fatigued layer for every 20 to 25 million gross ton of traffic. Other railways have experimented with over 50 million ton with good results. There is also high speed grinding that can be considered which may have a limited or no influence on traffic though it is required more frequently versus several passes of the conventional grinding machine at longer grinding cycles but with a larger influence on traffic and a full maintenance occupation. There are various variables that will influence the required grinding cycle, the type of machine to be used, the profiles to grind to, etc. Determining the grinding strategy for specific line requires specialised consultation combined with some experimentation but for economic reasons the limit for the amplitude of corrugations, for example, should not exceed 0,1 mm after which grinding becomes a priority to prevent damage to the rest of the track infrastructure. In other words, grinding is not purely a time based activity but condition should also be factored into the decision-making model.

Studies have shown that this strategy may extend the service life of the rail which will have a major influence on the lifecycle cost of the rail and the whole track infrastructure. The elimination of rail surface defects and the resulting dynamic forces will also lead to a 30 to 50% increase in the durability of the track geometry, further reducing the track lifecycle cost.

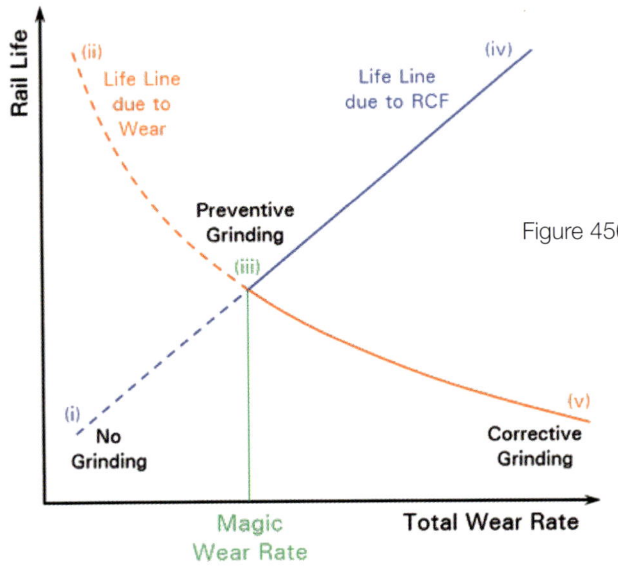

Figure 456: Position of the MWR

3. RAIL GRINDING

Rail grinding can be condition-based and planned when using the results from the Infrastructure Measuring and Recording Vehicle, if equipped with rail profile and/or corrugation measuring equipment. Refer to CHAPTER 9 paragraphs 4.2 and 4.4. The rail profile measurements can be enhanced with the use of additional Rail Grinding Analysis Software (see CHAPTER 9 paragraph 6.3.2) to determine the rail grinding programme.

Rail is ground to a specific profile template which was designed by considering contact dynamics to ensure ideal wheel and rail contact. The grinding template for the crown profile will also differ between tangent (straight) track, flat curves wit radii between 800 and 2000 metres and tight curves with smaller radii. The high rail and low rail profiles may also differ in the same curve.

There are generally two types of grinding machines used, machines with oscillating stones and machines with rotating grinding stones.

3.1. Grinding Machines with Rotating Stones

Speno International is the largest manufacturer of grinding machines using rotating stone technology. This is the most often applied method for preventive or corrective maintenance of the rail surface profile.

Depending on the model, these machines can be fitted with 16, 24, 32 or 48 driven grinding wheels, or even more with the combination of units in tandem, with which up to 0.13 mm of metal can be removed (48 stone machine) per grinding pass. To grind to a target rail profile the grinding wheels are set at different angles and facets.

A computer-aided control system presses the grinding wheels onto the rails at a set pressure to achieve optimum grinding at the minimum number of required passes to achieve the target profile.

The grinding speed per pass is up to 8 km/h reducing to 6 km/h where more corrective grinding is required.

In turnouts the grinding constraints are more complex than on plain track for which special turnout-grinding machines have been developed by Speno. Specially adapted grinding stones enable efficient grinding and unrestricted re-profiling even in guardrail zones.

Grinding generates sparks which poses a fire hazard, especially in arid areas. It is therefore not uncommon for grinding machines to be accompanied by a fire team with water hoses or for grinding to take place at night during which dew will reduce the likelihood of fires or when flames will be more easily noticed.

3.2. High Speed Grinding

High speed grinding (HSG) is a fairly new technology which is gaining acceptance in Europe and provided by Vossloh Rail Services.

The basic idea of HSG is to remove the worn and brittle surface layer (the rolling contact fatigue layer) of the rails which is the breeding ground for defects such as head checks, etc. The HSG machine removes approximately 0.03 to 0.05 mm of material per grinding pass and would therefore require two passes to remove the work-hardened layer approximately every 10 to 15 million gross ton traffic. A third smoothing pass is sometimes used to combat noise pollution.

Figure 457: Rail Grinding Machine

The process is based on 96 grinding stones mounted to 4 units which are towed along the rail under a certain lateral orientation angle thus inducing both a rotation and a grinding motion without additional power. The machine has a working speed of 60 to 100 km/h.

HSG maintains the rail's good surface condition but the rail profile will still deteriorate over time. Conventional grinding might therefore still be necessary to re-profile the rails.

Figure 459: GWM 250 Rail Grinding Machine

3.3. Grinding Machines with Oscillating Stones

This type of grinding machine uses oscillating grinding stones in the longitudinal direction of the rail in combination with the continuous forward motion of the machine and hydraulically adjustable vertical load.

The GWM series is equipped with either 6, 12 or 30 brick-shaped grinding stones per rail treating the rail crown surface with a working frequency of 4,5 Hz. The grinding speed is 1,2 km/h during which 0.05 to 0.07 mm material is removed per pass. The shape of the stones, however, does not allow for re-profiling of the crown.

With 6 stones rigidly mounted on one frame and a reference length of 2,400 mm, the contact surface between the grinding stones and rail guarantees that the grinding stones will not follow the corrugated or wavy longitudinal profile of the track.

The quality produced is exceptional and is the reason why this machine is often used for noise reduction. The high surface quality will also delay the formation of corrugations to a minimum.

The grinding units are able to grind plain track, track with guard rails, as well as turnouts.
This grinding system is the only system available which offers a spark-free operation utilising a water supply for lubrication of the grinding stones with the added advantage of dust elimination, which is an important factor in highly populated areas. This process also emits less noise than grinding machines utilising rotating stones.

Figure 458: High Speed Grinding Machine

4. RAIL PLANING

If the grinding programme was not maintained or is not effective, the crown profile can deteriorate beyond the economic intervention by a grinding machine due to the number of passes that would be required to reinstate the targeted profile. Short of rail replacement, the rail planing machine (Figure 460) can be used as a corrective measure to remove a considerable volume of material in just a few passes to ensure that some life can still be gained from the rail.

The SBM 250 rail planing machine (Figure 460) is made up of two sections, the four-axle planing machine and the single-axle swarf collector. Two planing units are positioned between the bogies, one per rail, and are supported on the rail by a number of tool guide rollers that are hydraulically pressed onto the rail. Each planing unit has a tool holder for securing the tool inserts. Accurate transverse positioning of the tools on the rail surface is achieved using datum rollers on the field side of the rails.

A testimony of the quality produced by the rail planing machine as a corrective activity is South Africa's Sishen to Saldanha iron ore line where extensive hunting damage was experienced in the 1990's. The SBM 140 rail planing machine was used to prevent premature rail replacement. At the time Transnet carried out suitability tests which showed that the planing machine provided a very accurate transverse position for the rail wheel contact area (Figure 461) with a consistent accuracy of ± 1 mm.

Figure 460: SBM 250 Rail Planing Machine

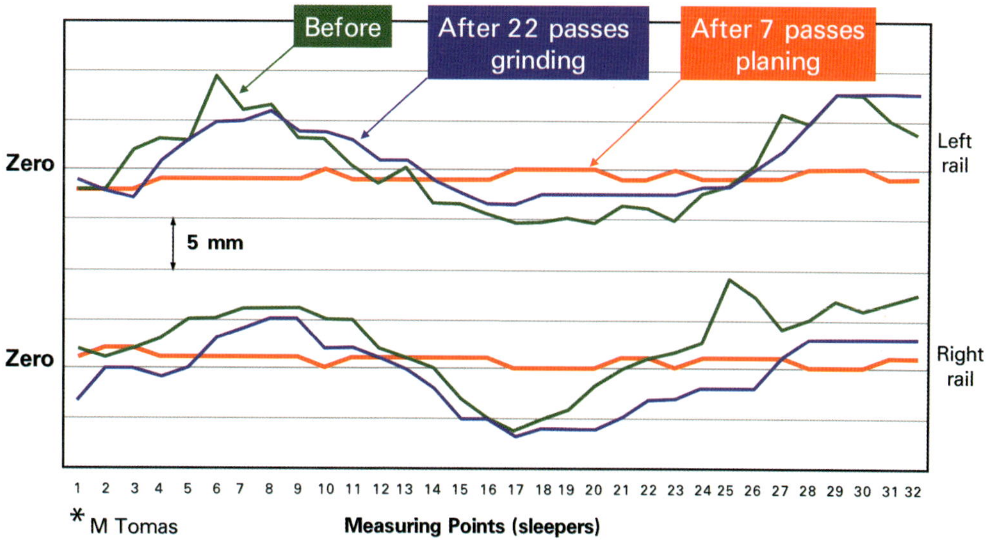

Figure 461: 1/20 Contact Point

Plasser South Africa experimented with various tool bits and cutting positions to achieve optimum production and quality for the specific problem at hand. The following illustrations show the cutting positions that were used and is only placed here for illustration purposes as it would not necessarily be the same for all rail conditions.

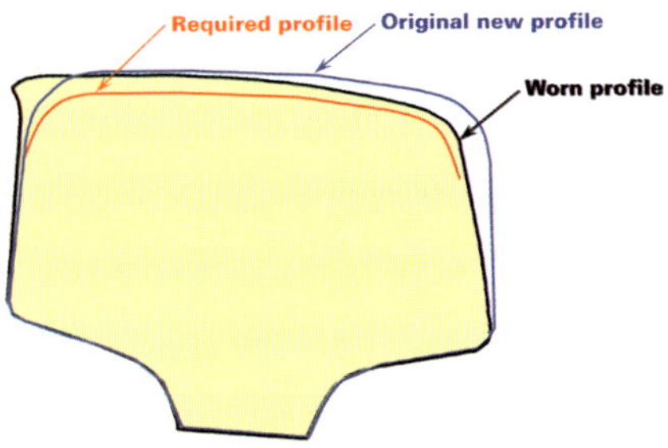

Figure 462: Illustration Showing the Original Profile, the Worn Profile and Required Profile after Planing (the Excessive Volume of Material to be Removed Clearly Visible)

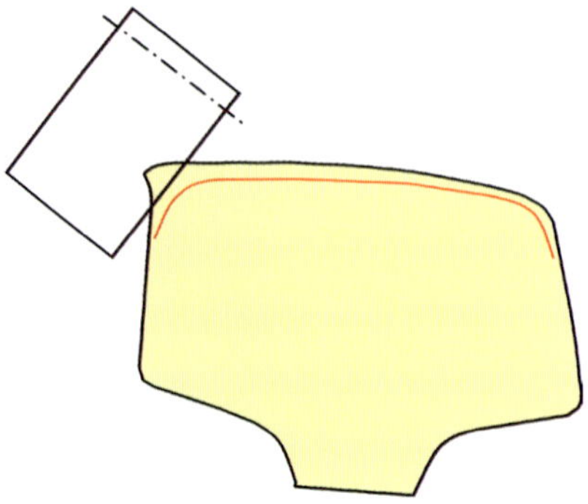

Figure 463: Burr Cutting Wheel is Applied First to Remove the Metal Flow on the Field Side

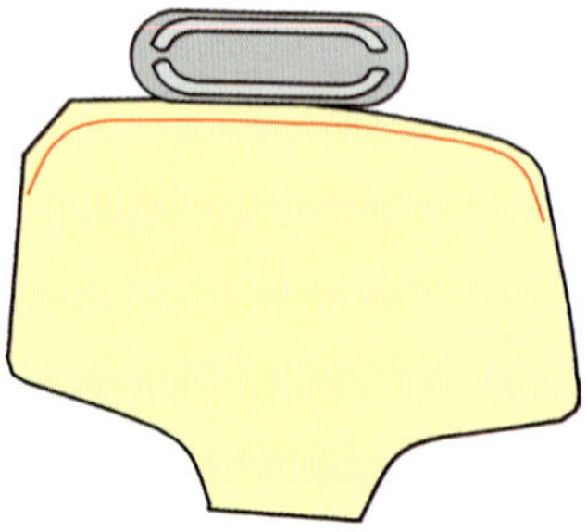

Figure 464: Centre Cut to Remove Corrugations and Create Smooth Running Surface for Guide Rollers

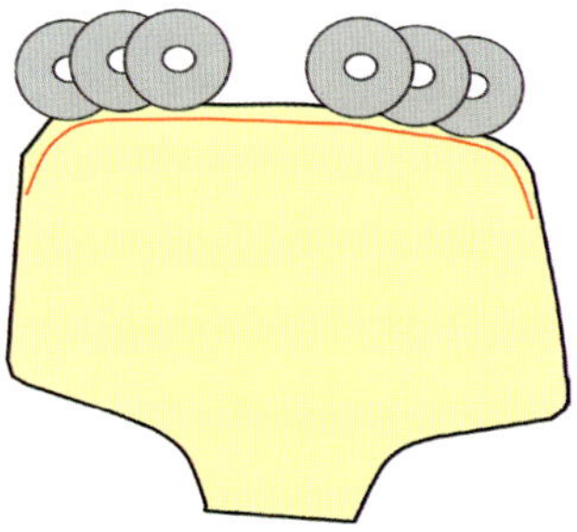

Figure 465: Button Tool Cut for Bulk Material Removal

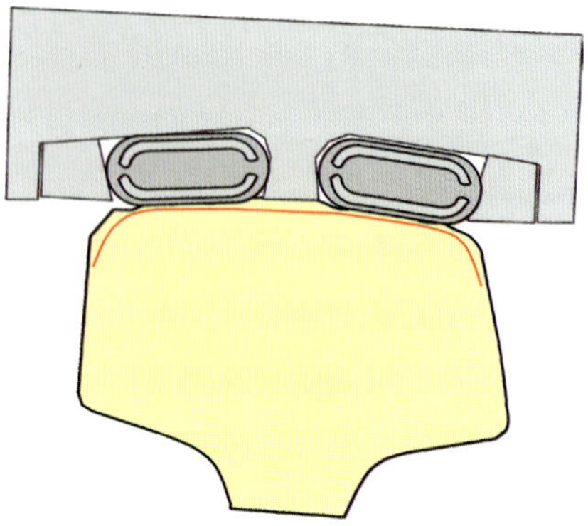

Figure 466: Flat Tool in Double 11° Arrangement

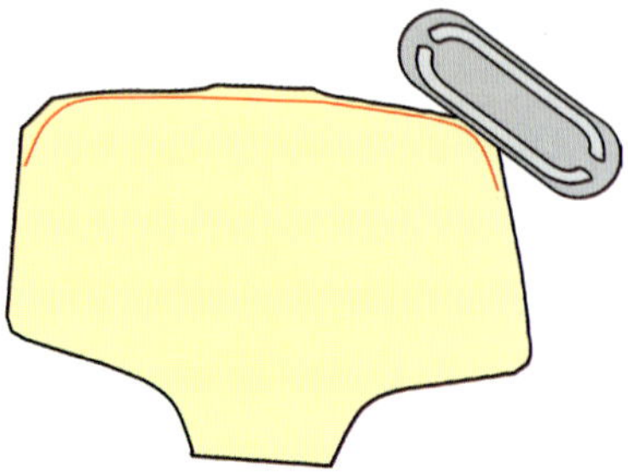

Figure 467: Flat Tool in 45° Arrangement on the Gauge Corner

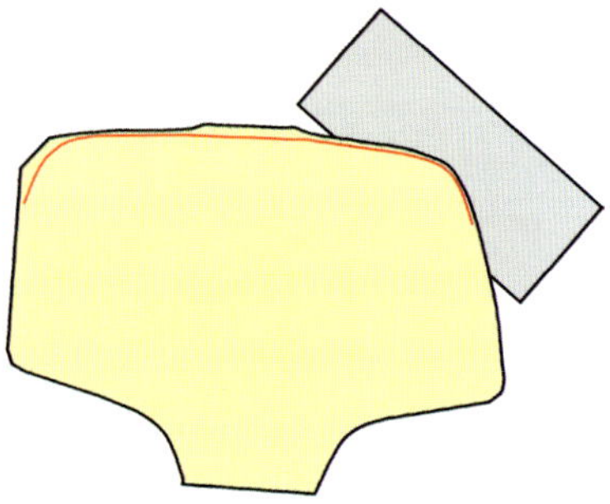

Figure 468: 15 mm Radial Tool for the Gauge Corner

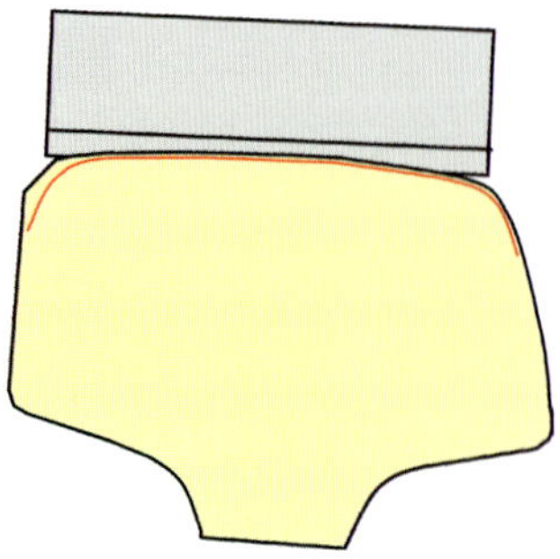

Figure 469: 300 mm Radial Tool for Contact Band

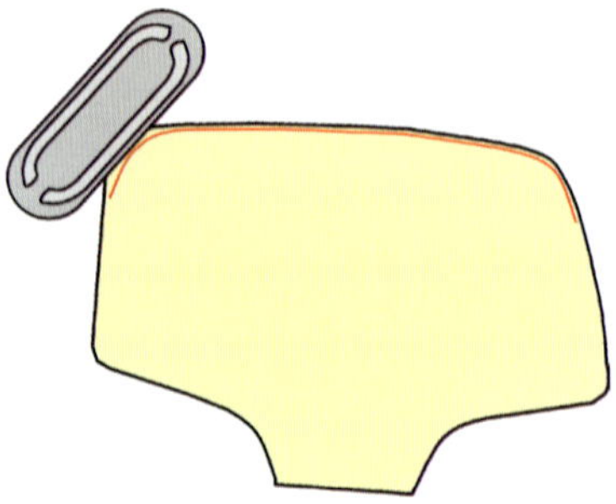

Figure 470: Flat Tool in 45° Arrangement on the Field Corner

To evaluate the effectiveness of rail planing a few track sections were selected during a study, all showing similar results. The sections were selected for their conformity to certain requirements to eliminate any other factors that could have contributed towards improvement of the lateral acceleration of the line except for planing and tamping. The graph in Figure 472 shows the comparative RMS runs for one of the sections which clearly show the improvement in lateral acceleration after planing.

5. RAIL MILLING

Rail milling as a mobile rail profiling method (Figure 473) has developed over the past 10 years and is therefore a relatively new technology. However, it is a very fascinating development as it is equally capable of preventive rail profiling competing with rotating grinding methods as it is ideal for corrective rail profiling competing with rail planing.

It does however have certain advantages over the other two processes under certain circumstances. The most important of which is that it can remove between 0.3 and 1.5 mm off the running surface and between 0.3 and 3.0 mm (partial 6.0 mm vertical measured is possible) off the gauge corner in one pass while at a longitudinal production speed of between 400 and 1800 m/h depending on the size and depth of faults in the transverse profile of the rail head. With the new finishing system of Maschinenfabrik Liezen und Giesserei GmbH (MFL) milling of 0.1 to 0.2 mm is possible with a surface quality of 3 µm. In one train these two systems together can therefore mill from 0.1 mm to 1.7 mm in one pass.

Figure 471: Very Smooth Final Finish Achieved with Rail Planing

Figure 473: Rail Milling Machine

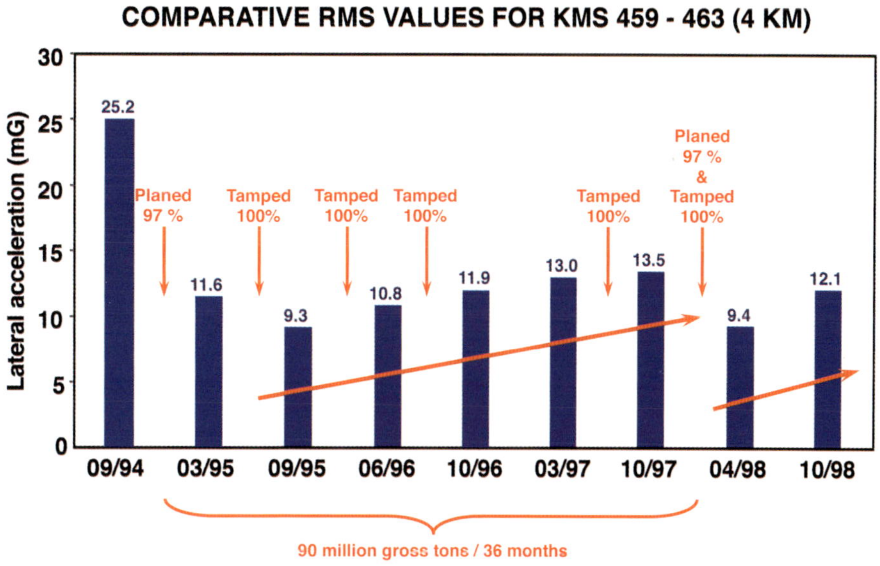

Figure 472: Comparative RMS Values for Kms 459–463

Other advantages of milling as a rail profiling method are that:

- it is environmentally friendly as –
  - it does not create sparks so no fire protection is required;
  - no dust is created which makes it more suitable in tunnels and bridges;
  - it does not require any chemicals that contaminates the ballast; and
  - 99% of the metal removal (chips) are picked up during the process which can be recycled;
- it does not change the microstructure of the steel due to the low heat generated;
- the process produces very little noise;
- production is not affected by the weather;
- there is no need to remove rail bonds;
- milling can continue through level crossings;
- sections with check rails can also be milled without restrictions;
- milling is not affected by tight curves or steep gradients (up or down hill); and
- the overheads does not need to be switched off.

Milling machines, such as those provided by MFL (Figure 473), consist of 2 semi-permanently coupled units. The one unit carries the milling heads and operating cab and the other the engine and swarf container.

The milling unit is equipped with 2 milling heads per rail and work in one direction only. The milling heads can be operated individually (left or right rail), in pairs or all together. Each milling head diameter is 1,400 mm and equipped with up to 60 cartridges with cutting tools.

Each cartridge is equipped with several straight and one or two radius cutting tools (Figure 474) which together profile the entire contour of the rail head.

The cutting tools produce a very precise transverse profile with a nominal deviation not worse than ±0.2 mm. All the working processes of the machine can be monitored on screens by the operators.

The surface ripple left by the milling tools is at right-angles to the length of the rail and is sufficiently small (within European requirements) due to the large diameter milling head. It therefore does not require subsequent finishing treatment though for a super fine finish (Figure 475) finishing can be incorporated in the process.

The chips are picked up with a vacuum system at the milling head and transported to the swarf container from where it can be emptied at the end of the shift.

Figure 475: Final Product with Finishing

Figure 474: Examples of Cartridges with Cutting Tools

6. SELECTING THE APPROPRIATE PROCESS

With the different rail profiling processes overlapping in capabilities and advantages it is not that clear which is the best process to use. It depends on the application (economics not considered):

- Light preventive profiling – Rotating grinding can take off very little material and if the profiling strategy is to carry out regular light grinding single passes, rotating grinding could be the best option. High speed grinding is also a strategy that can be considered.
- Preventive profiling – If rolling contact fatigue (RCF) is allowed to develop, i.e. a longer profiling interval strategy, the rotating grinding process will require more than one pass to remove the head checks and cracking that would have developed. Under these circumstances the milling process may be a better option.
- Corrective profiling – If severe RCF and deep head checks have developed, the number of passes using a rotating grinding machine may make it unsuitable for which milling or planing will be a better option.
- Noise reduction – The oscillating grinding process has shown the best results for smoothing the rail head to reduce noise. This machine can work in conjunction with rotating grinding or the milling process.
- Profiling new rails – It has been proven that grinding the rail head as quickly as possible after installation of new rails will provide better results. Rotary grinding will remove the least material and would be the better option.

It is therefore safe to say that the process of choice will depend on the maintenance strategy, on-track maintenance time available and number of machines available.

7. CONCLUSION

Rail profiling is regularly carried out by all modern railways today for the extension of rail life and reduction in lifecycle cost. A lack of rail maintenance is not only expensive in material terms but may also be expensive in terms of rolling stock loss in derailments and the cost of life.

If preventive rail profiling was neglected, corrective profiling will become necessary which will remove large volumes of material to reinstate the targeted profile with a defined contact band between rail and wheel and remove the surface cracks that may have grown to considerable depths. This should however be avoided due to the volume of material that must be removed and the resultant reduction in expected rail life.

Chapter 17
Rail Flash Butt Welding

1. INTRODUCTION

Fish-plated track joints have always been the area of concentrated maintenance effort and costs. Discontinuity of the track running surface produces dynamic impact loads which increases track settlement and results in an uneven track longitudinal profile. The pumping action at the joints also accelerates rail fatigue, rail damage, sleeper breakage, fouling of the ballast at the joint due to ballast crushing and increased stress on the formation. If the joint is poorly maintained as well (Figure 476), the track deterioration in the area of the joint is very fast and very destructive.

Great technical progress in rail manufacturing was made from the first cast-iron rail of 914 mm in length laid in 1798 to today's continuous welded rails (CWR) which eliminates the joint entirely (Figure 477). See also CHAPTER 3, paragraph 2.

The introduction of CWR has vastly improved the geometric stability of the track structure. Nevertheless, the localised change in the metallurgical character of the rail that can occur in the vicinity of thermite welds can ultimately lead to a loss of running smoothness which can initiate corrugations and promote the development of localised track geometry irregularities, pulverisation and movement of the ballast and damage to fastenings and sleepers.

This has, however, improved quite dramatically with the development of the flash butt welding process in 1930. This process with its limited change to the metallurgical character of the rail, reduces the possibility of loss of running smoothness, thus producing a very high quality continuous welded rail.

Flash butt welding charges the two rails to be welded at very high amperage at low voltage. The two rail ends are brought close to one another without touching, which produce an arcing between the rail ends, basically an electrical short circuit. The resistance of the current flow between the two rails causes rapid heating of the two rail ends (Joules Law) to melting temperature.

Based purely on a time-controlled and programmed basis for the specific rail type, the heated rail ends are forced together (butted) in an upset stroke at high pressure during which all impurities are forced out causing the ends to join. The upset metal is trimmed off immediately after the upset stroke.

Flash butt welding does not require any foreign filler material as the rail itself is used as a welding compound. The result is a near flawless weld without a significant change in the hardness curve or metallurgical character of the rail. The heat-affected zone (spheroidised zone) is limited to the smallest possible area (Figure 478) thereby reducing the possibility of loss of running smoothness dramatically. This is in stark contrast to thermite welding where the spheroidised zone can be between 12 and 22 mm, causing cupping at these softer areas.

Figure 476: Poorly Maintained Rail Joint

Figure 477: Flash Butt Welded Joint

Initially flash butt welding was limited to stationary welding depots which required the transportation of the long welded rails to site and thermite welding of the long rails into continuous welded rail, thereby losing some of the advantages offered by flash butt welding. Plasser & Theurer therefore developed the first rail-mounted self-propelled flash butt welding machine in 1973 (Figure 479) with an integrated K 355 welding head developed at the J O Paton Institute in Kiev, Russia. The mobile flash butt welding machine produced similar quality welds as found in a controlled factory environment.

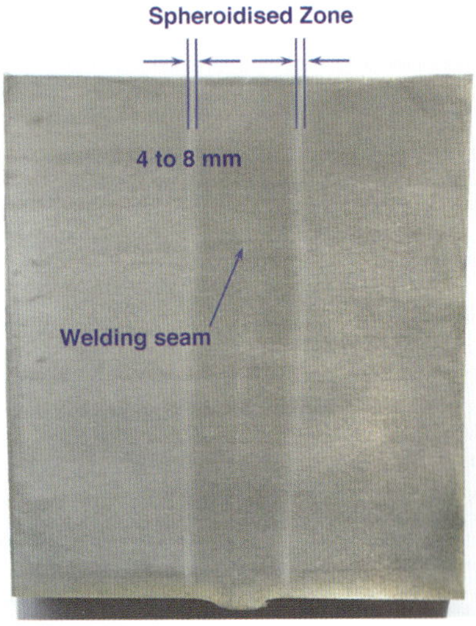

Figure 478: Cross-Section of a Flash Butt Welded Rail Joint

Figure 479: Flash Butt Welding Machine with the K355 Welding Head

Figure 480: The APT 1500 R Flash Butt Welding Machine

Over the next couple of decades the welding head remained basically unchanged and literally hundreds of machines using this welding head have been sold worldwide by various suppliers. Plasser & Theurer alone has sold 115 flash butt welding machines to 41 countries making it the most successful manufacturer of mobile flash butt welding machines in the world. The most important of changes were to the control system for the programming of the welding process which kept up with the advancement in computer technology and lessons learned in practice. However, a European welding standard, EN 14587-2 published in 2009, changed the status quo by specifying additional requirements from mobile flash butt welding machines.

Though the Russian welding head partly meets the new requirements, Plasser & Theurer subsequently set out to design and develop a new welding head introduced on the new APT 1500 R (Figure 487) welding machine to completely comply with these EN standards and to set new quality standards based on their extensive experience in this field. This was also an opportunity to build various features into the new head to improve productivity and quality, some of which was not available at all with the K355 head and had to be manually carried out.

2. FLASH BUTT WELDING MACHINES

2.1. APT 500 Flash Butt Welding Machine

There are many APT 500 flash butt welding machines working around the world still using the K355 welding head and a similar machine is still available. This machine is for all practical purposes still perfectly suitable for lines in developing countries where the EN standards are not enforced.

2.1.1. The Electrical Plant

The three phase 50 Hz alternator with a rated output of 150 kVA is designed as a revolving field generator i.e. with a fixed anchor and a revolving magnet wheel. The primary voltage is set between 340 and 380 volts depending upon the cross-section of the rail being welded.

2.1.2. Telescopic Crane

During travel the welding head is completely enclosed inside the machine. At the work site the cab is opened and the welding head lowered by means of a double arm telescopic crane from which it is suspended.
The cab in which the welding head is housed can also slide backwards to enable the crane to move through 90° to reach a distance of 3,250 mm from centre of track. The machine can therefore be applied either for welding rails in situ, for welding pre-deposited rails alongside or in the track centre or for setting up a stationary welding depot.

Figure 481: Cab Housing the Welding Unit Being Opened

2.1.3. The Welding Head

The welding head weighs 2,550 kg and consists of a mechanical lever system. All movements of the welding head are carried out hydraulically. Rails with a cross-section area of up to 10,000 mm^2 can be welded. S60 rails have a cross-section area of 7,702 mm^2.

Figure 482: Telescopic Crane Lowering the Welding Unit

2.1.4. Clamping Jaws

The welding sequence begins by clamping the rails in two pairs of contact jaws. The lever system applied achieves a clamping force of approximately 125 ton. During clamping the rail is thus adjusted accurately in terms of level and line. These copper jaws are also the electrodes which are water-cooled.

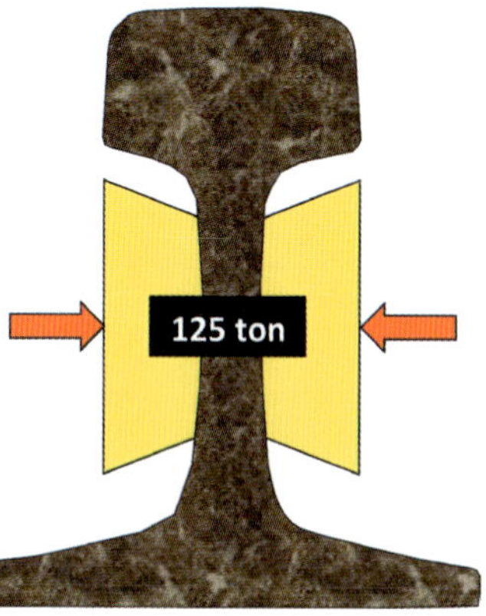

Figure 483: Clamping Jaws Gripping the Rail

2.1.5. Welding Transformers

The secondary circuit of the power plant is integrated in the welding head in order to transform the welding current from 380 volt to approximately 6 volt and the welding amperage from 395 ampere to more than 20,000 ampere. The welding transformers are also water-cooled.

2.1.6. Integrated Upset Trimming Tools

A hydraulic upset trimming device is integrated into the welding head to remove the welding upset immediately after the weld has been completed. This device perfectly follows the contours of the rail. It follows therefore that it has to be changed for every rail profile.

Figure 484: Trimming Tools Seen Between the Clamping Jaws

2.1.7. Program Unit

A PLC and an analogue controller controls the welding process automatically without any further operator input after the start button has been pressed.

Welding programs are pre-programmed for many different types of rails, are stored and can be extended at any time. User-defined settings can also be stored in addition to this.

2.1.8. Pulling Device

The machine is equipped with a hydraulic rail pulling device to bring the rail into the correct position directly before welding. This device is capable of pulling lengths of up to 216 metre 60 kg/m rails. The rails must, however, be placed on rollers.

2.1.9. Printed Record

In contrast to other welding methods, the electric flash butt welding machine can produce a report of each welding sequence which is checked by specially developed software. The recorded parameters are:

- welding current intensity
- compression force
- displacement (feed travel) – This recording shows the movement of the two halves of the welding unit relative to one another. This indicates the burn-off, the compression displacement, the total material consumption and the trimming process.

Figure 485: Hydraulic Pulling Device and Rail Fitting

2.2. APT 1500 R Flash Butt Welding Robot

Subsequent to the publication of the EN14587-2 flash butt welding standard in 2009, Plasser & Theurer designed and developed an all new welding head, the APT 1500 R welding robot (Figure 486) introduced on the APT 1500 RA (Figure 480) welding machine to completely comply with the EN standards and to set new quality standards.

Subsequently the APT 1500 R robot was also introduced in road/rail (APT 1500 RL) and container designs (APT 1500 RC) (Figure 487). The container design has further flexibility that it can be loaded either on a railway wagon or flatbed road trailer.

Some of the most important benefits and features of the APT 1500 R welding robot include:

- The process of aligning the rails has been included in the automatic work sequence to very high accuracy thereby eliminating the time-consuming manual alignment associated with the Russian welding head. Auxiliary clamping jaws raise the rails and automatically and accurately align the rail ends to required slight upward inclination using a total of eight distance transducers (Figure 488).

Figure 486: The APT 1500 Flash Butt Welding Robot

The rails are also horizontally aligned using a horizontal-action centring device. The upward inclination is required to ensure that the rail top remains within the specified fine tolerance in the longitudinal level after cooling down. Due to the shape of the rail and the irregular volumetric distribution of material over the cross section, the rail foot and base will cool down first while the thicker rail crown will take longer to cool. The natural consequence is a depression of the rail in the longitudinal plane over the weld which can be avoided by welding the rails at a slight inclination.

- The rail clamping jaws grip the rail with a 3,200 kN clamping force using four hydraulic cylinders. This allows high traction forces without slippage to weld closure welds.

- The design of the rail clamping jaws eliminates the need to remove rail markings from the rail web before welding, reducing preparation time.

- A unique design feature and a first for flash butt welding is the 1,000 Hz frequency used for the current supply from a 400 V alternating current generator. The transformers on the welding robot rectify the alternating current into direct current and transform the voltage down to low voltages at very high amperage. The high frequency produces a very low residual ripple of the electric welding voltage which has a very positive effect on the welding quality. The continuous welding current is 26,000 A with a maximum welding current of 75,000 A.

- During trimming of the excess upset (Figure 489) the weld is held in compression by all the rail clamping jaws which remain closed. The rail clamping jaws and trimming tools are positioned so that it is possible to trim the weld immediately after the upset stroke without having to open the rail clamping jaws. The trimming tools are also designed to apply the load on the welded joint in the compression direction. This fulfils one of the important requirements of the EN standard. The trimming tools perfectly follow the contours of the rail.

- An integrated no-contact temperature-measuring device allows monitoring of the rail cooling temperature after welding to ensure that high-alloy rails, such as chrome manganese rails, can be made to follow a specific time-temperature-transformation cooling curve after welding (see CHAPTER 3 paragraph 2.3), therefore ensuring the desired perlite crystalline structure of the rail material form unhindered.

No post heating material or blankets are required since the cooling rate is slowed down by the welding robot using current pulses.

Figure 487: The APT 1500 R Flash Butt Welding Robot in Road/Rail and Container Design

Figure 488: Transducer and Centring Device

Figure 489: Upset Removed by the Trimming Tools

- If required for the accelerated cooling of head-hardened rails, there is a device incorporated that supplies a stream of air directly to the area of the rail head (air quenching).

- Arguably the most important new and unique feature of the APT 1500 is its ability to perform closure welds (the last weld to connect two sections of continuous welded rail) without an additional pulling device, thereby eliminating a thermite weld. The pulling cylinders exerts a traction force of 1,500 N at stroke distance of up to 250 mm which allows welds to be made at rail temperatures up to 20°C below neutral rail temperature. After closure welding at below neutral temperatures, the rails are held by the rail clamping jaws until the weld can withstand the pulling forces in the rail. Since the rail clamping jaws and trimming tools work independent from one another, trimming can take place directly after the upset stroke without releasing the rail clamping jaws.

Figure 492: Welding Robot Retracted into the Machine for Travel

- A programmable control system is used to guide the process which meets the stricter hardware requirements (extended temperature and oscillation resistance ranges, etc.) in compliance with EN 50155. Welding programs which include all parameters for fully automatic welding are pre-programmed for many different types of rails and can be extended at any time. Flash butt welding is therefore not subject-

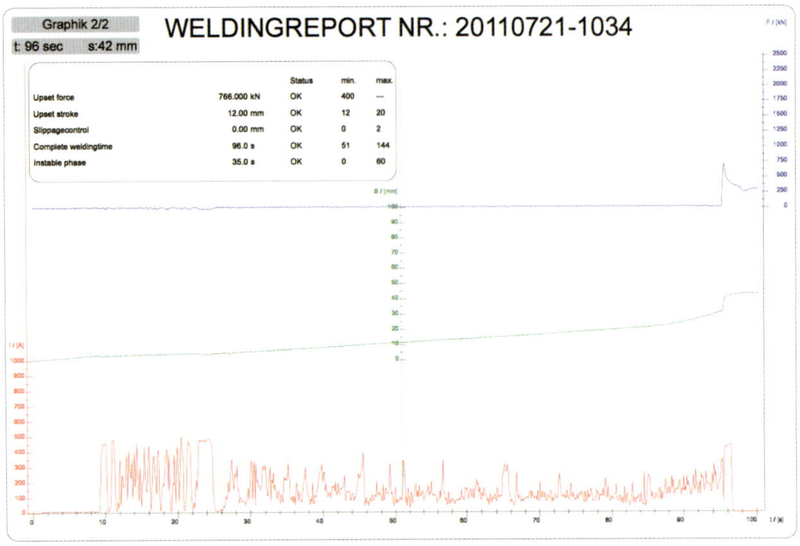

Figure 490: Welding Report

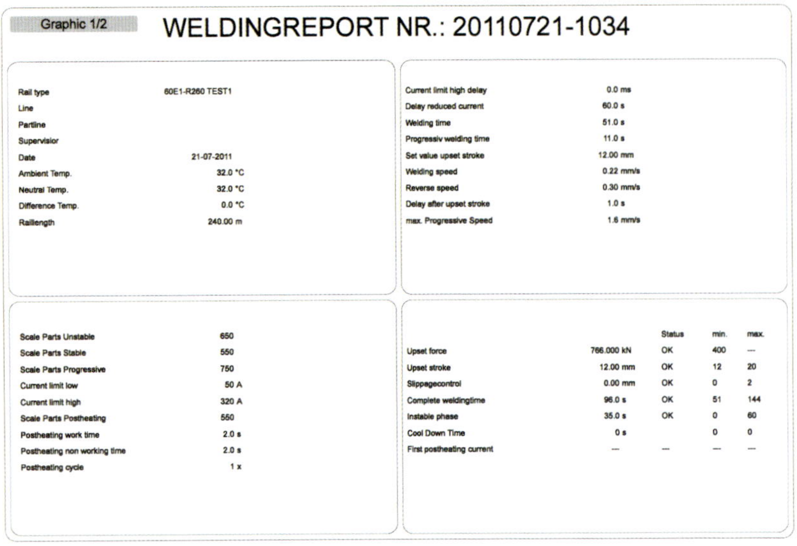
Graphic 1/2

WELDINGREPORT NR.: 20110721-1034

Rail type	60E1-R260 TEST1
Line	
Partline	
Supervisior	
Date	21-07-2011
Ambient Temp.	32.0 °C
Neutral Temp.	32.0 °C
Difference Temp.	0.0 °C
Raillength	240.00 m

Current limit high delay	0.0 ms
Delay reduced current	60.0 s
Welding time	51.0 s
Progressiv welding time	11.0 s
Set value upset stroke	12.00 mm
Welding speed	0.22 mm/s
Reverse speed	0.30 mm/s
Delay after upset stroke	1.0 s
max. Progressive Speed	1.6 mm/s

Scale Parts Unstable	650
Scale Parts Stable	550
Scale Parts Progressive	750
Current limit low	50 A
Current limit high	320 A
Scale Parts Postheating	550
Postheating work time	2.0 s
Postheating non working time	2.0 s
Postheating cycle	1 x

		Status	min.	max.
Upset force	766.000 kN	OK	400	—
Upset stroke	12.00 mm	OK	12	20
Slippagecontrol	0.00 mm	OK	0	2
Complete weldingtime	96.0 s	OK	51	144
Instable phase	35.0 s	OK	0	60
Cool Down Time	0 s		0	0
First postheating current	—	—	—	—

Figure 491: Welding Report

ed to operator errors. During welding the parameters are monitored constantly and if the specified tolerance thresholds and quality values are exceeded, welding is aborted automatically.

- All data relevant to the weld, such as operator, line, rail type, temperature, etc., are entered using the keypad and are stored simultaneously in the log in the same way as the machine parameters. The logs contain the text section and a recording of the weld in graph form (Figure 490 and Figure 491). This data is available at any time and continuous monitoring is carried out during the welding process. The software compares the actual data with the target data and shows the operator the state of the weld. The log is protected from manipulation and can be used on standard computers without additional troubleshooting software.

- During travel the welding robot is retracted into an enclosed compartment.

3. THE FLASH BUTT WELDING PROCESS

3.1. Arrival of the Machine on Site

The self-contained machine arrives on site with the welding robot completely enclosed to protect it against inclement weather and damage during transfer. At the working site the welding robot is extended using the crane specifically designed for this purpose.

3.2. Rail Cutting and Cropping

If new rails are welded, cutting and cropping would generally not be required but when rail defects such as cracks are repaired a minimum of 100 mm section from the extremities on either side of the defect (Figure 493) must be cut out.

If a thermite weld is removed, a minimum of 150 mm section from the extremities on either side of the thermite weld is cut out due to the heat-affected zone of the thermite weld.

When battered fish plated rail joints are welded, the battered tip must be removed. For flash butt welding at least 70 mm must still be left between the rail face and the 1st fish plate hole (Figure 494). If not, the rail must be cut behind the last fish plate hole.

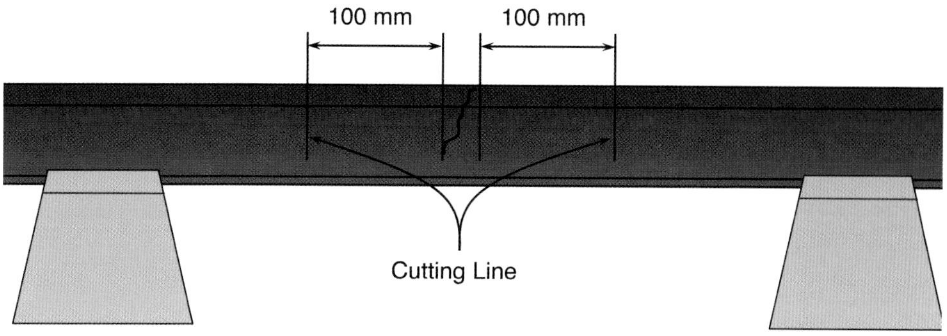

Figure 493: Cutting Out Rail Defects

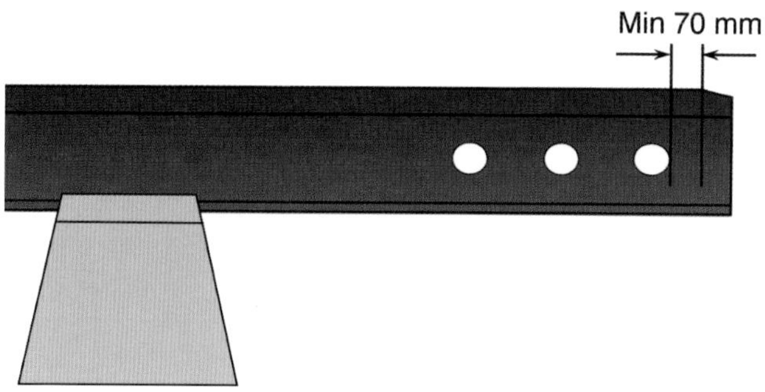

Figure 494: Material Required Between Rail Face and Hole

3.3. Removing Rail Fastenings

During welding there is a consumption of material due to the burn-off and compression of the rails. The consumption of material amounts to approximately 30 mm or less for the APT 1500. In order that the welding head can move the rails without hindrance, rail fastenings of the entire section have to be removed and the rail laid on rollers (Figure 495 and Figure 496) so that the frictional resistance between the rail base and the sleeper can be reduced to a minimum.

Figure 495: Robel Rollers

Figure 496: Vortok Stressing Rollers

Figure 497: Integrated Lifting Jacks (APT 1500 RA)

These rollers are inserted approximately every 10 to 20 sleepers. Vortok International developed a roller which does not require the rail to be lifted with the aid of a jack since the lifting process is incorporated in the roller (Figure 496). These rollers fit inside the clip shoulder, one on each side of the rail. The lifting process is therefore very quick and easy.

3.4. Preparing The Rail

Preparation of the weld involves grinding the rail ends (face) clean of any corrosion as well as the areas on the rail where the electrodes will make contact with the rail. Any arcing between the rail and the clamping jaws will prevent effective heating of the butting rail ends.

Any rail identification markings on the rail web must be removed to prevent damage to the copper jaws and to ensure good electrical contact. A minimum of 500 mm from the rail ends are ground. Using the APT 1500 welding robot there is no need to remove the identification markings on the rail. Slight grinding of the rail web is required to ensure electrical conductivity. The clamping jaws are universal and can be used for more than one rail profile.

3.5. Rail Alignment

Depending on the direction from which the rail will be pulled during welding, the machine wheels can be lifted clear off the rail using the integrated lifting jacks (Figure 497). The machine is capable of pulling from either side.

The welding robot is lowered over the two rails to be welded. The rails should be no more than approximately 250 mm apart (the maximum stroke of the cylinders of the APT 1500) for the machine to bring the two rails together within the defined gap of approximately 3 to 5 mm.

With the APT 500 welding robot the rails must be aligned manually using a straight edge. Good alignment is crucial for a quality weld since, if the rails are aligned at an angle, uneven preheating will result.

It is much quicker, however, with the APT 1500 since the auxiliary clamping jaws raise the rails and automatically and accurately align the rail ends to required slight upward inclination using the eight distance transducers.

The lateral alignment is the last step. The rail clamping jaws will hold the rails in this position which means that welds can even be made over sleepers.

3.6. Welding Phases

Once the welding process has been initiated, the welding procedure will follow the programme totally free from any operator input along very distinct phases which are:

3.6.1. Phase 1 – Preheating

The welding starts with the discontinuous preheating phase. At the start of this phase the welding head will bring the rail together to close the gap during alignment. The secondary welding current is controlled automatically, preventing local overheating of the ends.

Discontinuous preheating more than 50 seconds is not permissible. The duration of discontinuous preheating is not controlled, but regulates itself via resistance and temperature.

After sufficient preheating, continuous flashing is performed, which will start the time program after 8 seconds. The time setting is according to a predetermined program.

The preheating phase is composed of repeated pulses of high amperage. According to Joules Law, the maximum heat is produced on the points of maximal resistance. Maximum heat is therefore produced where the high current flows through a small area of contact when the rails are brought together.

3.6.2.Phase 2 – Flash Burning (Progressive Flashing)

During this phase the rails are pulled together at a constant rate and a continuous input of current for a uniform burn-off.

3.6.3. Phase 3 – Upset

The third phase is the upset phase. Once the required upset temperature of approximately 1,100°C has been reached, the rail ends are pulled and butted together with a force of 1,500 kN. The temperature is, however, not the trigger for the upset phase to start. It starts as the last phase of the fully automatic timed sequence.

The upsetting distance should be sufficiently high to prevent micro-porosity and non-metallic inclusions in the welding seam. The upsetting distance is, however, not set in terms of distance but rather in terms of pressure. Upsetting will continue until the pressure reaches its maximum when all the pasty metal consisting of slag and impurities has been squeezed out. The upsetting distance can be as much as 15 mm.

In this regard correct alignment plays a major role. As said earlier, alignment errors cause uneven preheating. If one part of the cross-section area – the crown for example – was at a lower temperature, it would absorb the whole upset force, leaving the other part – in this case the foot, inadequately butted.

3.7. Trimming

Directly after welding and before the welding head is released, the trimming device removes the welding seam with the aid of compression cylinders. This ends the welding procedure; the clamping jaws are released and the welding head lifted.

Figure 498: Welding Upset Visible Shortly After Trimming

3.8. Post-Welding Treatment of Chromium Manganese Rails

Standard rail types do not require any heat treatment after welding as their cooling rate between 800°C and 500°C is sufficiently slow to allow complete formation of a pearlitic microstructure.

However, alloy rails reduce the rate at which pearlite will form. To avoid the formation of martensite a controlled slower cooling cycle must be initiated. This process is automatically carried out by the APT 1500 R welding robot through continuous pulses of high amperage.

The APT 500 welding robot, however, does not have this feature and post-heating must therefore commence within 30 seconds after the weld. Exothermic heating powder is poured into a steel clinker former and ignited using ignition blocs. The powder burns at a temperature of between 370°C and 430°C. The clinker former must stay on for approximately 30 minutes, covered with an asbestos blanket. This will slow down the cooling sufficiently for complete pearlite formation.

The welded joint may not to be subjected to longitudinal or lateral stresses for a period of 12 minutes. Post-heating means that the machine cannot move forward as is usually the case. The hydraulic lifting cylinders will therefore lift the machine clear of the track to relieve the rail of the machine's load to allow the machine to pull the rail from the rear so as not to pull against the post-heated weld. After the weld the machine will move backwards to the next weld.

3.9. Finalising

Behind the welding machine the rail joints have to be ground according to the rail profile, true to shape and line within set standards.

When the sides of the rails are welded out of alignment up to a maximum of 0.8 mm, the joint is ground back at a slant not steeper than 1:500 on the running edge of the rail and to 1:50 on the field side. Welded joints which are out of alignment in excess of 0.8 mm in the crown are rejected, cut out and re-welded.

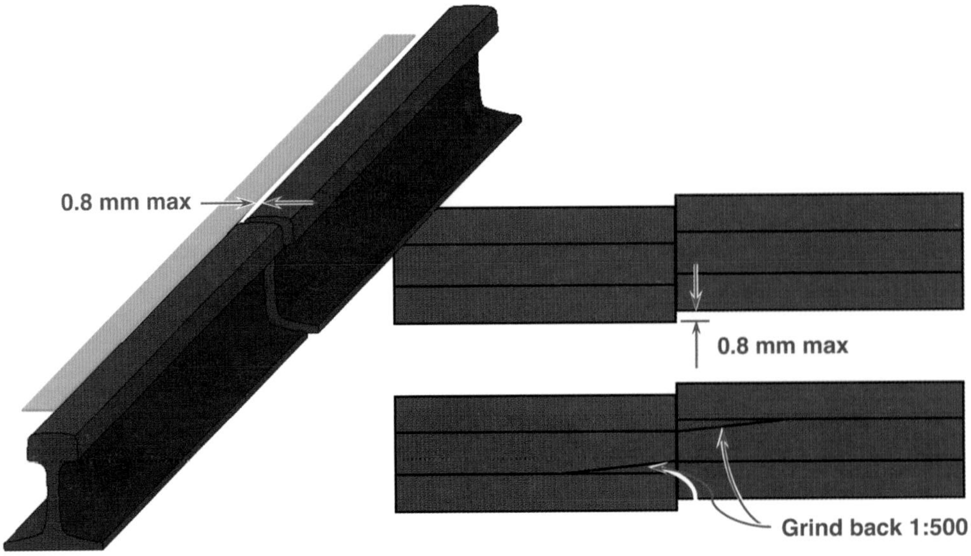

Figure 499: Rail Alignment Standards (Transnet)

3.10. Check Alignment

After grinding the alignment of each weld is checked using a 1 metre straight edge placed on top and on the running edge of the crown for straightness in the vertical and horizontal planes. A 0.2 mm feeler gauge may not enter at any point along the 1.0 metre straight edge placed centrally over the joint.

3.11. De-Stressing

The track must be de-stressed at the end of the process to prevent rail breaks or kick-outs. De-stressing is accomplished by replacing all sleeper fastenings whilst the rail temperature is within the de-stressing range as specified for the track section (see CHAPTER 3 paragraph 2.12.).

3.12. Thermite Weld Last Weld

At the end of the section to be welded, a closure rail (minimum length of 4.2 metres) is inserted to cover the gap created by the rail consumption during the previous welds. The last weld to complete the continuous welding process will be a thermite weld if the APT 500 welding robot is used. As explained earlier, the APT 1500 R is capable of flash butt welding the last weld, eliminating the need for a thermite weld.

4. THE CHARACTERISTICS OF FLASH BUTT WELDING

4.1. Static Bending

Static bending tests (deflection) should be carried out in accordance with set standards of the railway concerned. The Transnet standard, for example, requires a static bending test to be performed by their metallurgical laboratory every 500 welds. Two 600 mm lengths of rail are welded together for a test piece which must show a specified minimum deflection before fracture, see Figure 501.

Experience in South Africa was that the rails usually failed clear of both the heat-affected zone and the weld seam in the parent material confirming the higher strength of the weld. Fatigue tests done by the former British Rail showed similar failure results. Under cyclic load, the fatigue life of flash butt welds is generally greater than those of the parent material.

Type of Rail	Rolling Mark	Minimum Deflection (mm) at Breaking
40 kg	HCOB	40
48 kg	HCOB	40
48 kg	Grade A	35
48 kg	Grade B	30
57 kg	Grade A	30
57 kg	Grade B	25
60 kg	HH	20
60 kg	CrMn	20

Figure 501: Minimum Deflection Values of Flash Butt Welds

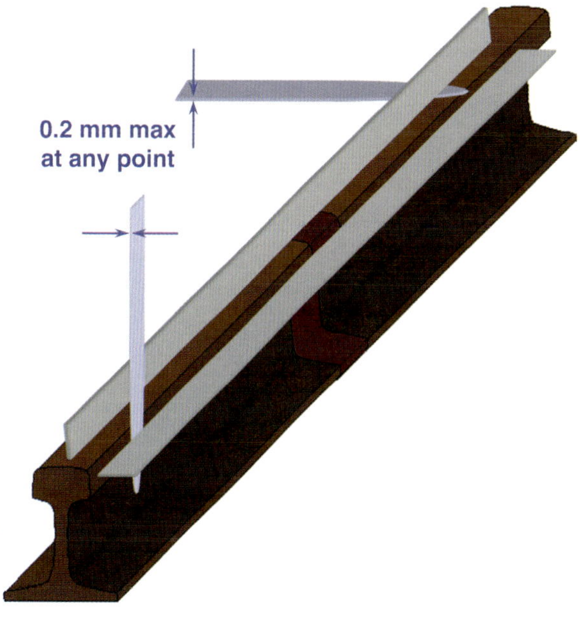

Figure 500: Checking Alignment after Grinding

3. TURNOUT TRANSPORTATION UNITS

A moving vehicle structure gauge (Figure 505 illustrates the South African gauge) is generally set by railways to ensure that trains can safely pass one another on double lines and in loops, safely cross bridges and tunnels and fit within the electrical overhead structure where applicable.

Due to the length of the long sleepers in a turnout (Figure 507), it will exceed the moving vehicle structure gauge if transported in an assembled state in the horizontal position on normal flat wagons (Figure 506).

To transport preassembled turnouts, Turnout Transportation Units (Figure 508) are required which hydraulically tilts the sections with long sleepers to fit within the moving vehicle structure gauge (Figure 506). Flat wagons are also used to transport the panels with normal sleepers.

4. TURNOUT INSTALLATION USING A TRACKLAYING MACHINE

Once a pre-assembled turnout has arrived at the installation site, specialised machinery is once again required to lift the turnout from the transportation unit and manoeuvre it accurately into position without exceeding the bending limitations of the turnout.

Figure 508: Turnout Transportation Units

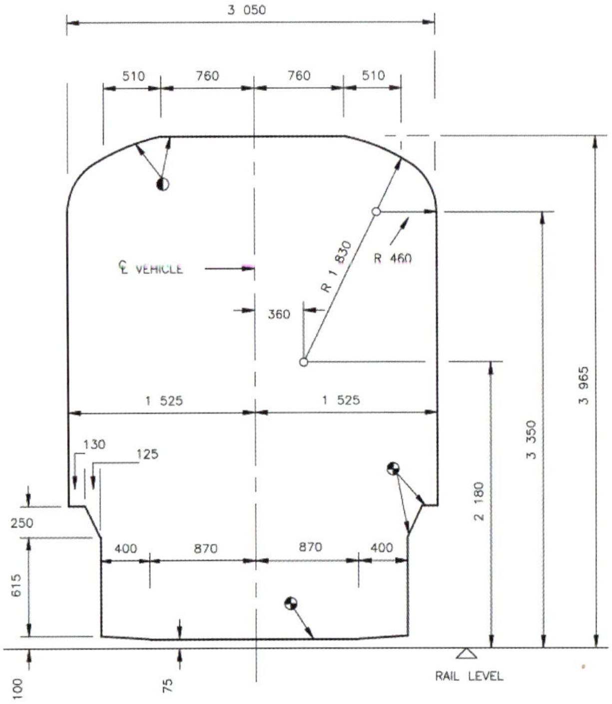

Figure 505: Moving Vehicle Structure Gauge for 1067 mm Gauge in RSA

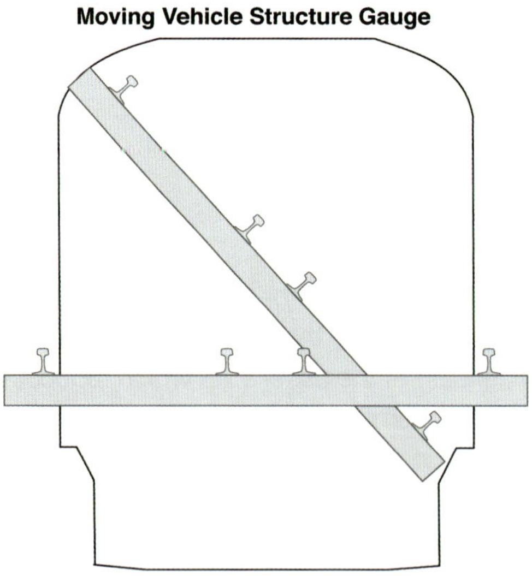

Figure 506: Transporting Pre-Assembled Turnouts by Rail

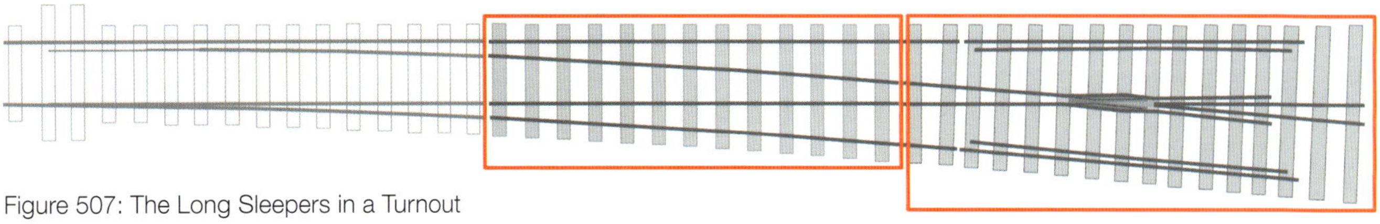

Figure 507: The Long Sleepers in a Turnout

Turnout installation machines (Figure 509) are designed to carry the turnout in the minimum number of sections into the installation gap with the least amount of deflection on the turnout to ensure its integrity and long life.

Turnout installation machines can be used for renewal and laying of turnouts as well as track panels. Turnouts up to a length of 36 metres and a maximum weight of 40 tonnes are relayed in one piece. Longer and heavier turnouts are handled in two or more sections.

4.1. Main Components of Turnout Installation Machines

The following is a description of one such machine, the Plasser & Theurer turnout installation machine.

4.1.1. Longitudinal Main Frame

The telescopic arms with crawler tracks, lifting beams and main engine are mounted to the longitudinal main frame. See Figure 510.

4.1.2. Lifting Beams

Four lifting beams are attached to the longitudinal frame on which the lifting bars are mounted. They may be folded in to be within the moving structure gauge during transfer travel. Each of the lifting bars is equipped with either self-securing rail tongs or hooks that can be laterally displaced along the entire length of the lifting bars.

The lifting beams are hydraulically adjustable with two lifting cylinders to ensure exact positioning of the lifting bars. The rail tongs or hooks can grip the turnout panel at any required place to ensure that it remains straight without bending or sagging. In relaying longer switch parts than the main frame, the protruding parts are held in horizontal position by additional rail-tongs, cables and hydraulic cylinders.

4.1.3. Horizontally and Vertically Extendable Telescopic Arms with Crawler Tracks

Four telescopic arms are mounted onto the longitudinal frame by horizontal guiding columns. Each telescopic arm is fitted with two crawler tracks. See Figure 510.

The distance between the telescopic arms can be adjusted according to the width of the switch by horizontal hydraulic cylinders. Each hydraulic cylinder has a maximum stroke which is in accordance with customer needs.

Figure 509: Turnout Installation Machine

Figure 510: Main Components of Turnout Installation Machines

The crosswise stepping movement is achieved by repeated lateral displacement of the lifting rams to the left or right and lifting and lowering of the longitudinal frame or load.

The vertical clearance height of the longitudinal frame can be adjusted.

4.2. Travelling

4.2.1. Moving on Site

Each of the 8 crawler tracks (2 per telescopic arm) is hydrostatically driven and provides good traction and stability. The crawler tracks can be steered by hydraulic cylinders to a minimum radius of 35 m (± 10°).

The crawler tracks are specially designed in length (1,400 mm) and width (400 mm) to ensure minimum surface pressure on the ground.

The drive is by radio remote-control and the speed is infinitely adjustable up to a speed of 4 km/h (maximum 1 km/h when a turnout is carried). For safety reasons in case of radio interference a second control unit with a cable link is mounted on the machine.

Figure 511: Self-Loading

Figure 512: The Machine is Retracted to its Minimum Size

4.2.2. Travelling Between Work Sites

The Turnout Installation Machine can load itself onto a railway wagon for transportation between worksites.

The machine's horizontally and vertically extending telescopic arms are retracted to its minimum size to fit within the moving vehicle structure gauge.

4.3. Removing the Old Turnout

The old turnout can either be cut up in manageable sizes in track for road transportation or, if the turnout is reusable, the turnout installation machine can be used to lift the turnout from its position and move it to the opposite side of the track from where the new turnout will arrive.

4.4. Formation Rehabilitation

While the formation is exposed formation rehabilitation can take place with conventional methods using earthmoving equipment. Formation rehabilitation is often required at the same time as turnout replacement due to the higher dynamic loading associated with turnouts. Neglecting to carry out formation rehabilitation could reduce the life expectancy of the turnout dramatically. Refer to CHAPTER 21, paragraph 4.1.

4.5. Manoeuvring the New Turnout into Position

The turnout installation machine will straddle the turnout transportation unit to lift the turnout panels one by one from the wagons. The panels on the tilt unit will be rotated to the horizontal position.

Once secured underneath the turnout installation machine, the machine will move turnout panels accurately into position.

5. CONCLUSION

The maximum life and minimum lifecycle cost can be achieved from turnouts if the pre-assembled turnout is transported to the installation site using turnout transportation units and if turnout installation machines are used to manoeuvre the turnout into position in the minimum number of sections and minimal deflection of the turnout.

Chapter 19
Rail Handling and Transportation

1. INTRODUCTION

Rail is a sensitive component which can easily be bent beyond its permissible rail stress during handling. Any over-stressing, nicks or marks on the rail may not be visible at the time but could be the cause of a rail break later on in the life of the rail with possible disastrous consequences.

There are many reasons why the rail must be handled. Rail pads may have a shorter life than rail, which requires the rail to be lifted to replace the pad, the rail may have to be lifted to remove a broken sleeper, the rail may have to me threaded to the side of the track if large scale sleepers replacement takes place, new rails may have to be transported, offloaded and threaded into position, etc. Each of these activities has the potential to damage the rail.

2. RAIL LIFTING

Figure 513 shows a couple of rail lifting jacks and tongs available. This equipment is purpose-designed and forms a very important part of track maintenance equipment that should be available to maintenance teams.

3. RAIL THREADING

Rail threading is an extremely dangerous and arduous function when done manually during re-railing or other maintenance activity where the rail must be moved into or out of the track. Specialised equipment is available to make this activity safe with the least amount of stress on the rail. See also CHAPTER 20 on track renewal.

Figure 513: Rail Lifting and Clamping Devices

Figure 514: Rail Threading Equipment

4. RAIL TRANSPORTATION

Rail transportation methods depend on the length of the rail to be transported, which is often dependent on the rail welding equipment available or strategy being followed for rail welding during re-railing or track renewal. See more about this topic under CHAPTER 17, Rail Welding and CHAPTER 20, Track Renewal.

4.1. Using Standard Railway Wagons

Shorter rails that fit within the length of a railway wagon are far easier to transport than long-welded rails but must still be picked up onto the wagon in a safe manner without over bending the rail.

Rail loading cranes are designed to be fitted to a standard rail wagon to safely lift rail onto the wagon. A corresponding number of rail loading cranes can be used for the length of rail to be transported.

4.2. Rail Train

If long welded rails are to be transported, specialised equipment is required. The Robel Rail Train is one such system and is an efficient and safe rail loading, transportation and unloading method for new and used rails of all profiles without unduly stressing the rail. During loading and unloading neither will the process, nor will the machinery foul the adjacent line. If re-railing is required on a regular basis or major track rehabilitation is required, the Rail Train will bring about vast savings in transportation, together with safety to personnel and track material.

Rails can be loaded and unloaded which are located between or outside of an existing track. The equipment for loading and unloading is designed for rails up to a length of maximum 500 m.

Figure 515: Rail Loading Cranes

Figure 516 : Rail Train

The rail carrying train is suitable for a transport speed of up to 100 km/h (track conditions allowing) therefore making short turnaround times possible. The maximum permissible deceleration rate is 2 m/s^2.

The smallest permissible radius for the loaded train to travel is 120 metres, however, for a radius smaller than 180 metres, the travelling speed must be reduced accordingly and for a radius smaller than 135 metres, the maximum speed is 30 km/h. Should radii smaller than 120 metres be absolutely necessary, small changes can be made to the train before the curve, which will allow the train to pass at 5 km/h.

Figure 518: Bi-Directional Chute Wagon

4.2.1. Functional Components

Refer to Figure 517 for the functional units of the rail train and the corresponding paragraphs below for a detailed description of the functions of these units.

4.2.1.1. Bi-Directional Chute Wagons

Two bi-directional chute wagons guide the rails during the loading and unloading process onto and from the transport unit and are located at the beginning of the rail train. No rails are transported on these wagons.

The chute wagons contain universal rollers and chutes which prevent the rails from tilting. The universal roller heads are suitable for all common rail profiles. The roller heads are mounted on the first third of the chute wagon and can be repositioned by the manipulator to the other end depending on the working direction of the unit.

Figure 519: Universal Roller Head on the Left and Chute Rollers on the Right Prevent the Rails from Tilting During Loading/Unloading

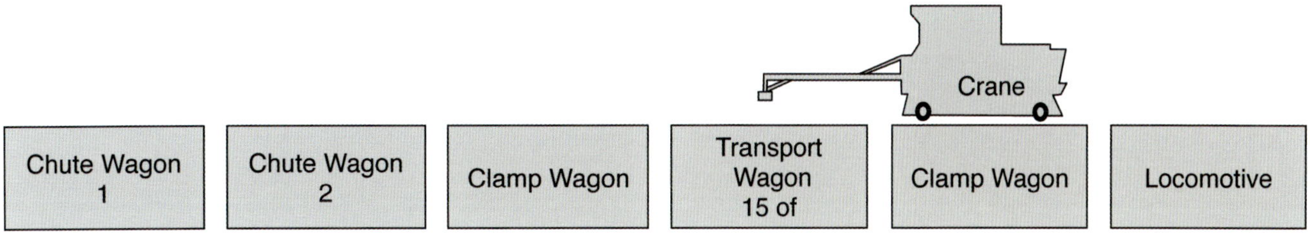

Figure 517 : Rail Train Consists for Bi-Directional Rail Handling

Figure 520 : Rails Being Guided into the Universal Roller Heads

Figure 521 : Guiding Rollers

The rails can either be loaded from or unloaded to the sleeper heads via the lateral chutes or from/to the centre of the track via the two middle chutes.

The rollers can accommodate rails with thermite weld beads on the rail foot of up to 20 mm. Bigger weld beads must be removed before loading the rails.

An electronic scanning device at the entrance of the chute will detect any oversized welds and automatically shut down the operation to prevent any damage or jamming of the rails in the chute.

The design of the chute wagons minimises the bending of the rail to a minimum as illustrated below, therefore avoiding any stress risers in rail steel.

Note the low height above track level of the point of exit or entry of the rail into the special chutes (Figure 524). No whip-lashing of rails is likely which may foul the clearance to the overhead electrification.

For the unloading process, folding spring-loaded chutes (Figure 525) are fitted to the end of the chute wagon to guide the rail end down to ground without falling, avoiding any possible damage to the rail or the sleepers.

Figure 522: Very Small Bending Of The Rail on the Robel Rail Train

The universal roller heads on the chute wagons are arranged symmetrically, thus allowing loading and unloading of the rails from both ends (no time-intensive shunting movements required). One of the chute wagons is equipped with a turning device for the manipulator.

A diesel engine is mounted on a chute wagon to supply power to the turning device for the rail manipulator and for the electricity supply.

After lifting the manipulator, it can be manually turned through 180°. The lifting of the turning device is hydraulically controlled via the operating box of the diesel power pack. Should the engine break down, the turning device can be manually operated via a hand pump.

A diesel tank with 500 litre capacity is installed on the frame of the chute wagon with a fuel indicator and filling device on both sides. An electrical pump with nozzle and tube is provided for refuelling the manipulator.

4.2.1.2. Rail Manipulator

The self-propelled, diesel-hydrostatic rail manipulator travels on running rails along the whole length of the rail carrying train and serves for pulling the rails on and off the rail carrying train (Figure 526). The driving speed is infinitely variable from 0 to 10 km/h in both directions. Full traction force is already achieved at minimum speed. The gantry crane weighs approximately 12 ton.

To prevent wheel-slip in damp weather conditions, sanding equipment is provided on all four wheels, front or rear, depending on working direction. The sanding equipment is activated by a joystick button from either working stations.

Hydrostatic service brakes act on 4 wheels and hydraulic disk brakes (for the parking brakes) act on 2 wheels. The parking brakes activate automatically when the diesel engine is switched off.

Figure 523: Excessive Bending of the Rail with Rail Bolsters

Figure 524: Low Entry Point for Loading of Rails

Figure 525: Folding Spring-Loaded Chutes for Off-Loading

Figure 526: Rail Manipulator

Figure 527 : Running Rails for the Manipulator

Figure 528: Rail Tongs

Figure 530: Closing and Locking of the Two Halves of the Support Blocks to Allow Loading of the Next Level

The track for the manipulator is fixed over the total length of the transport, clamp and chute wagons. During travelling of the rail train, the bridging rails between the wagons are swivelled out of the way (Figure 527). Slewing crane buffers are mounted at both ends of the train for limiting the travel of the rail manipulator.

The operation of two slewing arms with rail tongs (Figure 528) is effected independently by two machine operators from the cabin and can reach 2.3 metres laterally from the centre of the track or even more with an optional positioning device (see paragraph 4.2.1.5). They have a lifting force of 15,000 N at the maximum 3.57 metre extension. The maximum slewing force is 10,000 N. Special rail tongs, suitable for all rail profiles, are provided to prevent the rail from tilting during the loading and unloading operation.

The articulated rail tongs clamp the rail under the lower edge of the rail crown. For safety of operation the tongs are closed by spring force and opened hydraulically.

The manipulator is equipped with an electric signalling horn and a rear view video camera for the safety of staff working on and around the rail train.

Figure 529: The Upper Supports of the Support Blocks can be Swivelled Out of the Way

4.2.1.3. Transport Wagons

The number of transport wagons required depends on the length of rail to be transported and the type of rail wagon, i.e. the length of the available wagons if supplied by the railway. Custom wagons can also be manufactured by the supplier. Two of the wagons, one at each end, will be equipped with clamping equipment (clamp wagons) whereas all the other wagons will be equipped with stacking support blocks.

Two support blocks are mounted on every transport wagon. The rails are loaded in three to five layers onto the stacking equipment depending on the height of the structure gauge, the rail profiles to be transported, permissible axle loading, etc. Each layer takes approximately 10 rails (30 to 50 in total can be loaded depending on the number of layers). The supports of the support blocks are divided in the middle and can be swivelled to the side (Figure 529 and Figure 530) to allow loading of the layers.

The support blocks are equipped with carrying rollers with anti-friction bearings. This allows easy movement of the rails during the loading and unloading procedure as well as length compensation during travel through curves to prevent any danger of derailment.

4.2.1.4. Clamp Wagons

To secure the rails against the longitudinal forces resulting from acceleration, deceleration or length compensation in curves, each rail is clamped down on the rear most wagon on an anchoring scaffold (clamping equipment).

Fixing blocks are integrated in the anchoring scaffold and is in design similar to the support blocks, however, it does not contain carrying rollers but slots for the rail clamps that secure the rails individually.

Pulling and pushing rods are supplied to absorb all the forces transferred during transportation.

At the front and the rear end of the clamp wagons, pivotable rebound stops are mounted, which prevent movement of the rail in an accident or unintended loosening and consequently slipping of a rail. This is an important safety feature, which makes the need for a dedicated person to watch over the stacked rails during transit unnecessary. They further provide an appropriate safety zone to the operator on the clamping wagon.

Figure 531: Each Rail is Individually Clamped on the Rear Most Wagon

Figure 532: Pulling and Pushing Rods

4.2.1.5. Positioning Device

As an optional extra, a rail guiding device can be provided at each end of the chute wagons, which provides the exact guiding and positioning of the rails during the loading and unloading process. This will increase the loading and unloading distance to 2,500 mm from track centre in either direction (Figure 534).

The spring-loaded chutes usually attached to the chute wagon can be attached to the positioning device to still guarantee a gentle landing of the rail and to ensure that no sleepers will be damaged in the process.

4.2.2. Loading and Unloading Process

4.2.2.1. Loading from the Track Centre or Outside of the Track

Rail-loading is usually carried out in pairs, either from the centre of the track or from the outside of the track. Loading starts by gripping the ends of the rails with the crane tongs, raising them and guiding the rails through either the middle or lateral chutes of the chute wagon (Figure 526).

The operators in the manipulator then instruct the locomotive driver to move the rail train underneath the lifted rails.

The slewing arms are controlled from the cabin of the manipulator to lay the rails between guide rollers at the front end of the chute wagon, and in the universal roller heads on top of the second chute wagon. The rails are held on the rail head and foot by the rollers to secure the rails against tilting under their own weight.

The rail train continues to move forward underneath the rails, while the manipulator uses its own power to pull the rails onto the wagons by moving with a synchronic speed into the opposite direction of the locomotive.

Figure 533: Slewing Rebounding Walls in the Closed Position

Figure 534: Positioning Device

Immediate application of rail clamps on the fixing block of the anchoring scaffold secures the long rails against sliding due to longitudinal forces.

The manipulator runs back to the chute wagon in order to load the next pair of rails.

4.2.2.2. Unloading to the Track Centre or Outside of the Track

Unloading is also carried out in pairs. The ends of both rails are gripped by the manipulator slewing arms, pulled towards the chute wagons, and guided via the rollers into either the middle or lateral chutes. The manipulator runs back to the other end of the rails, grips the ends and pushes the rails from the train until their ends touch the ground. The rail ends can be very accurately aligned with the previously unloaded rails. If required in small curves, the rail ends can be fastened to the track with clamping heads and short chains.

The locomotive then moves from underneath the rails being unloaded, while the manipulator guides the rails off the train by moving with a synchronic speed into the opposite direction of the locomotive. As soon as a sufficient length of rails (approximately 50 metre) is placed on the ballast bed or on the sleepers, the friction force is big enough and the manipulator can release the rails and already grip the next pair of rails for unloading.

The rail ends will slide down the spring-loaded chutes, which are fitted to the end of the chute wagon to guide the rail end down to the ground without falling, avoiding any possible damage to the rail or the sleepers.

4.2.2.3. Loading and Unloading Limitations

- Loading the Rail Train at the Welding Factory

  The rail train is not designed to load rails from a stock pile, nor is it suitable to accept rails pushed on rollers from a welding plant. Gantry cranes at the welding plant are therefore required to load rails on the rail carrying wagons.

- Loading and Unloading Distance From Track Centre

  Rails must be within 1.9 metres from the track centre and in the upright position to be lifted and guided through the chutes without tilting. If the rails are lying down the embankment or on its side, additional labour will be required.

  A ballast profile within standard is also required as this may also influence the ability of the machine to reach the rail for loading. For unloading purposes on a ballast bed that does not conform to standards (excessive ballast depth), a stable "step" may have to be made in the ballast bed.

- Working under Live Overhead Electrification (OHE)

  Staff standing on top of the wagons during the loading and unloading process, especially when fully loaded, may breach the safety distance from the OHE with arms extended upward. The rail train is therefore not recommended to work under live OHE conditions unless strict safety discipline is provided when clamping and unclamping the rails on the wagons and when the support blocks (stacking racks) are opened or closed.

4.2.3. Conclusion

The Robel Rail Train has the following advantages:

- The equipment for loading and offloading of rails using a rail carrying train requires a crew of only four people, minimising the risk of injury.
- Very short loading and offloading times, requiring short occupations and faster turnaround times for the rail train.
- Rails can be loaded and unloaded which are either located in-between the rails or outside on the sleeper head or ballast formation of an existing track.
- Thirty rails of 240 metre can be transported on three layers.
- No holes are required in the rail for loading or offloading.
- The rails are not unduly stressed during transport, loading or unloading.
- The rails are individually clamped in such a way to prevent any stress risers and any longitudinal movement during travelling.
- Rollers prevent the rails from tilting during offloading or loading, giving huge savings in time and labour.
- Short preparation time on site.
- The process is very safe because handling of the rail is minimised.
- The operation does not foul the adjacent track.
- The rail train can work under live overhead electrical systems depending the on the structure gauge and height of the contact wire.

Chapter 20
Track Renewal

1. INTRODUCTION

All the track components have different life expectancies and will be replaced at different cycles. Regardless of the component to be replaced, it is an involved process which will see line closure for an extended period. The components can either all be replaced at once if the track has been neglected and requires complete renewal or it can be due to a strategy to replace all the components even if some will be replaced prematurely when the costs of occupations and installation are considered. Single major components such as sleepers or rails only can also be replaced. The processes are similar and the same methods and machinery can be used. However, in this chapter track renewal does not include formation rehabilitation. Refer to CHAPTER 21 for formation rehabilitation processes and methods.

Depending on the availability of capital and/or machinery, track renewal can be carried out with semi-mechanised or fully mechanised methods.

Figure 535: Offload New Rails at Ballast Toe

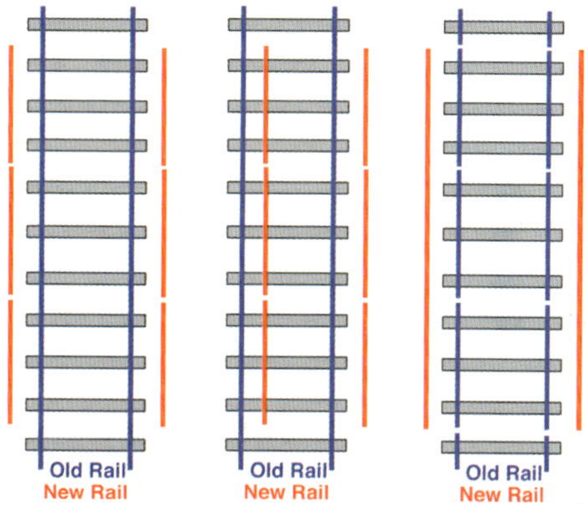

Figure 536: Threading and Weldin

Figure 537: Rail Train

2. SEMI-MECHANISED TRACK RENEWAL METHOD

Semi-mechanised methods using a combination of machines and large labour gangs have proven to be very cost-effective in Southern African countries. There are many possible variations for achieving the same results. Following is only one possible process in a step by step explanation which has been used and proved to be very effective. This same method can also be used if only the rails or sleepers are replaced.

Step 1
Offloading New Rails

The process starts by offloading new rails at the toe of the ballast (Figure 535). If a rail carrying train (Figure 537 and CHAPTER 19) is available, the relaying process can be sped up by first welding the rails up to 240 metres in length in a stationary welding depot near the renewal site (Figure 538) using a flash butt welding machine (CHAPTER 17). The long welded rails can then be transported to the relay section and offloaded at the toe of the ballast.

Figure 538: Flash Butt Welding Machine in Stationary Deport Setup

However, in many countries a rail train is not available, in which case short length rails of 12 to 18 metres must be loaded on rail wagons and offloaded at the ballast toe of the relay section.

Welding must start long in advance of the actual relaying process due to the potential delay that so many welds may have on the process. The welding can take place between trains. The short rails of the one leg are threaded to the centre of the track (Figure 536) where a mobile flash butt welding machine welds the rails into long welded rails of up to 500 metres. The long welded rail is threaded back to the toe of the ballast and the other leg threaded in for welding.

The new long welded rails are supported in the upright position as they will become the rails for portal units (Figure 539 and Figure 540).

Figure 539: Typical Portal Unit

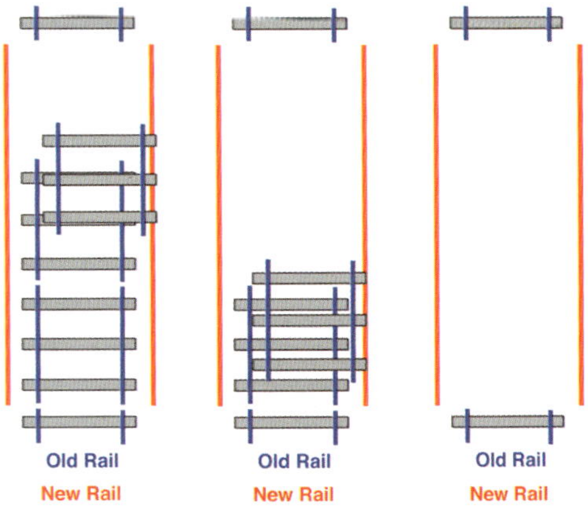

Figure 540: The 13 Metre Panels are Lifted Using a Portal Unit

Step 2

Cutting and Removing the Old Track Skeleton

A portal unit (Figure 539) is used to lift track panels up to a maximum length of approximately 13 metres, depending on the model used. The old rails are therefore cut into 13 metre lengths (refer back to Figure 536, illustration on the right) using a disc cutting machine. The rail-to-sleeper fastenings are not removed. The 13 metre panels are lifted one by one using the portal crane (Figure 540) until a construction gap of 4 to 6 panels is open.

The panels are placed on flat railway wagons (Figure 541) ahead of the construction gap. The ballast is levelled to fill the holes where the old sleepers were (Figure 542) using hand labour with ballast forks.

If the ballast depth underneath the sleeper was inadequate and severely fouled, the ballast can be compacted with a road roller into a sub-ballast layer. Alternatively, the levelled ballast bed can be left as is, to be screened later on in the relaying process.

Figure 541: The Panels are Placed on Flat Railway Wagons

Figure 542: The Ballast Bed is Levelled and the Holes Filled Where the Old Sleepers were

Step 3

Laying of the New Sleepers

To lay the sleepers, a continuous sleeper laying unit can be attached to Plasser & Theurer's PA 1-20 portal unit. This unit consists essentially of a lifting bridge combined with a continuous action sequence controlled automatic device for laying individual sleepers. The unit picks up a maximum of 20 concrete sleepers (24 timber sleepers) from flat railway wagons (Figure 543). The PA 1-20 ES travels on the new rails lying at the ballast toe to the relay area and places the sleepers at the correct sleeper spacing. The sleeper spacing is programmed using an electronic distance wheel and each sleeper is laid accurately (Figure 544). The electronic control ensures an automatic and continuous work process. The laying unit is able to lay 20 sleepers in 100 seconds.

Gantry cranes with a spreader been can also be used.

The portal unit returns to the wagons carrying the new sleepers, picks up another load and places the next load of sleepers. This process continues until the entire construction gap is filled with sleepers at the required final spacing.

Step 4

Placing Rail Pads, Threading in Rails and Attaching Fastenings

The rail pads are placed and the fastenings distributed (Figure 545). A rail threader is used to thread the rails into position (Figure 546) for safety reasons, though this can be done by a labour gang as well. The rail fastenings are attached which completes the construction of the track skeleton.

Figure 543: The PA 1-20 ES Picks Up New Sleepers from Flat Railway Wagons

Figure 545: Rail Pads are Placed

Figure 544: The PA-1-20 ES Places Sleepers at the Required Spacing

Figure 546: Rail is Threaded in

Step 5
Screening and Regulating the Ballast Bed

A ballast regulating machine is used to plough in all the ballast material spread over a wide area towards the shoulder of the sleeper (refer to CHAPTER 13). If there was adequate ballast to start off with, a ballast cleaning machine (Figure 547) is brought in to screen the ballast while the material replacement carries on ahead. If the ballast was compacted into a sub-ballast layer at the start of the relaying process, ballast regulating and cleaning would not be required.

Step 6
Offloading and Regulating New Ballast

New ballast is offloaded (Figure 549), distributed and regulated to correct the ballast profile using a ballast regulating machine. If the track skeleton lies on the sub-ballast layer, an HGR track lifting machine (Figure 548) may be required to pre-lift the track ahead of the tamping machine or the ballast cleaning machine if the ballast is screened.

The track is lifted, levelled, aligned and tamped in several passes (CHAPTER 11) to produce a totally relayed track of high initial quality. Dynamic stabilisation in layers is preferred in order to open the track at normal track speed without a speed restriction.

Figure 547: Ballast Cleaning

Figure 548: Pre-Lifting of the Track (Photo for Illustration Purposes Only)

Figure 549: New Ballast is Offloaded

3. TRACK RENEWAL USING MECHANISED METHODS

The Plasser & Theurer SMD 80 Track Renewal and Track Laying Machine (Figure 550) and the SVM 1000 Track Laying Machine are two mechanised systems that have been in operation in many countries around the world and have proven its reliability and cost efficiency many times over.

3.1. SMD 80 Track Renewal and Track Laying machine

This machine offers technology which can be used both for track renewal and for laying new track.

The SMD 80 removes old rails and sleepers, grades the ballast bed and lays new sleepers and rails. The SMD 80 is capable of laying and picking up all types of wooden, concrete or steel sleepers and all rail profiles.

Refer to Figure 551 for the working principle of the SMD 80 and the components referred to in the following paragraphs. A locomotive is used to take the SMD 80 to the worksite, together with conventional standard flat-bed wagons (A) loaded with new sleepers. The wagons are fitted with crane tracks for the gantry cranes (B). The gantry cranes serve to supply the laying unit with new sleepers and to take away the old sleepers.

When starting to work, the machine spreads the old rails, removes a few sleepers, and a crawler track (C) is lowered onto the ballast which supports the rear end of the machine and drives the machine forward during operation. The crawler track guarantees high tractive power even in wet weather and on gradients. During operating the locomotive is therefore not required. The machine pushes the sleeper wagons ahead of itself.

The time required to set up and close down the machine is only 15 minutes. Rail fastenings are loosened ahead of the machine. For safety, every 10th sleeper can remain fastened. These are loosened in the course of track relaying under the power unit (E). If the rails are to be replaced during the same operation, the new rails must also be laid out at the worksite before work begins (it is possible to exchange the rails or to reinstall the old rails again).

The old and the new rails are picked up by adjustable rail guiding clamps and rollers (D). The old rails are lifted behind the rear bogie of the power unit (E), spread out and deposited at the side. The new rails are guided and installed on the base plates directly after the new sleepers are laid.

Figure 550:
Plasser & Theurer SMD 80 for Track Renewal and Track Laying

Once the old rails have been spread, the old sleepers are lifted out of the encrusted ballast bed individually by the prongs of the sleeper pick-up unit (F) consisting of vertical chain conveyors and transferred to the transfer platform (G) by a conveyor belt where they are picked up for removal to the sleeper wagons by the gantry crane.

The machine can also be equipped with a sleeper expulsion unit so that badly damaged or non-transportable sleepers can be removed to the side of the track.

Behind the sleeper pick-up unit (F) a regulating plough and a grading chain (I), which is similar to the excavating chain of a ballast cleaning machine, ensures a smooth ballast formation for laying the new sleepers. The ballast is picked up by the grading chain and taken to the side by two transfer conveyor belts. The grading chain guide is constructed so that the ballast is graded deeper in the middle of the track in order to prevent the new sleeper from becoming centre-bound.

By this time the gantry crane has already picked up a layer of new sleepers from the sleeper wagons and transported them to the transfer platform for new sleepers (J) on the relaying machine. On the gantry crane's return journey it will transport a load of old sleepers to the sleeper wagons.

Fig B - Crosswise Loading Of the Sleepers

Fig C - Crawler Track

Fig H - Guiding the new Rails Onto the Rail Pads

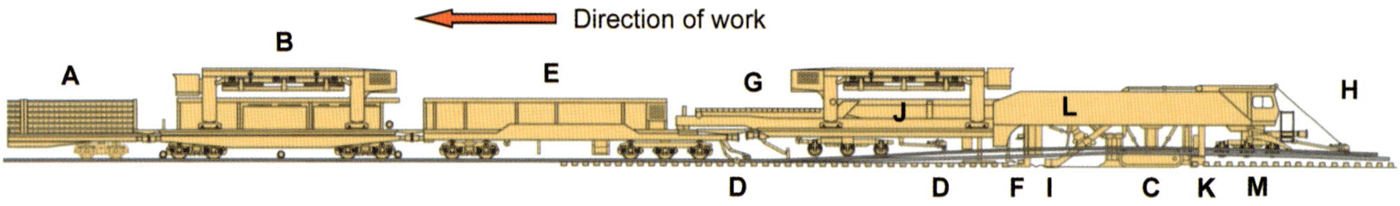

Fig F - Removal of Old Sleepers

Fig I - Grading Chain and Plough

Fig K - Laying New Sleepers

A - Flat bed sleeper wagons
B - Gantry crane
C - Driven crawler track
D - Rail guiding clamps and rollers
E - Power unit
F - Sleeper pick-up unit
G - Transfer Platform Unit for old sleepers
H - Guiding the rail
I - Regulating plough and grading chain
J - Transfer platform for new sleepers
K - Sleeper laying unit
L - Auxiliary chassis
M - Rear bogie

Figure 551: Working Principle of the SMD 80 Track Renewal and Track Laying Machine

With the help of conveyor belts, the new sleepers are passed from the transfer platform (J) on the relaying machine to the sleeper laying unit (K). The unit centres the sleepers and lays them on the ballast formation in the correct spacing and alignment. In curves the sleepers are laid radially, therefore no follow-up work is necessary. The laying process is monitored by an operator sitting in the cabin next to the work unit.

In no phase of the relaying are the rails subjected to inadmissible bending stresses. The rail fastenings are mounted behind the machine after the rails have been laid on the base plates.

At the end of the worksite the sleeper pick-up unit lifts the last old sleeper out of the ballast bed. The grading unit also stops its work. Both work units are placed and fixed hydraulically in the transfer travel position. The auxiliary chassis (L) is lowered as soon as it stands over the track that no longer has to be renewed. Now the crawler track can be lifted. When the last new sleeper has been laid, the laying unit is fixed in the transfer travel position. The machine moves forward until the rear bogie (M) is also over the track that no longer has to be renewed. The auxiliary chassis is retracted and the bogie is lowered onto the track again. The SMD 80 and the wagons loaded with dismantled old sleepers leave the worksite.

If the ballast is fouled, a ballast cleaning machine (CHAPTER 14) can be used ahead of the SMD 80 to screen the ballast and remove the fines with MFS material conveying systems (CHAPTER 15) but the offloading of new ballast would not be advisable until after the material replacement process.

If the ballast is not fouled, the ballast that has been graded to the side can be profiled back using a ballast regulating machine (CHAPTER 13). After any ballast maintenance or disturbance, the track must be lifted, levelled, aligned, tamped and dynamically stabilised. Refer to CHAPTER 11 and CHAPTER 12.

Figure 552: SVM 1000 R Track Laying Machine with New Sleepers and Rails

Chapter 21
Formation Rehabilitation

1. INTRODUCTION

Refer to CHAPTER 3, Paragraph 6 for a detailed description of the formation as a track component. In CHAPTER 20 track renewal processes were discussed in a scenario where formation rehabilitation was not necessary. Also, in CHAPTER 18, paragraph 4.4 it was suggested that formation rehabilitation would often be required when turnout replacement is done. Formation rehabilitation is a very involved process and will cause line closures for extended periods over and above the time required for track renewal or turnout replacement, hence the importance of a proper preventative maintenance programme for all the track components to preserve the formation life for as long as possible. This chapter will look at formation rehabilitation methods in isolation.

Formation rehabilitation can also not be described as a single method or design. There are various different methods ranging from the use of conventional earth moving machines and manual labour to fully mechanised on-track methods. The formation rehabilitation design may also vary extensively depending on the type and axle loading of the track, the extent of the formation failure, the prevailing type and quality of the in-situ material, etc.

2. FORMATION REHABILITATION DESIGNS AND MATERIALS

Formation rehabilitation is required where the subgrade (the platform on which the track is constructed) is not able to support the track structure. The reasons for this lack of support could be that (see also CHAPTER 3, paragraph 6.2):

- the in-situ soil consists of clays or silts which cause poor drainage and an unstable platform; and
- the presence of water in the formation due to:
  - subsurface seepage; or
  - saturation of the formation due to fouled ballast retaining rain water or surface flow water.

Formation rehabilitation could be required over relatively short distances, often referred to as a mudhole, or can be over very long distances where continuous formation failure is encountered.

"Formation rehabilitation" refers to a combination of removal and replacement of poor or damaged soils with graded backfill material, strengthening of the formation using geo-synthetic materials and the installation of surface and subsurface drainage systems.

A minimum California Bearing Ratio (CBR) value for the in-situ material would generally be prescribed by railway earthworks standards as the minimum at which to re-construct the formation to provide the required stable platform. The depth below ballast level where the required CBR value is encountered will determine the depth of the excavation, the type of backfill material used, as well as the need and type of geo-synthetic material. The presence of subsurface water will determine the type and use of subsurface drains.

As an example of the extent to which formation rehabilitation can be designed, Figure 556 illustrates an extensive formation rehabilitation design for heavy haul line applications used in South Africa, according to the Transnet Specification for Railway Earthworks S410. Formation rehabilitation could also be as simple as removal of 100 mm of formation material and replacing it with geo-textiles as a separation level and 100 mm of graded backfill material.

Figure 556: An Extensive Formation Rehabilitation Design in Accordance with the Transnet Specification for Railway Earthworks S410 on Heavy Haul Lines

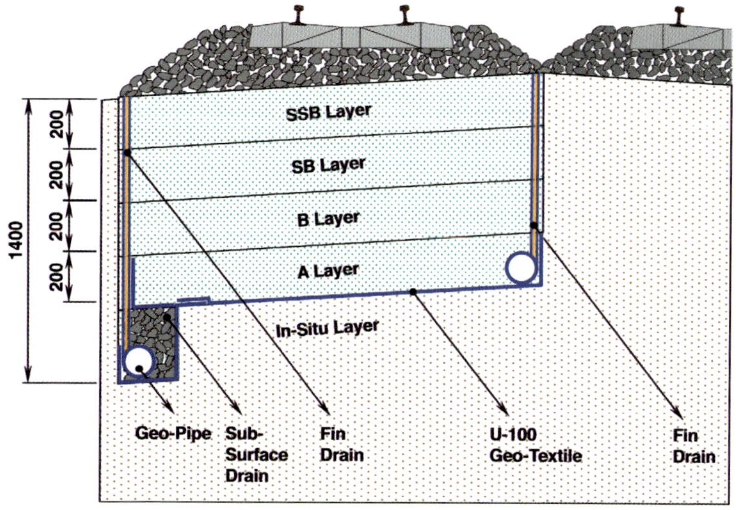

The different formation rehabilitation materials are discussed in the following paragraphs.

2.1. Subsurface Drains and Geo-Pipes

Subsurface drains, as with all formation rehabilitation activities, must be properly designed by a professional in the field. Subsurface drains can also have various designs and depths, depending on the subsurface flow characteristics. To facilitate the effective flow of water in the drain, these drains are often lined with geo-textiles, filled with ballast and may also contain a geo-pipe.

2.2. Geo-Synthetic Materials

"Geo-synthetics" is a generic name for a variety of synthetic materials used in the field of soil mechanics. The most common geo-synthetic materials are:

2.2.1. Geo-Textiles

Geo-textiles are made of polymers such as polyester or polypropylene and are either used as a filter or as a separation layer between two different soil types, thereby maintaining the integrity and functionality of the soils.

Non-woven geo-textile fabrics are used for their permeability which allows the passage of water but prevents the passage of granular material, whereas woven geo-textiles are less permeable and will avoid the passage of water.

Geo-textiles are rolled out on top a soft subgrade to prevent the intermixing of the subgrade material and the placed backfill material, especially where a high water table or subsurface seepage is encountered. Geo-textiles should not be directly underneath the ballast since the stones will puncture holes in the material and when the track has to be ballast cleaned, the geotextiles will interfere with the cutting chain.

Figure 557: Geo-Textiles Being Rolled Out over the Prepared Formation

Figure 558: Geo-Grid Being Rolled Out over the Prepared Formation

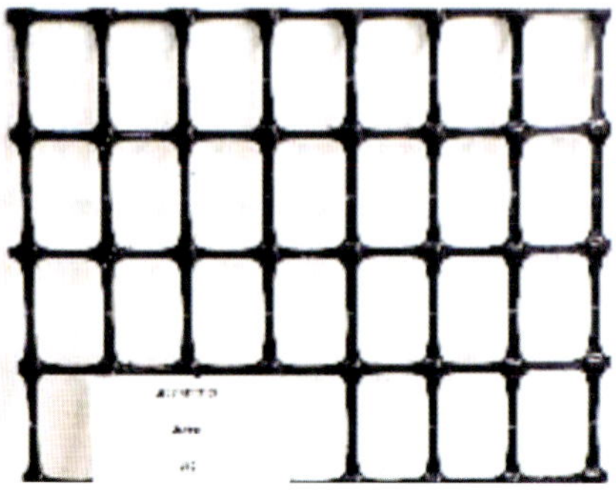

2.2.2. Geo-Grid

Geo-grid is a mesh like polymer structure which is rolled out on top of a weak subgrade to strengthen it. It is also used in fin drains.

2.2.3. Geo-Cells

Geo-cells are honeycomb-shaped cells that are filled with backfill material to create the structural strength required on which to construct the new formation. The strength that the cellular confinement provides may reduce the depth of the required excavation.

2.3. Fin Drains

A fin drain is a prefabricated filter comprising a geo-grid core sandwiched by geo-textiles. The core acts as a drainage conduit along which water flows freely after it has been filtered by the geo-textile skin. This is referred to as a geo-composite for it combines the use of two different types of geo-synthetic materials to benefit from the features of both.

2.4. Backfill Material

The number of layers, layer thickness and selection of backfill material type and grading will depend on the formation rehabilitation design which is, to a large degree, based on the axle loading and traffic density of the line and the condition of the formation. The layers are, depending on the specification, graded according to the following:

- Plasticity
- Ballast grading
- Minimum compaction percentage of modified AASHTO density (American Association of State Highway and Transportation Officials)
- Minimum CBR strength after compaction

Each layer is compacted to a specified density. The specified moisture content must also be observed. Geo-textiles will generally be used together with layer work to prevent intermixing of the in-situ material and the layer work.

2.5. The Formation Protective Layer (FPL)

If the formation does not meet requirements, it may be necessary to install an appropriately thick formation protective layer (FPL) consisting of a graded gravel-sand mixture (backfill material). This can be reinforced by adding a membrane of geo-synthetic material.

The installation of a formation protective layer is an effective and well-proven method of raising the bearing strength of the substructure as it reduces the soil pressure tensions. Consequently, this constructional measure brings an enormous reduction of the costs for track maintenance. The installation of an FPL is associated mostly with mechanised methods of formation rehabilitation. See Figure 594.

The thickness of the formation protective layer depends above all on the bearing strength of the earth formation. On the other hand, the required bearing strength of the FPL depends on the maximum line speed. Normally, an FPL with a thickness of 30–50 cm is installed.

Figure 559: Geo-Cells Being Used Together With Geo-Textiles and Filled with Backfill Material

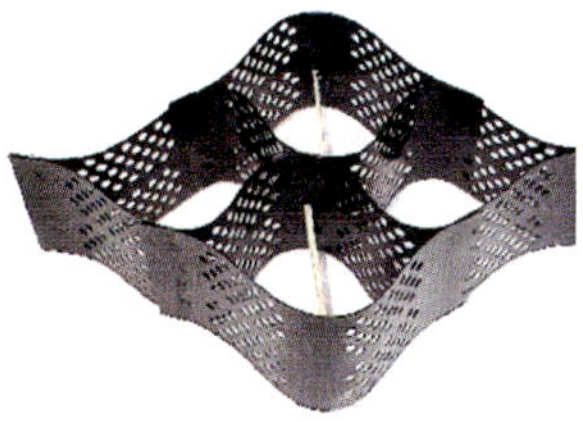

Figure 561: Cutting the Rails

Figure 562: Excavating the Ballast

Figure 563: Excavating Damaged Formation

Figure 564: Compacting the Formation

Step 1
Dismantling and Removing the Track Skeleton

The first step is to dismantle and remove the track skeleton to the side of the track by cutting the rail. If the rail is going to be re-used, it must be cut in the longest possible sections of 240 metres or even longer where practically possible, to reduce the number of welds that would be required during the reconstruction phase. If the rail is going to be replaced with new rail that has been delivered to the side of the track before the rehabilitation started, the old rail can be cut in short sections for ease of removal, if transposing thereof is not required (Figure 561).

Step 2
Excavating the Ballast

The second step is to excavate the ballast down to formation level using an excavator. Tipper trucks are used to transport the ballast to an environmentally-suited and prepared washing plant. At the washing plant the ballast that fits in the reusable envelope is recovered and stockpiled, to be returned to the track during the reconstruction phase (Figure 562).

Step 3
Excavating the Damaged Formation

The third step is to excavate the fatigued formation to a depth in line with the selected formation rehabilitation design. The excavated material is removed to a spoil site using tipper trucks. Exits out of the excavated area are made at various points for the tipper trucks to improve the turnaround times for the trucks at deep excavations. At shallow excavations this is however not a problem (Figure 563).

Once the required depth has been reached, the formation is compacted using a vibratory roller. The level below top of rail is measured at regular intervals throughout the process.

At deep excavations there is generally not enough space in the 4 metre wide excavation for the trucks to pass one another. This poses a problem where the excavation is in a long cutting or on a long bank. The incoming tipper truck will have to wait for the outgoing truck to leave the excavation first. In long cuttings/banks this slows down the process (Figure 564).

Step 12
Lifting, Aligning, Levelling, Tamping and Stabilising the Ballast Bed

The track is lifted, aligned, levelled and tamped to the required specification. There is no compromise here since hand tamping cannot provide the precision that is required to ensure a good track profile. The use of a dynamic track stabilising machine will further enhance the stability and durability of the track.

Advantages of the Conventional Formation Rehabilitation Process

- Formation rehabilitation can be carried out simultaneously at various sites or fronts on the same site.
- The high availability of commercial moving plant.
- Resources can be shared between working sites.

Disadvantages of the Conventional Formation Rehabilitation Process

- The conventional method is slow and logistically complex.
- The line is completely disassembled by cutting the rail and moving all the material to the side of the track. The line will therefore be closed for traffic during the entire extended period.
- In long rehabilitation sections, where the excavation is deep, on a bank or in a cutting, there is not enough space for tipper trucks to pass one another. The incoming truck must wait until the outgoing truck is out. The excavator's productivity is therefore reduced during this period, delaying the entire rehabilitation process.
- Quarries for graded material are generally very far, which requires a large number of trucks to achieve a reasonable turnaround time. Alternatively, a ballast stockpile area must be created, which still requires a large number of truck trips and double handling of material.
- Due to the very deep excavation, large volumes of material must be removed to suitable spoiling sites which can also be a long distance away.
- Danger of collapsing of the excavation wall adjacent to the open running line, due to the excessive depth of the excavation and the extended time that the excavation wall is exposed to the stresses from passing trains.

Step 4
Constructing the Subsurface Drain

After the excavation, subsurface drains are constructed using a tractor loader backhoe (TLB), if applicable, according to the formation rehabilitation design.

Step 5
Laying Geo-Synthetic Material and Fin Drains

The subsurface drain and formation is lined with geo-synthetic material. Various types of materials are available including geo-textiles, geo-grid or geo-cells and are used depending again on the design.

A fin drain and geo-pipes may also be installed on the sides of the excavation and the subsurface drain filled with ballast stones (Figure 565).

Figure 565: Rolling Out Geo-Textiles

Step 6
Importing Backfill Material in Layers

Graded backfill material is brought in by tipper trucks, levelled and compacted in layers. The thickness of the layers and the grading of the backfill material depend on the formation design specification.

If geo-cells are used, back tipping is required, which requires the backfill material to be levelled before the next load can be offloaded, since the empty geo-cells cannot support the weight of trucks (Figure 566).

Figure 566: Back Tipping of Backfill Material

If space is available, frontend loaders can offload the backfill material from the side. This process is uneconomically slow for long sections. A vibratory roller is used to compact the backfill material to specification. A density gauge is used to measure the compaction and moisture content of the backfill material in several places.

Step 7
Re-Ballasting

There are various options for re-ballasting. If the excavated ballast was reusable, tipper trucks can be used to bring the screened ballast back. Alternatively, new ballast can be brought in with tipper trucks or if an adjacent line is available, ballast can be brought in with ballast wagons and levelled and distributed using skid steer loaders (Figure 567).

Figure 567: Re-Ballasting

Figure 568: Reconstructing the Track Skeleton

Figure 569: Thermite Welding

Figure 570: Grinding

Figure 571: Ballast Offloading

4.2. Semi-Mechanised Methods using a Variety of On-Track Machinery

The following is a step-by-step explanation of semi-mechanised formation rehabilitation using a combination of various on-track machines. Semi-mechanised methods and the conventional methods can be combined depending on the availability of on-track machinery. The same steps as above will be followed. However, the order of the steps may change.

Figure 575: Ballast Cleaning Machine

Figure 576: MFS Material Conveying Systems Discharging

Figure 577: Rail Threader

Step 1
Excavating the Ballast

Instead of using excavators and tipper trucks to remove the ballast bed, a ballast cleaning machine in conjunction with MFS Material Conveying Systems can be used. The ballast will not be screened during this process. The ballast cleaning machine will excavate the ballast bed down to a predetermined depth and transfer it to MFS wagons in front of the ballast cleaning machine.

The MFS wagons will take the excavated ballast to an environmentally approved and prepared site where the ballast can be screened for re-use. Refer to CHAPTER 15 for the working principle of the MFS systems, which is designed to ensure that the process can carry on continuously without any stoppages.

The recovered ballast can be dealt with in one of two ways:

- The ballast can be offloaded at an established ballast washing plant. The disadvantages of establishing a washing plant are the cumbersome environmental legislation that may have to be followed and rounded ballast will be returned to the track.
- Due to the versatility of the MFS wagons, the recovered ballast can be returned to the quarry for reprocessing or for alternative use. In the past, with the use of tipper trucks, the distance to the quarry made this possibility uneconomical. However, using MFS wagons with their large carrying capacity and travel by rail does make this economically possible. At the quarry the fast offloading feature of the MFS wagons will make turnaround times of the wagons short.

Step 2
Dismantling and Removing the Track Skeleton

As before, the length the rail sections will be cut into will depend on the re-use or replacement of the rail. The efficiency and safety of moving the rail to the side of the track will be vastly improved if the rail threader is used. The sleepers can be moved individually to the side of the track by hand or, if they will be disposed of, by using the frontend loader.

If track renewal takes place simultaneously with the formation rehabilitation process, the track can be cut into 13 metre sections.

Portal Units are designed to lift track panels of varying lengths depending on the system used and run on the new rail that must be delivered before the formation rehabilitation process starts.

The track panels are then lifted by the portal unit and loaded onto flat railway wagons to be removed from site. These are very efficient units for high production and safety for the removal of material.

Step 3
Excavating the Damaged Formation

The third step is to excavate the fatigued formation to a depth in line with the selected formation rehabilitation design.

This will still be done using excavators but the alternative to using tipper trucks to remove the excavated material is using MFS wagons fitted with crawler tracks in addition to bogies (the MFS 40 ZW).

Once the required depth has been reached, the formation is compacted using a vibratory road roller.

Step 4
Constructing the Subsurface Drain

This process is the same as with the conventional method.

Step 5
Laying Geo-Synthetic Material and Fin Drains

Once again, this process is the same as with the conventional method.

Step 6
Importing Backfill Material in Layers

On single lines the graded backfill material can be brought in by tipper trucks or the MFS 40 ZW wagons can be used in reverse to the process explained in step 3. On double lines MFS wagons can be used to offload backfill material which will bring about huge savings in time, due to the even distribution made possible by MFS wagons and the volume of material that can be offloaded in a very short time.

Refer to CHAPTER 15, paragraph 3.1 for a detailed explanation of the process.

As mentioned before, using geo-cells for strengthening the formation was perhaps not considered in the past due to the slow process of back tipping and levelling of the backfill material before the next load could be offloaded. The MFS wagons have the advantage that huge volumes of backfill material can be offloaded straight into the cells.

Figure 578: PA 1-20 Portal Unit

Figure 579: MFS 40 ZW Material Conveying Systems with Crawler Tracks

Figure 580: Using MFS's to Offload Backfill Material

Figure 581: Loading Station Required to Load New Material

Figure 582: The PA 1-20 ES with Single Sleeper Laying Unit

Figure 583: Flash Butt Welding

Figure 584: Rail Train

The use of geo-cells will also reduce the depth of the excavation and the volume of backfill material required by at least half. The process of compacting is the same as with the conventional method by using a vibratory roller.

Step 7
Re-Ballasting

The alternative to using tipper trucks or ballast wagons (if an adjacent line is available); MFS wagons can be filled with ballast stone (using a loading station) and the same offloading process as described in CHAPTER 15, paragraph 3.2 followed, to quickly and evenly spread the 1st ballast stone layer of 100 to 200 mm (as specified for the type of line for ballast depth under the sleeper). The ballast layer is again levelled and compacted.

Step 8
Reconstructing the Track Skeleton

The track skeleton can be re-constructed on top of the compacted ballast layer using a sleeper laying unit to improve production. Refer also to the use of the SVM track laying machine described in CHAPTER 20 paragraph 3.2.

The rails can be threaded in using a rail threader (Figure 577), fastenings attached and the rails jointed using fishplates and joggle or Robel jaw-clamps.

Step 9
Rail Welding

The flash butt welding process is superior to any other for the welding of rail. The flash butt welding machine can either be used to weld rails together in-situ or the machine can be setup in a welding depot near a port, station or in a material yard.

Refer to CHAPTER 17 for the flash butt welding process.

If re-railing takes place at the same time as the formation rehabilitation and the new rail is only available in short sections (e.g. 12 or 18 m), it would be advisable to weld them into long rails in a depot, since in-situ welding will be too slow and may delay the project.

The disadvantage, however, is that the long welded rails would require a specialised rail train to transport and offload the rails. See also CHAPTER 19.

Another alternative is to offload the short rail sections long before the actual rehabilitation process starts and for the flash butt welding machine to weld the rails in-between train traffic.

Step 10
Offloading of the Top Ballast Layer

Offloading ballast in the conventional way using ballast wagons is still the most cost-effective method. More sophisticated ballast wagons are, however, available on the international market.

Step 11
Regulating the Ballast Bed to the Required Profile

Ballast regulating machines are equipped with different design shoulder ploughs and transfer ploughs, depending on the model, to restore the required cross-sectional ballast profile and redistribute ballast quantities across and longitudinally to the track (Figure 585).

Ballast regulating machines are discussed in CHAPTER 13.

Step 12
Lifting, Aligning, Levelling, Tamping and Stabilising the Ballast Bed

As with the conventional method, the lifting, aligning, levelling and tamping of the track after formation rehabilitation cannot be compromised for any other method but to use a tamping machine of high precision. Refer to CHAPTER 11.

The lack of precision associated with any other tamping process will increase the load on the new formation due to dynamic loading created by an uneven vertical alignment (Figure 586).

The use of dynamic track stabilisation directly after tamping is also highly advisable to consolidate the ballast, to increase the track's resistance to displacement, increase the durability of tamping and to open the track at normal track speed directly after maintenance. Refer to CHAPTER 12.

Advantages Associated with Semi-Mechanised Formation Rehabilitation

- Every step described in the semi-mechanised method does not have to be carried out using mechanised machinery. A combination of the conventional and semi-mechanised methods can be used, depending on the availability of machinery.
- All the advantages of the conventional method are retained despite using mechanised on-track machinery.
- Mechanise machinery will reduce the total time that the track is closed for traffic by a large percentage.
- A very large volume of backfill material can be brought in and distributed evenly along the length of the excavation in a very short time by using the MFS material conveying system where an adjacent line is available.

Figure 585: Ballast Regulating Machine

Figure 586: Tamping Machine with Dynamic Stabilising Machine Behind

Step 4

Ballast Offloading and Finalising the Track

Ballast wagons closely follow the SVV 100 to offload ballast. It may be advisable to offload the ballast in layers in order for the tamping machine to lift level and align, and the stabiliser to stabilise the ballast in layers not exceeding 100 mm (Figure 591).

Alternatively, the SVV 100 can make a second pass with ballast in the trailing material conveyor wagons. The same process as offloading and compacting the FPL material is used to distribute and compact the bottom layer of ballast under the sleepers.

Advantages Associated with Fully Mechanised Formation Rehabilitation using the SVV 100 Sand Distribution and Compacting Machine

- Insertions of a formation protective layer underneath the existing skeleton track
- Laying of geo-textiles at the same time
- Insertion of formation protective layer up to 50 cm thick in one pass
- Working width infinitely variable from 4,000 to 6,000 mm
- No need to lay out material beforehand
- Supply and removal of material only on the track under repair, therefore no occupation of the adjacent line
- Environmentally friendly, no ground damage caused by lorry and construction machine traffic
- Insertion and compaction of a ballast layer, also under the existing skeleton track
- High output for both FPL and ballast insertion
- Utilisation of existing machines (ballast cleaning machine and material/ballast wagons)

Disadvantages Associated with Fully Mechanised Formation Rehabilitation using the SVV 100 Sand Distribution and Compacting Machine

- Ballast cleaning machines may experience difficulty when excavating ballast and or formation material with a high moisture content and/or plasticity.
- The maximum excavating depth when using ballast cleaning machines is 1,200 mm below top of rail (RM 74/76 and RM 80 respectively)

4.3.2. AHM 800 Formation Rehabilitation Machine

Working Process Using the AHM 800 Formation Rehabilitation Machine

The AHM 800 combines various steps, mentioned in the conventional and semi-mechanised formation rehabilitation methods, into one operation without removing the track skeleton. This includes excavating the ballast and damaged formation material, laying of geo-textiles, installation and compaction of backfill material (also referred to as the formation protection layer or FPL) and the transportation of all the material by rail on the same line under repair. The newly produced formation is therefore not subjected to loads by construction machines or trucks. Refer to Figure 593 for an illustration of the machine's working method.

The machine was, however, not designed to repair formations that have been neglected to the extent where very deep excavations are required as illustrated in Figure 556. The maximum excavating depth of the AHM 800 is 1,200 mm below top of rail. Also the maximum thickness of the protective layer inserted in one operation is 500 mm. Figure 594 shows a typical formation repair design using the fully mechanised AHM 800 formation rehabilitation machine.

Figure 592: AHM 800 R Formation Rehabilitation Machine

Figure 591: Ballast Offloading

The smaller of two excavating chains [(4) in Figure 593] removes the upper layer of 200 to 250 mm of ballast over a width of 4,000 mm. This material is transported on conveyor belts (the blue line) to a crushing unit in the front section of the machine (2) to recycle the ballast material. The result is a substantial reduction in costs for backfill material and associated transportation.

Before the material enters the crushing unit, a magnetic separator removes all iron parts. The crushing unit breaks the ballast to a grain size of 0 to 35 mm. The crushed stone is conveyed to a mixer unit towards the back of the machine (the green line in Figure 593) where it is mixed with fresh material graded to specification, brought in from behind (the purple line) using skips or preferably MFS material conveyor units, to produce a sand-gravel mixture for the backfill material (FPL). The moisture content of the FPL material is measured continually and maintained at the required value by the controlled addition of water. This assures one precondition for the ideal degree of consolidation of the installed FPL.

The track is cut in the beginning and the end of the process for the installation and removal of the second, larger excavating chain [(3) in Figure 593]. This chain excavates the remaining ballast bed material and the upper layer of the formation. The working width can be adjusted from 4,050 to 6,000 mm. The excavated material is conveyed to the front of the machine and onto MFS material conveyor units to be removed from site without any impact on the operations of the AHM 800. Refer to CHAPTER 15, paragraph 2.1 for the working principle of MFS conveyor systems to remove spoil material.

If required, geo-synthetics can be rolled out onto the formation. The prepared sand-gravel mixture is laid on top of this using the slewing conveyor belt to distribute it under the track skeleton over the entire width. The regulating unit distributes the material evenly and the six vibrating plates following behind compact it to form the finished formation protective layer. The track skeleton is laid down on top.

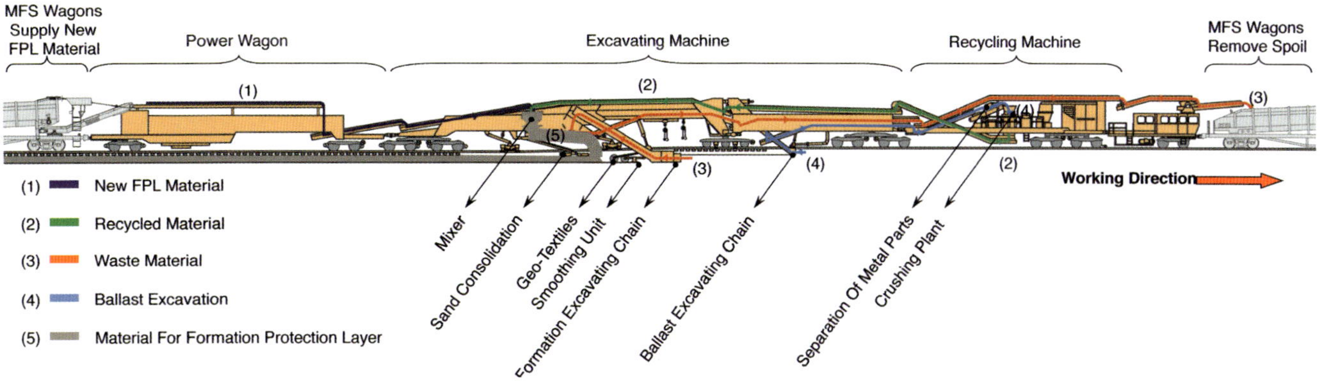

Figure 593: AHM 800 R Formation Rehabilitation Machine

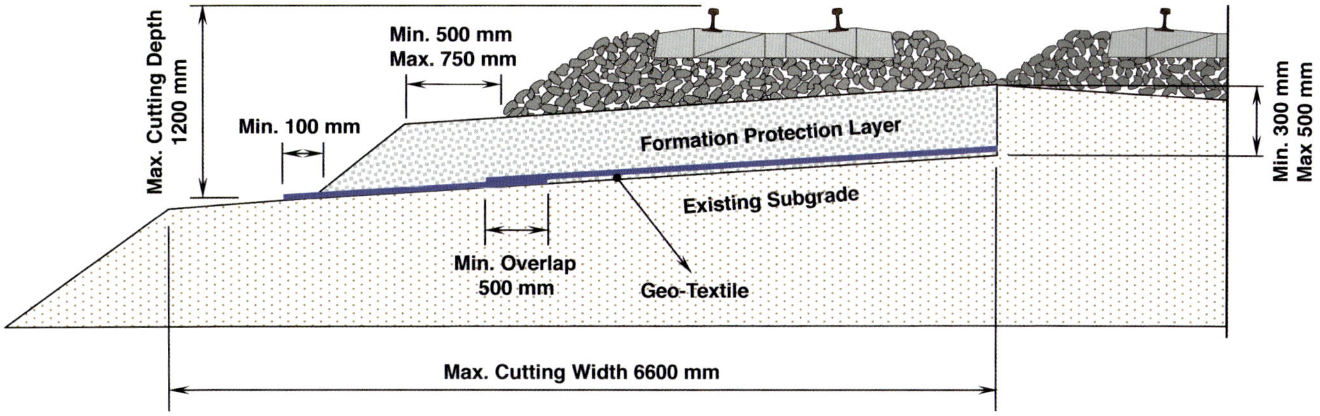

Figure 594: Formation Repair Design Using Fully Mechanised Methods for Formation Rehabilitation

The track is finalised by offloading new ballast using ballast wagons, profiling the ballast bed with a ballast regulating machine, lifting, levelling and aligning the track with a ballast tamping machine and dynamically stabilising the track (steps 10, 11 and 12 in paragraph 4.2).

Advantages Associated with the AHM 800 Fully Mechanised Formation Rehabilitation Machine

- The track can be opened for traffic at the end of a working day
- Cost savings due to reduced expense for material and transport
- Optimum moisture content of the FPL material, achieving excellent consolidation
- Installation of an FPL up to 50 cm thick in one operation
- Concurrent installation of protective materials such as geo-textiles, Styrofoam, plastic sheeting or geo-grid
- No removal of the track skeleton
- No pre-depositing of material
- No ground damage caused by machine operation because no traffic routes have to be made for construction machines or lorries
- No hindrance of rail traffic on the adjacent track
- No strain on the formation during work
- Easy operation even on difficult line conditions (long cuttings, high embankments)
- High production

5. CONCLUSION

Formation rehabilitation is a very extensive activity for which the track may be completely closed for a very long time. If it is a single line, this will prevent any traffic whatsoever, which may come at a very high cost.

It therefore pays to carry out regular maintenance on all the other track components to extend the life of the formation for as long as possible. But since the formation also has a finite life, the time will come when formation rehabilitation will be required. The earlier the rehabilitation is done, the less extensive it will be and the more options are available for the method.

The cost of the line closure should also be considered when the budget is prepared for the repair work. It may pay to use fully mechanised methods for the rehabilitation which have the advantage of high production, and opening the line for traffic at the end of a working day.

Chapter 22
Overhead Electrification Equipment Maintenance

1. INTRODUCTION

The track which comprises the rails, sleepers, fastenings, ballast and formation, lends itself to full mechanisation using highly sophisticated and efficient machinery, such as tamping machines, ballast regulators and ballast cleaning machines.

The overhead electrification system (OHE), also referred to as overhead traction equipment or overhead track equipment (OHTE), on the other hand consists of a large number of small and large components which do not lend itself to full mechanisation. Labour is still required to inspect, adjust, repair, clean and replace these components.

Maintenance of the OHE must be the most tedious and ergonomically challenged of all railway maintenance (Figure 595). It certainly is the most dangerous of all railway track maintenance as well.

Some attempts with varying success have been made by railways in the past to make OHE maintenance safer, more ergonomic and more efficient, by converting on-track trolleys with platforms (Figure 596) for easy access to the overhead structure. These machines were, however, rudimentary, unreliable and lacked all the features that would truly improve OHE maintenance.

Maintenance mechanisation of the OHE only refers to improving the access to the OHE by replacing the ladder with a hydraulically operated elevating platform. Other equipment and small tools can then be added to the machine to further improve safety and productivity.

Today there are many different models of machines available, equipped with various features such as elevating or aerial platforms, high pressure washing equipment for cleaning insulators, AC power supply and air supply for electrical or pneumatic tools, lighting, cranes, mobile earthing points, storage space and a host of other features to make OHE maintenance vastly more efficient, reliable and safe.

It would, however, not be economically sensible to have all these components on one machine since there are different types of maintenance tasks which may only use some of the components some of the time. The addition of each component has an effect on the size and cost of the machine. A trade-off has to be made regarding the use of the component and the additional cost of the machine. It is therefore essential that the most efficient machine be selected for the type of work it will be doing.

Figure 595: Conventional Overhead Electrification Maintenance from a Ladder

Figure 596: Trolleys and Other Vehicles Converted by Railway Depots with Scissors Type Platforms to Improve Overhead Electrification Maintenance in South Africa

2.3. Machine Features

2.3.1. Elevating Platform

Various types of elevating platforms are in use, including a scissors type solid platform, column type (Figure 597 above) which can swivel to the side, purpose designed freely moving elevating platform (Figure 598 above), a loading crane with a work basket (Figure 600 below), etc.

Purpose designed freely moving elevating platforms can be designed to reach any height or position that will be encountered on a railway electrification system (Figure 601). Solid platforms on the other hand are limited in reach but have other advantages.

The fact that the purpose designed freely moving elevating platforms can reach any position on the OHE does not automatically mean that this type of platform should be favoured under all circumstances.

An analysis of the percentage of time that work is required at different positions on the OHE should be carried out, since a solid elevating platform does have certain advantages such as a more stable and larger area for ease of manoeuvring and placing of components and tools. An analysis that was done at Metrorail in South Africa is illustrated in Figure 603 for DC and AC lines.

Figure 600: Loading Crane with a Work Basket

Figure 601: Purpose Designed Freely Moving Elevating Platforms can be Made to Reach any Position on the OHE

Figure 602: Different Types of Platforms in use

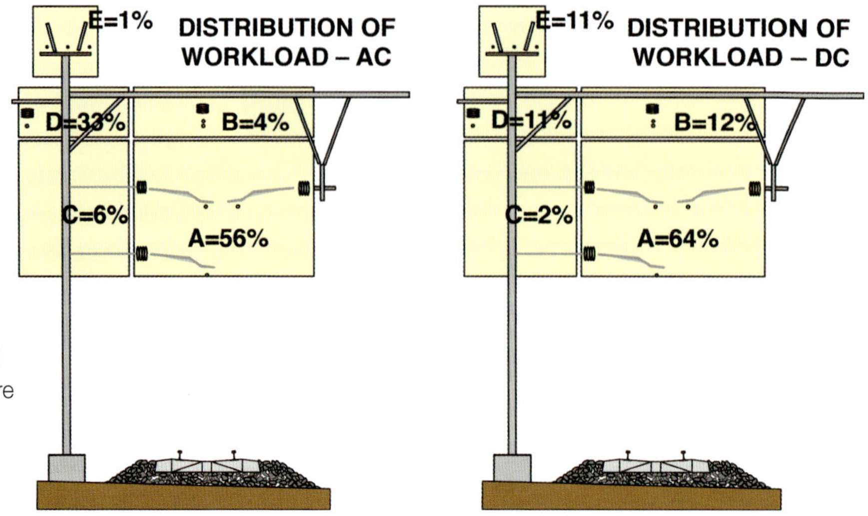

Figure 603: Distribution of Total Maintenance Time per Kilometre per Year

The percentage shown is the share of the total maintenance time per kilometre per year spent on that specific section of the OHE. This workload distribution must then further be broken down into the type of maintenance such as illustrated in Figure 604.

The solid-type platform may have difficulty in reaching the suspension insulator area (Block B), the feeder area (Block D) and may not be able to reach the signalling lines (Block E) at all. For corrective maintenance it means that 60 % of the maintenance time would prove difficult with a solid platform. A purpose designed freely moving elevating platforms type platform will, however, not have any limitations.

It is also clear from the illustration that the bulk of the routine maintenance workload (83%) is easily accessible with a solid platform. Even though routine maintenance comprises 57% (Figure 599) of the total working time, the work is light in nature, for which a simpler and smaller machine will suffice.

These statistics are already guiding towards the design of the machine for Metrorail in South Africa. It confirms the suggestion in paragraph 2.2.1.1 that the light duty machine would be an ideal machine for routine maintenance but that a larger machine with a purpose designed freely moving elevating platforms would be required for corrective maintenance.

A railway having to prepare a machine specification must go through the same process to establish the platform-type best suited for their needs.

2.3.2. Overhead Wire Support Mast

The overhead wire support mast (Figure 605) is a very useful component, which will dramatically increase the productivity when, for example, insulators are replaced. It is used to grip the contact and/or carrying wire (depending on whether one or two units are fitted on the machine) to remove the tension off the insulators, which was traditionally done with cumbersome chain blocks. In Figure 605 two support masts can be seen in operation.

Figure 605: Overhead Wire Support Mast

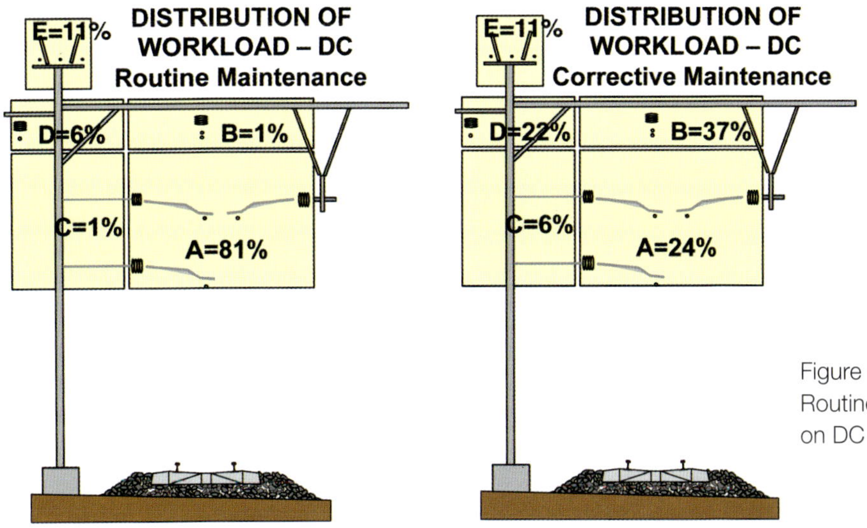

Figure 604: Distribution of Routine vs Corrective Maintenance on DC Lines

2.3.3. Earthing and Measuring Pantographs

Safety regulations in South Africa stipulate that OHE maintenance must be performed between two earthing points. Fixing and moving earthing points is a cumbersome process and would generally require an additional road vehicle to affix and move the earthing points.

OHE maintenance machines are therefore generally fitted with two standard locomotive pantographs with a number of small modifications to improve safety. These improvements may include alarms if the pan loses contact with contact wire, turning the contact force system around to have spring load up and air pressure down, etc.

One pantograph may be required to be insulated for live geometry measurements with either a printed scale on the pantograph pan for stagger and height measurements or with electronic proximity switches for electronic measurement displays. With electronic displays the stagger and height can also be recorded if a distance encoding wheel is used.

The addition of pantographs to a machine will have a significant effect on the size and cost of the machine and the need and advantage of the pantographs must therefore be carefully considered.

2.3.4. Crane

A crane is often specified on a machine for use during heavy maintenance such as replacing mast poles, contact wire or mast foot insulation. In most cases a 7 ton metre crane is adequate. On machines fitted with a solid platform, a basket may be supplied with the crane to reach the few areas that cannot be reached from the solid platform (Figure 600).

2.3.5. Machine Controls on the Platform

For repetitive type work from mast pole to mast pole, the machine and platform are required to be controllable from the machine platform cradle. This enhances productivity and avoids communication problems from the cradle to a driver in the cab and is advisable for all scissors and purpose designed freely moving elevating type platforms.

2.3.6. Power Tools

Air pressure power tools enhance productivity, especially in the coastal areas where rust is common.
The Metrorail study revealed that a large percentage of corrective maintenance activities will benefit from the availability of power tools.

2.3.7. High Pressure Washing Equipment

Cleaning of insulators is a major maintenance activity, especially in coastal areas. The high pressure washing equipment is relatively cheap. What should, however, be kept in mind is the demand on the 220 V system and the water tank which, when filled, can add another 600 kg to the machine. Space must also be provided for the water tank. The Metrorail study once again revealed that a very large percentage of corrective maintenance activities involves the use of high pressure washing equipment but will differ between inland and coastal use.

2.3.8. Electrical Generator

If electrically operated high pressure washing equipment is used, a generator is required. This can also be used to drive other small electrical tools.

3. CONCLUSION

This chapter only mentioned at a small number of decision-making criteria for OHE maintenance machines. A railway needing to specify a machine needs to carefully consider all the operational requirements to ensure the most cost-efficient machine is procured.

Figure 606: Measuring and Earthing Pantograph

Chapter 23
Overhead Electrification System Renewal

1. INTRODUCTION

The overhead electrification system (OHE) generally has a very long life, limited mostly by corrosion, vandalism and theft and is not as much determined by traffic volumes and wear. The OHE components will generally be replaced as part of normal corrective maintenance activities, as discussed in CHAPTER 22. The contact wire is, however, the only wear item and also the most difficult and time-consuming component to replace.

Similar to the homemade trolleys with platforms of CHAPTER 22, railways in the past developed equipment for contact wire replacement as well (Figure 607 and Figure 608). The equipment has served the railways well over the years but it does have many disadvantages, the least of which is that a locomotive would be required in all cases for motive power, which is an expensive and scarce resource. Safety of the homemade equipment was also of concern.

Contact wire theft for the scrap metal value of the copper is on the increase world-wide and has already reached problematic proportions in some countries. Employing an overhead wire installation machine (Figure 609) can reduce the financial implications and traffic disruption substantially for replacing stolen or worn contact wire. Using high end machines that install the contact wire at the final tension and stagger, the work can be completed safely and in a far shorter time. In Europe, occupation times have been cut by 80%, even though they were already using mechanised methods.

Figure 607: Overhead Maintenance and Installation

Figure 608: Rudimentary Drum Holder With Friction Braking System

Figure 609: Mechanised OHE Wire Installation Machine

2. MECHANISED OVERHEAD WIRE INSTALLATION MACHINES

Over time several incarnations of OHE wire installation machines have been developed to suit specific requirements of the customers, OHE designs and budgets. The main working unit of these machines however, is the winch unit (Figure 610) which consists of the following components:

(i) Hydraulically-powered storage drum holder.

(ii) The tensioning device consisting of four friction winch reels (two in pairs) which are electronically controlled to install the contact wire at the final operating tension. Three of the four friction winch reels have a separate drive system. The actual wire tension is measured on one reel using a wire resistance strain gauge.

(iii) A telescopic lifting mast with a guiding roller head lifts the cable to the final height of the installation.

(iv) An auxiliary cable winch is provided to feed the contact wire through the friction winch reels and to pull the contact wire towards the attachment points on the mast pole at the beginning of the process.

The winch unit can be mounted to a slewing and tilting table (model depending) which provides the lateral reach for the contact wire to instate the stagger.

The winch unit can be installed in various configurations. For example, the Plasser & Theurer CEM100.121 (Figure 611) is equipped with one tilting winch unit and an elevating work platform with operating controls to provide access to the overhead structure for attaching the contact wire to the mast poles and to hang temporary suspension wires to keep the contact wire in position. This elevating platform can swing freely in both directions. A driver's cab at the one end provides seating for all the staff, the driving and operating controls and the central on-board computer, which controls all the systems.

Figure 611: CEM100.121 Wire Installation Machine with Elevating Platform

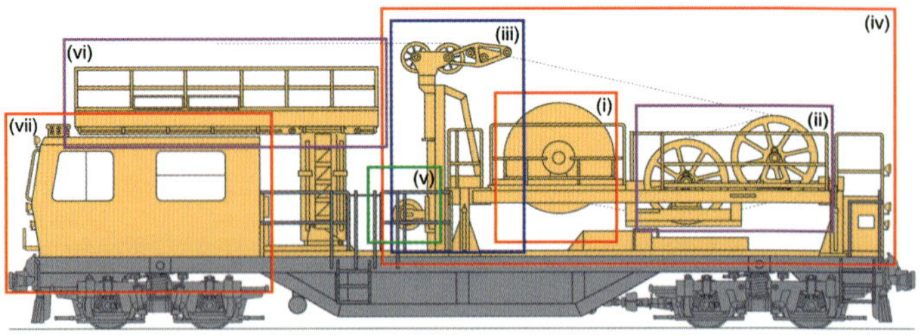

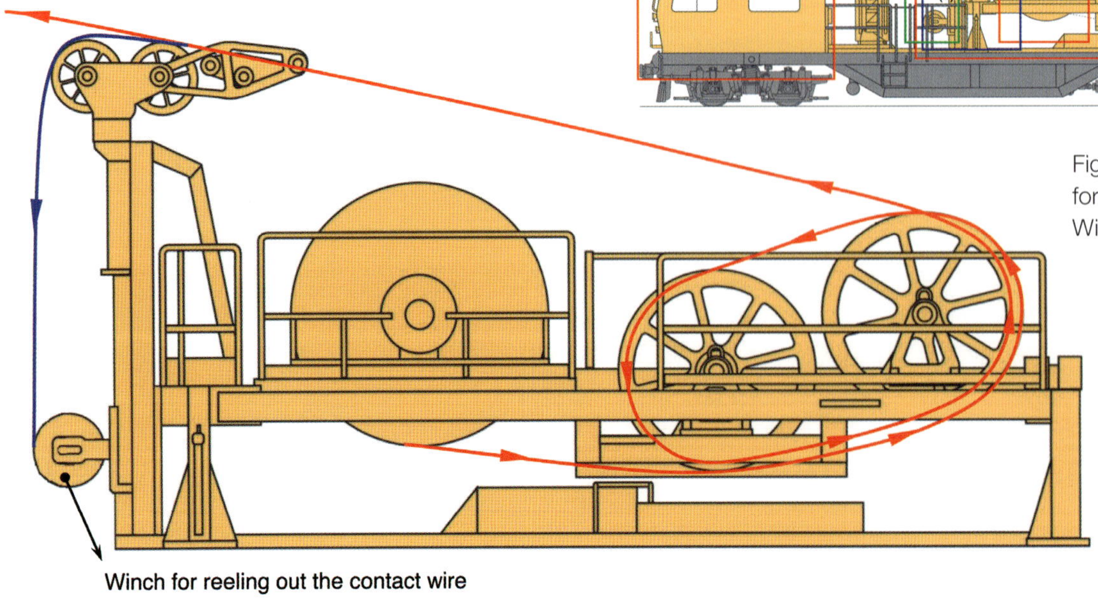

Figure 610: Winch Unit for Installation of OHE Wires at Final Tension

Larger machines such as the FUM100.128 (Figure 609) is equipped with two tilting winch units for simultaneous installation or removal of two contact wires or one contact wire and one carrying cable. The even larger FUM100.051 (Figure 612) is equipped with three tilting winch units for simultaneous installation of one carrying cable and two contact wires.

A further development is the installation of the winch units in a container-like setup on an 18.5 metre railway wagon (the VTC system in Figure 613). The winch units are mounted to a separate frame which makes it possible to lift it off for storage or fit the system on another wagon. The control desk/cabin from where all the working units can be controlled and the power unit are both also in a container-like layout. This system is therefore not self-propelled but must be towed to the worksite and hauled along during work by another machine such as an elevating platform which is anyway required in the renewal process.

The VTC system is fitted with three storage drum units and a separate tensioning unit, which is of different design as described above, and can therefore also perform a variation of installation or removal of 3 cables simultaneously i.e.:

- Install two contact wires and one carrying cable; or
- Install one contact wire, one carrying cable and to collect a third.

Unlike the swivelling winch units, the drum storage units cannot tilt which requires an overhead maintenance machine fitted with an overhead wire support mast (see Figure 605) to position the cable at its mounting points.

Figure 612: FUM100.051 OHE Wire Installation Machine (in Travel Mode)

Figure 613: VTC OHE Wire Installation Machine

3. WORKING METHOD

Once the full drum of contact wire is in place in the drum holder, the auxiliary cable winch is used, to reel the wire through the four friction winch reels and through the rollers of the telescopic lifting mast (Figure 610).

The new contact wire can now be spliced with the end of the existing contact wire or the auxiliary cable winch can again be used to pull the contact wire towards the mast pole where it can be affixed to the tensioning device. The telescopic mast is extended to find the required height of the contact wire.

The target data such as the desired contact wire tension is entered into the central computer using an interactive monitor with touch screen.

The actual contact wire tension is measured at the one reel with the strain gauge and compared with the desired tension. The central computer will produce a hydraulic pressure correction on the individual drive systems of the three friction winch reels. This enables a very sensitive controlled build-up of the wire tension from reel to reel in an adjustment range from 3 kN to 30 kN for the final wire tension. The set target value and the measured actual value can be monitored continuously on the computer screen in the cabin. All the values can be stored as a record of the work performed.

The most critical period for causing over- or under-tensioning of the contact wire is when the machine starts moving. This could lead to damage to the contact wire or even wire breakage. Wire Installation machines should therefore have an electronically controlled starting sequence, which is automatically initiated when the machine operator selects the forward motion of the machine. With the machine's brakes still engaged, the hydraulic pressure in the drives of the friction winches begins to build up until the required wire tension forces is reached. When this point is reached, the machine's brakes, as well as the brakes of the friction winches, are released automatically.

The combined assembly of storage drum holder, friction winch reels and lifting mast on the winch table enables synchronous slewing movements for the installation of the contact wire stagger. It rules out any bending or twisting of the wire.

As the machine moves forward at a constant speed, staff on the elevating work platform can attach the temporary suspension wires to keep the contact wire in position.

Once the machine reaches the end of the span, the contact wire can be attached to the contact wire tensioning device of the OHE. A simple self-propelled elevating platform is required to follow close behind the installation machine to attach the droppers (suspension wires for the contact wire) behind the machine. Once the installation is complete, the line can be opened at normal operating line speed.

4. CONCLUSION

A mechanised overhead wire installation machine will certainly bring about a huge improvement in the safety, efficiency and costs of contact wire replacement.

Chapter 24
Glossary of Track Terminology

abrasion plate	A special plate used to support the rail and prevent abrasion to the **sleeper**.	*Par 4.2.3 Page 43* *Par 4.2.4 Page 43*
abutment	A substructure composed of stone, concrete, brick, or timber supporting the end of a bridge span.	
AC electrified section	A section of railway line where the **overhead** electrification system provides **alternating current**.	
adhesion	The **friction** between wheel and rail. When the friction is exceeded during acceleration or braking, wheel slipping or sliding takes place.	*Par 4.2 Page 80*
affordability	A measure of cost relative to the amount that the purchaser is able to spend.	*Par 2.3 Page 88*
alloy steel	Steel to which elements such as silicon, manganese, nickel, etc. has been added or removed to change the steel properties for a particular use.	*Par 2.3.2 Page 13*
alternating current (AC)	An electric current that reverses its direction at regular intervals, usually abbreviated as 'AC'.	
alumina-thermic weld	See **thermite weld**.	
angle bar	One of two bars used to join two rail ends together to form continuous track. Also known as a **joint bar.**	
anti-creeper	A device firmly attached to the base of a rail and bearing against a **sleeper,** to keep the rail from moving longitudinally under traffic. Also called a **rail anchor.**	*Par 2.2 Page 84*
availability	The ability of a product to be in a state to perform a required function under given conditions at a given instant of time or over a given time interval assuming that the required external resources are provided.	*Par 2.3 Page 88*
axle counter	Track mounted equipment that counts the number of axles entering and leaving a track section to determine if it is occupied.	
axle load	The axle load refers to the total weight of a wagon divide by the number of axles. It normally refers to the maximum weight permitted on a single axle.	
balance speed	The speed through a **curve** at which the wheel loads are equal on the **high** and **low rails.**	*Par 3.1.3 Page 125*
ballast	Ballast forms the track bed on which **sleepers** and track are laid.	*Par 5 Page 45*
ballast cleaning	The process during which the fine material that does not conform to specification is removed from the **ballast** bed.	*Par 5.2.3 Page 49*
ballast cleaning machine	An on-track machine specifically designed to excavate **ballast** down to **formation** level, to separate usable ballast from fine material that does not conform to specification and to return reusable ballast to the track while **spoil**ing the unusable material.	*CHAPTER 14 Page 199*
ballast curb	Retaining material along the outer edge of the floor on **ballast**ed bridges to retain the ballast.	
ballast depth	The depth measurement from the bottom of the sleeper to the top of the formation.	*Par 5.1.2 Page 45* *Par 5.3 Page 52*
ballast regulating	Redistribution of the ballast to reinstate the required **ballast** bed cross sectional profile.	*CHAPTER 13 Page 193*
ballast regulating machine	An on-track machine equipped with various ploughs to redistribute and shape the **ballast** bed cross sectional profile to specification.	*CHAPTER 13 Page 193*
ballast shoulder	The section of the **ballast** bed between the end of the **sleeper** to where the ballast slopes down to the toe of the ballast bed.	*Par 5.3 Page 52*
ballast stabilising	Reinstating of the track's resistance to **lateral forces** by introducing a horizontal vibration and vertical force into the **ballast** bed which rearranges the ballast stones in relation to one another to find a stable position.	*CHAPTER 12 Page 185*
ballast stabilising machine	An on-track machine which sets the track in a horizontal vibration whilst applying a controlled **vertical load** to rearrange the **ballast** stones to find a stable position where it will have an increased resistance to **lateral forces.**	*CHAPTER 12 Page 185*
ballast sweeper	Also called a broom which sweeps ballast off the top of the **sleepers.** Can also be used to pick up excess **ballast** into a hopper on **ballast regulating machines** equipped for this purpose.	*Par 7.4 Page 198*
ballasted track	Track of which the track skeleton consisting of rails fastened to **sleepers** are laid in a bed of graded **ballast** stones which conforms to a specified cross sectional profile.	*Figure 12 Page 9*
base plate	An iron or steel plate used to spread the weight of rail over a larger area of the **sleeper** and facilitate a secure, low maintenance, **fastening** with bolts or clips. It derives from the former rail chairs.	*Par 4.2.3 Page 43*
berm	An earthwork used to intercept or divert surface water from running down the **cutting** wall towards the track.	*Figure 1 Page 1*

dynamic stabilisation	Consolidation of the **ballast** bed following ballast maintenance to restore track's resistance to **lateral forces** (displacement) using an on-track **dynamic track stabilising machine.**	*CHAPTER 12 Page 185*
dynamic track stabilising machine	A self-propelled on-track machine which consolidates track **ballast** after ballast maintenance by inducing a horizontal vibration whilst applying a controlled vertical load.	*CHAPTER 12 Page 185*
easement	A gradual transition between **curves** or **gradients.**	*Par 2.2 Page 116* *Par 2.3 Page 121*
EC	Abbreviation for **eddy current.**	*Par 3 Page 156*
eddy current (EC)	A local current in an iron core caused by the electromotive force generated by moving the core in a magnetic field.	*Par 3 Page 156*
embankment	Where it is necessary for the track to run above ground level to provide a gently graded route for the railway.	*Figure 1 Page 1*
exothermic weld	See **thermite weld.**	
expansion joint	See **rail stress** and **stress.**	
expansion shim	Spacer inserted between ends of abutting rails at **joints** while track is being laid to provide allowance for expansion of steel when temperature changes.	
false flange	The convex radius that occurs as a reverse of the concave hollowing of the wheel tread and can form either on the flange side or the end of the wheel tread.	*Figure 187 Page 79*
fastenings	Bolts, clips and spikes used to fasten the rail to **sleepers.**	*Par 4 Page 42*
fatigue	Fatigue is the progressive weakening of a material resulting in structural damage that occurs due to the cyclic loading. See also **rolling contact fatigue.**	*Par 2.4 Page 85* *Par 2.3.1 Page 12* *Par 2.1 Page 83*
field corner / side	The corner of the **rail head** facing away from the track.	*Figure 13 Page 9* *Figure 176 Page 75*
fish bolt-hole crack	See **bolt-hole crack.**	
fishing area / surface	The area at the bottom of the **rail head** where it meets the **web** that makes contact with **fishplates / joint bars.**	*Figure 13 Page 9*
fishplate	A short metal or wooden plate bolted to the sides of two abutting rails to join them. The use of long welded rail reduces the need for fishplates.	*Par 2.4 Page 16*
fishplate bond	See **bond.**	
flaking	Small chippings and cavities on the rail running surface. Flaking should not be confused with shelling, as the flaking takes place only on the running surface near the gauge corner of the rail and is not as deep as shelling.	*Par 2.19.1 Page 29*
flangeway	The open way through a track structure that provides a passageway for wheel flanges.	
flangeway guard	A protective device that keeps the wheels in proper alignment on a **turnout.** See **check rail.**	
flash butt welded joint	A weld made by electrically heating two abutting rail ends. When the steel reaches the proper temperature, the rail ends are compressed together eliminating the need for **joint bars.**	*Par 2.8.2 Page 20* *CHAPTER 17 Page 225*
flat spot / wheel flats	A flat area on a wheel tread caused by the wheel locking and skidding during braking.	
footplate	The cab of a steam locomotive, or the driving compartment from which a diesel locomotive, electric locomotive, electric motor coach or electric driving trailer, is being controlled.	
formation	The levelled and graded base onto which the track is constructed.	*Figure 12 Page 9* *Par 6 Page 54*
fouled ballast	**Ballast** that has become contaminated by fine material that inhibits the drainage of water through the ballast.	Par 5.2.1.3 Page 47 *Par 5.2.2 Page 48*
fouling track	Obstruction of a running line by objects or encroaching vegetation in such a way that it could be struck by a passing train. See **gauge, structures and vehicles.**	
friction	Friction is a force in the opposite direction to the direction of motion that one object encounters when moving over or against another object.	*Par 4.5 Page 81*
friction modifier	An engineered product that is formulated to be applied in a thin film at the rail/wheel interface to alter the composition of the naturally occurring third body layer between the wheel and the rail, providing **friction** levels within specified limits.	
frog	A casting with „X" shaped grooves used in **switches** and **crossovers.** A track structure used at the intersection of two running rails to provide support for wheels and passageways for their flanges, thus permitting wheels on either rail to cross the other. Also called a **crossing.**	Par *7.3.4 Page 62*
frog angle	The angle formed by the intersecting gauge lines of a **frog.**	*Par 7.3.2 Page 61*
gauge	Distance between the running faces of the two rails, measured a nominal distance below the tops of the **rail heads.**	*Par 3.1.4 Page 126*
gauge corner	The top corner of the **rail head** facing the centre of the track.	*Figure 176 Page 75*

gauge corner cracking	**Gauge corner cracking** of rails is a type of **rolling contact fatigue.** See also **head checks.** It is the result of excessive contact stresses at the wheel/rail interface, which in turn produces deformities in the structure of the rail near its surface.	*Par 2.19 Page 29*
gauge, structures and vehicles	1) Vehicle structure gauge – It forms the 'envelope' within which train profiles must fit. 2) Static vehicle profile – The profile formed by the maximum permitted cross-sectional dimensions of vehicles and, where applicable, their loads when at rest on straight and level track. It thus defines the maximum profile to which railway vehicles can be built in terms of dimensions of height and width, and then loaded, without risk of striking a **wayside** structure. 3) Dynamic/moving vehicle profile – This is the static profile enlarged to allow for the maximum possible displacement of the vehicle at rest or in motion, with respect to the rails on straight track. It should take into account vehicle suspension characteristics including body tilting where provided, and allowances for tolerances in the maintenance of vehicles, including **wear.** The effects of end-throw and centre-throw of vehicles on **curved** track are not included and are disregarded in the development of the dynamic vehicle profile. 4) Kinematic envelope – The increasing speed of trains over the years has resulted in the need for a more advanced concept of the space which a train occupies. This is the kinematic envelope, defined as the volume of space swept by a train in motion. The dynamic vehicle profile is enlarged to allow for the permitted tolerances in **track gauge,** alignment, **level** and **cross level** and the dynamic and static effects of track **wear.** The effects of end-throw and centre-throw of vehicles on **curved** track are not included, and are disregarded in the development of the kinematic envelope. 5) Swept envelope – The swept envelope is the kinematic envelope enlarged to allow for the effects of vertical and horizontal curvature, including centre- and end-throw of vehicles, and the super-elevation applied to the track. The swept envelope may be defined separately for each structure or for sections of the route and should take account of all railway vehicles using the line.	*Par 3.4 Page 131*
geometry	See **track geometry** and **overhead traction equipment geometry.**	*CHAPTER 8 Page 115*
geotextile	A material designed to permit the passage of water through it, but not particles of soil carried by the water and often used during **formation** rehabilitation for various applications.	*Par 2.2.1 Page 262*
gradient	The rate at which a railway track inclines or declines expressed either as a fraction or percentile. For example 1 in 500 indicates a vertical change in the level of one metre along a horizontal stretch of 500 metres. Expressed as a percentage, this is 1/500*100% = 0.2%. To eliminate the decimal point usually associated with railway radients, this may be expressed as 'per thousand' rather than 'percent'. Thus, 1 in 500 becomes 2%. Also referred to as a vertical curve.	*Par 2.3 Page 121*
greasing	See **lubrication.**	
grinding	See **rail grinding** and **turnout grinding.**	
gross weight / gross tons	The total combined weight of the wagon and its contents.	
guardrail	See **check rail.**	
head and web separation / crack	A progressive fracture, longitudinally separating the head from the web of the rail at the head fillet area.	*Figure 53 Page 27* *Par 2.19.2.6 Page 32*
head check	A rail defect consisting of shallow surface or hairline cracks that appear in the **gauge corner** of the **rail head.** They generally run at a 45-degree angle to the axis of the rail and usually occur on the **high rail** of **curves.**	*Figure 53 Page 27* *Par 2.19.1.1 Page 30* *Par 4.3 Page 137*
heat treatment	The process of altering the properties of a steel by heating and cooling in such a way as to obtain desired conditions or properties.	Par 2.3.3 Page 14
Hertzian contact stress	Refers to the theory of Heinrich Hertz who in 1882 solved the contact problem of two elastic bodies with curved surfaces. The Herzian contact stress is a description of the **stress** within mating parts and refers to the stress close to the area of contact between two spheres of different radii.	*Par 4.1 Page 80*
high rail	The upper or outside rail in a **curve.**	*Figure 269 Page 125*
hi-rail (noun)	A road vehicle with retractable flanged wheels so it may be used on either road or rail. Also called hi-railer or **road/rail** vehicle.	
hockey stick	See **steady arm.**	
hollow wheels	A concave **wear** on the wheel tread.	*Par 3.3 Page 79*
horizontal alignment	A deviation from the geometric longitudinal line.	*Par 3.1.1 Page 124*
horizontal curve	A bend in a track normally simply referred to as a curve. Curves connect two sections of tangent (straight) track to go around obstacles or change direction. Curves are further identified by its radius, **transition curve** and the **circular curve.**	*Par 2.2 Page 116*
horizontal split head	A horizontal defect originating below the running surface and progressing horizontally along the **rail head.** Visible as a flat spot on the running surface. Also called **longitudinal horizontal crack.** See **compound fissure.**	*Figure 53 Page 27* *Par 2.19.2.5 Page 32*
horizontal plane	Horizontal along the rail.	*Figure 14 Page 9*

rail seat abrasion	The abrasion action that occurs between the rail pad and the top of the concrete **sleeper** (the rail seat).	
rail stress	Rail have inherent residual **stress** after manufacturing and also as a result of installation outside of the **neutral temperature** and subsequent temperature variances, compressive stress at the bottom of long **gradients** due to braking of long trains, tensile stresses at the crest of the grade, etc. To avoid especially **continuous welded rail** from breaking under tensile stress or kickout/buckle under compressive stress the rail stress must be measured at regular intervals and destressed if required within the neutral temperature.	*Par 2.12 Page 22*
rail type	Generally measured in kg/m and cross sectional profile.	*Par 2.2 Page 10*
rail web	See **web.**	
rail weight	The weight in kilograms of a 1 meter long piece of new rail.	*Par 2.2 Page 10*
rail welding	There are two main types of welding used for welding rail into long welded rails: **thermite welding** and **flash butt welding.**	*Par 2.8 Page 18* *CHAPTER 17 Page 225*
railbound	Refers to any equipment, vehicle or rolling stock that can travel on railway track only.	
railbound manganese frog	A **frog** consisting of a manganese steel casting inserted between two rolled rails, held together by frog bolts.	*Par 7.3.4 Page 68*
RAMS	A European standard (EN 50126) which provides a process for the implementation of a system for the management of **Reliability, Availability, Maintainability** and **Safety.**	*Par 2.3 Page 88*
RAMAS	The abbreviation for **Reliability, Availability, Maintainability, Affordability** and **Safety** which are outputs required from a maintenance strategy adapted from the European standard (EN 50126) RAMS.	*Par 2.3 Page 88*
ratcheting	The incremental plastic flow of surface or very near surface material due to excessive contact **stresses,** often leading to the formation of **rolling contact fatigue** (RCF).	
RCF	Abbreviation for **rolling contact fatigue.**	
RCM	Abbreviation for reliability cantered maintenance.	
re-laying	Renewal of some, or maybe all, of the components of a section of railway track.	*CHAPTER 20 Page 251*
reliability	The ability of an asset or a system to perform a required function under stated conditions for a stated period of time or at a given point in time.	*Par 2.3 Page 88*
reliability centered maintenance (RCM)	A structured process, originally developed in the airline industry, but now commonly used in all industries to determine the equipment **maintenance strategies** required for any physical asset to ensure that it continues to fulfil its intended functions in its present operating context	
renewal	Replacement of track components, usually **sleepers** or rail on a larger scale over a track section.	*CHAPTER 18 Page 237* *CHAPTER 20 Page 251* *CHAPTER 21 Page 261*
repair	A **corrective maintenance** activity which returns the functionality of an asset that has failed.	
rerailer	A portable device designed so that when placed next to the rail, a derailed wagon or locomotive wheel will be guided back onto the rail when rolled over it.	
reverse curve	A **curve** which changes from a clockwise direction to an anticlockwise direction, or vice versa – an 'S' shape, without any intervening straight track.	*Par 2.2.2.2 Page 118*
right-of-way	The strip of land on which a railway track is built.	
roadbed	US term for **formation.**	
Road/rail vehicles	A motor vehicle which is equipped both with flanged steel wheels for rail use, and with pneumatic tyres for road use. One set is retracted when the other is being used.	
Rolling Contact Fatigue (RCF)	RCF is caused by repeated over**stress**ing of the rail running surface causing crack initiation and propagation on the surface or subsurface. Crack initiation and propagation is influenced by the rail/wheel interface, **curves, gradient,** the use of lubrication in curves, rail metallurgy, **track geometry** errors, rail **maintenance** practices, vehicle dynamics etc.	*Par 4.3 Page 80*
rolling stock	Any vehicle that operates on or uses a railway track, excluding a vehicle designed for both on- and off-track use when not operating on the track.	
route kilometres	The distance between two given points along the length of the railway, irrespective of the number of tracks.	
routine maintenance task	Any **maintenance** task that is performed at a regular, predefined time based interval.	*Par 5 Page 92*
ruling gradient	The ruling **gradient** is the steepest on a given section of line. This in turn determines the greatest load which can be worked over that line by a given locomotive at a given speed.	*Par 2.3 Page 121*
run-away points	Refer to **derailing devices.**	
running surface	Refers to the area on the **rail head** which is in contact with the wheel. Also called contact area or **contact band.** This is a fairly narrow band of as small as 20 mm but can increase as the rail profile deteriorates due to lack of rail profiling and wheel maintenance.	*Figure 176 Page 75*

runoff (curve)	The profile through which the **superelevation** of a **curve** is brought to the level of the **tangent,** or through which different elevations on a **compound curve** are connected.	*Par 2.2 Page 116*
safety	The lack of railway occurrences, fatalities, injuries or damage within railway operations.	*Par 2.3 Page 88*
schedule compliance	One of the key performance indicators often used to monitor and control **maintenance.** It is defined as the number of scheduled **work orders** completed in a given time period (normally one week), divided by the total number of scheduled work orders that should have been completed during that period, according to the approved **maintenance schedule** for that period. It is normally expressed as a percentage, and will always be less than or equal to 100%. The closer to 100%, the better the performance for that time period.	
scheduled maintenance	Any maintenance work that has been planned and included on an approved **maintenance schedule.**	
scissors crossover	See **crossover.**	
scotch	A wooden wedge placed on the rail and under the wheel of a stationary vehicle to prevent it from moving. See **point clips** and **scotch**es and **derailing devices.**	
screening machine	See **ballast cleaning machine.**	
screw spike	A threaded screw designed to be turned with a wrench into holes bored in **sleepers** to secure rails or to act as a sleeper-plate holder in sleeper plates.	
secondary damage	Any additional damage to an asset over and above the initial defect that occurs as a direct consequence of the initial defect.	
secondary line	Connect two destinations on lines of lessor importance and/or lower traffic density. Also referred to as a **branch line.**	*Par 2.1 Page 115*
section insulator	A device for dividing a **contact wire** and catenary wire into electrical sections while maintaining mechanical continuity and a continuous path for the pantographs.	
section switch	A switch mounted on a structure between two sections of **overhead line equipment** arranged to disconnect one from the other, but not designed to break current.	
sectioning	The dividing of an electrical distribution system or network into electrical sections.	
self-guarded frog (flange frog)	Usually a manganese steel casting with raised wings that contact the rim of a passing wheel to guide the wheel's flange past the frog point. This feature eliminates the need for a separate **guardrail.**	
self-normalising points	An alternative term for **self-restored points.**	
self-restored points	Power operated points which are automatically returned by the interlocking to the normal position to provide protection after a movement via the points reverse. Normalisation occurs after the route has been released and the track section has been clear for a predetermined time.	
set (points)	Often used as collective word for **turnouts,** slips and **crossing**s.	*Par 7 Page 59*
shatter crack	Minute, shallow cracks that occur in the rail after exposure to intense heat caused by locomotive wheel slipping.	*Par 2.20.4 Page 35*
shelling	See **spalling / shelling.**	
shoulder ballast	That portion of the **ballast** between the end of the sleeper and the toe of the ballast slope.	*Figure 116 Page 52*
shutdown	A planned period during which intensive maintenance work can be carried out on a specific track section. All traffic comes to a complete stop during this period and would normally stretch over a couple of days. It forms part of the maintenance strategy and is ideal for carrying out maintenance activity that cannot be done in the available short **occupations.**	
side drain	See **drain.**	
side wear	The **wear** on the **rail head,** reducing its width, caused by flange contact from the wheels of trains as they round a **curve.**	*Par 2.18.2 Page 28*
siding	A section of track off the **main line.** Sidings are often used for storing rolling stock or freight. A siding is also used as a form of rail access for warehouses and other businesses. See **private siding.**	*Par 2.1 Page 115*
siding extension	That portion of a **private siding** which is situated on land not belonging to the railway.	
single slip	A combination of a diamond **crossing** with two point ends without the need of separate crossings **(points).**	*Par 7.2.2 Page 59*
skate	A metal skid or chock (wedge) placed on a rail to stop the movement of rolling stock.	
slack	Defect of the track top (vertical alignment) on both rails.	*Par 3.1.2 Page 124*
sleeper	A beam of wood, concrete or steel on which the rails are fastened, that keeps the track **gauge** and spreads the wheel load on the **ballast.**	*Par 3 Page 37*
sleeper crib	The area between two **sleepers.**	*Figure 12 Page 9*

EN 50122-1	Fixed installations – Electrical safety, earthing and the return circuit – Protective provisions against electric shock
EN 50122-2	Fixed installations – Electrical safety, earthing and the return circuit – Provisions against the effects of stray currents caused by d.c. traction systems
EN 50122-3	Fixed installations – Electrical safety, earthing and the return circuit – Mutual Interaction of a.c. and d.c. traction systems
EN 50123	Fixed installations – D.C. switchgear
EN 50124-1	Insulation coordination – Basic requirements – Clearances and creepage distances for all electrical and electronic equipment
EN 50124-2	Insulation coordination – Overvoltages and related protection
EN 50126-1	The specification and demonstration of Reliability, Availability, Maintainability and Safety (RAMS) – Basic requirements and generic process
EN 50151	Fixed installations – Electric traction – Special requirements for composite insulators
EN 50163	Supply voltages of traction systems

References

1. Australian Rail Track Corporation Ltd: Engineering Practice Manual – Rail Defects Handbook (RC2400)
2. Becker J: Basic Metallurgy, Modes of Failure and Welding Metallurgy – Program for Fundamentals of Track Structure Design and Maintenance
3. BS EN 50126-1:1999: Railway Applications — The specification and demonstration of Reliability, Availability, Maintainability and Safety (RAMS) — Part 1: Basic requirements and generic process
4. Chair In Railway Engineering: University Of Pretoria Course Proceedings – Introduction To Multi-Disciplinary Concepts In Railway Engineering
5. Coetzee J, (Plasser South Africa) & Swarts E (Spoornet): Analysis of track geometry data to highlight problem areas: A simple approach
6. Corvino L & Intrieri T: Electrical Flash Butt Welding of Rails in Situ
7. Ebersöhn W & Rupert C: Erstellung Einer Gleisdatenbank Und Eines Instandhaltungs Systems Unter Verwendung Von GPS-, LiDAR-, Und Video-Technologien: Der Eisenbahn Ingenieur, Heft 8/2001
8. Ebersöhn W (University Of Pretoria Chair In Railway Engineering): Course Proceedings, Introduction To Multi-Disciplinary Concepts In Railway Engineering
9. Erasmus P: Efficient and Safe Rail for Heavy Haul in South Africa
10. Esveld C: Modern Railway Track, Second Edition
11. Ferguson H: British Rail Looks To Business Motive – New Civil Engineer (8 October 1987)
12. Fischer J: Influence of Frequency and Amplitude on the Stabilization of Track Ballast, June 1983
13. Fortin J: Dynamic Stabilisation To Follow Track Renewals – International Railway Journal (September 1978)
14. Franz Plasser Research & Testing Department: DGS The Dynamic Track Stabiliser, Principles, Tests And Operating Results (Issue 1991)
15. Glawischnig W & Riessberger K: Dynamic Track Stabilizer – Strength Through Vibrations – Internationales Verkwhrswesen (1/79)
16. Gräbe P & Maree J: Use Of A Dynamic Track Stabiliser To Improve Track Maintenance And Optimisation Of Track Tamping – RTR 4 (1997)
17. Haunold F (Austrian Federal Railways): Mobile Flash Butt Welding, With An Eye To The Future
18. Heckel T, Thomas H-M, Kreutzbruck M and Rühe S: High Speed Non-Destructive Rail Testing with Advanced Ultrasound and Eddy-Current Testing Techniques, December 2009, National Seminar & Exhibition of Non-Destructive Evaluation
19. Heller W, Koerfer E & Schmedders H: Naturally Hard Special Grade Rails for Heavy Duty Transportation
20. Hering H: Application And Testing Of The Dynamic Track Stabiliser On New Lines – Rail Engineering International (Edition 1989 Number 2)
21. Höne H (Plasser South Africa): Provisional paper – Rail performance : Rectification Of Rail Profiles On The Sishen-Saldanha Iron-Ore Export Line With The Plasser SBM 140 Rail Planing Machine (July 1999)
22. International Heavy Haul Association: Guidelines to Best Practice for Heavy Haul Railway Operations – Management of the Wheel and Rail Interface (2015)
23. Klotzinger E: Teaching Aid – Track (on CD)
24. Kluger K: Consideration on Stabilizing of Track, June 1977, Presentation in Budapest
25. Kretzmann A (Transnet): An overview Of The Multidisciplinary Rail/Wheel Interaction Plan Developed For The Sishen-Saldanha Line (Paper to IHHA Conference April 1996)
26. Kronenberg G: Study on the Compaction and Working Behaviour of a Single Mass Compactor on Ballast and Fine Gravel for Establishing the Leading Influence Factors Riddle Compaction, 1961
27. Li D, Cooke J & Shust B: The Effect Of Track Maintenance On Lateral Strength – AAR Research, Test & Development Report – Railway Track And Structures (August 1997)
28. Lichtberger B (Plasser & Theurer): International Edition 2001 Number 2
29. Lichtberger B (Plasser & Theurer): The Homogenisation And Stabilisation Of The Ballast Bed – Rail Engineering International (Edition 1993 Number 1)
30. Lichtberger B (Plasser & Theurer): Combatting Rolling Noise Of Trains By Means Of Rail Grinding – Rail Engineering International (Edition 2005 Number 3)
31. Lichtberger B: Track Compendium, 2nd Edition
32. Lugg P: BR Experience With Rail Welding - 2nd International Heavy Haul Railway Conference 1982
33. Maartens S & Stevens D: Wiel/spoor interaksie, Meting Van Dinamiese Gedrag Van 'N Gelaaide Trok Se Draaistel Om Sodoende Slingervoorkoms, Soos Ervaar Deur Die Gelaaide Trok, Op Die Ertslyn Te Bepaal (5 May 1998)
34. Mason A: Evaluation of the Plasser & Theurer K355PT Mobile Flash Welder – December 1978 – British Railways Board, Research and Development Division, Track Group, Technical Note TN MET 16
35. Mulder J, Joubert S, Maartens S & Theunissen C (Spoornet): Rail/wheel interaction, Sishen-Saldanha line, The Evaluation Of Acceptable Operating Speeds Over Track Under Different Conditions Of Slinger And The Determination Of The Advantages Of Limiting Hollow Wear To 2mm (3 June 1994)

36. Mülke F: Storm Water Drainage of Rail Track

37. Petty J: 60 Ton Giant Gives Slow Sunday Trains A Shake-Up – The Daily Telegraph (Monday 14 March 1988)

38. Plasser & Theurer: Various Product Brochures, Videos And Reports

39. Plasser American: PTS-90, Dynamic Track Stabilizer (18 January 1999)

40. Plasser South Africa: Regional Track Condition Assessment Project

41. Plasser South Africa: Various Internal Research Reports

42. Pretorius F: The State of the art of Ballast Fouling and Screening Maintenance Systems –
Section A - Technical Report on the Concept of Ballast Fouling (Jan 1993)
Section B - Manual on Practical Evaluation Techniques for Measuring And Classifying In-Track Ballast Layer Conditions (Feb 1993)
Section C - Screening Maintenance Systems (May 1993)

43. Pretorius G & Maartens S (Spoornet): Wiel/Spoor Interaksie, Meting Van Dinamiese Gedrag Van 'N Gelaaide Trok Se Draaistel Om Sodoende Slingervoorkoms, Soos Ervaar Deur Die Gelaaide Trok, Op Die Ertslyn Te Bepaal (30 March 1995)

44. Railtrack PLC: Rolling Contact Fatigue in Rails; a Guide to Current Understanding and Practice

45. Report BR: Research Derby Sectional Note No. 378 (10 March 1977)

46. Report CENELEC R009003 1998: Railway Applications – Guide to the specification of a guided transport system

47. Report NS: No Abt. 8e (November 1976)

48. Report SNCF: No R 3260-76-06

49. Riesberger K: Measurements of DB/Asynchronous Tamping, 1972

50. Riessberger K & Wenty R: Track Quality, Key To Load Bearing Capacity And Efficient Maintenance – IHHA Conference 1993

51. Riessberger K: Towards A More Stable Bed – Railway Gazette (March 1977)

52. Sari G: The Influence Of The Dynamic Track Stabiliser On Track Geometry – Transport International (Nr. 1 June 1981)

53. Schubert E: Contribution To The Stabilisation Of Track Geometry

54. Selig E & Waters J: Track Geotechnology and Substructure Management

55. Selig E: – 6th International Heavy Haul Conference – Characteristics Of Ballast Shoulder Cleaning And Ditching

56. Simons E: An outline of Metallurgy – 1968

57. Steele R & Reiff R: Rails, Its Behaviour and Relationship to Total System Wear - 2nd International Heavy Haul Railway Conference 1982

58. Swarts E (Spoornet): Measurement of Railway Track and Overhead Catenary with an On-track Machine

59. Tomas M (Spoornet): Fifth Interim Report – Km 209 Sishen-Saldanha line hunting and bad bogies (30 May 1996)

60. Tomas M (Spoornet): Second Interim Report - Strain measurement on track with hunting on the Sishen-Saldanha line at km 209 (25 February 1994)

61. Tomas M (Spoornet): Third Interim Report – Strain measurement on track with hunting on the Sishen-Saldanha line at km 209 (27 May 1994)

62. Transnet: Ballast Specification S406

63. Transnet: Manual For Track Maintenance (2000)

64. Transnet: Specification For Railway Earthworks S410

65. VAE Rail Brochures

66. Veit P (Technical University Graz): OVG Conference September 2004 Outsourcing – In Track Maintenance

67. Venter C: The Little Green Book